CLINICAL AND EXPERIMENTAL IMMUNOREPRODUCTION 1

IMMUNOBIOLOGY OF TROPHOBLAST

CLINICAL AND EXPERIMENTAL IMMUNOREPRODUCTION

EDITORS

R. G. EDWARDS

Department of Physiology, Cambridge

M. H. JOHNSON

Department of Anatomy, Cambridge U.

---

1. Immunobiology of Trophoblast

EDITED BY R. G. EDWARDS, C. W. S. HOWE AND M. H. JOHNSON

2. Maternofoetal Transmission of Immunoglobulins

EDITED BY W. A. HEMMINGS

IMMUNOBIOLOGY OF TROPHOBLAST

EDITED BY

R. G. EDWARDS, C. W. S. HOWE &

M. H. JOHNSON

The Department of Physiology, Cambridge University

CAMBRIDGE UNIVERSITY PRESS

CAMBRIDGE

LONDON · NEW YORK · MELBOURNE

Published by the Syndics of the Cambridge University Press
The Pitt Building, Trumpington Street, Cambridge CB2 1RP
Bentley House, 200 Euston Road, London NW1 2DB
32 East 57th Street, New York, NY 10022, USA
296 Beaconsfield Parade, Middle Park, Melbourne 3206, Australia

Library of Congress Catalogue Card Number: 74-31800

ISBN: 0 521 20636 7

First published 1975

Photoset and printed in Malta by St Paul's Press Ltd

CONTENTS

PREFACE

Interest in the immunology of reproduction has expanded rapidly during the last decade. Relevant studies in this subject demand an understanding of phenomena involved in reproductive biology, in both its clinical and scientific aspects, and in cellular and humoral immunology. The topics under investigation thus cover fundamental events of reproduction or immunology, the development of highly sensitive technical methods such as immunoassays, and clinical methods for purposes such as oncology, contraception or mitigating infertility.

This publication is the first of a series of meetings in which specific topics in immunoreproduction will be debated. The meetings are held in association with the activities of the International Co-ordination Committee for the Immunology of Reproduction, which has already published three volumes of more general scope. Two were general symposia entitled the *Immunology of Reproduction*, and were published in 1968 and 1971 respectively by the Bulgarian Academy of Science. The third meeting, entitled *Immunology and Reproduction*, was published in 1969, by the International Planned Parenthood Federation, London. Another related conference on the *Immunology of Obstetrics and Gynaecology* is currently in press with Excerpta Medica. The present publication initiates a series of reports of smaller meetings.

In the present series, a separate volume will be devoted to selected topics, and small groups of participants will be invited to present their data, discuss the results, and review the area generally. The Proceedings will be presented in full, and the meetings will be held in various parts of the world under the auspices of specific editors who undertake to organise the meeting. The presentations will be aimed at discussing the scientific and clinical aspects of immunoreproduction, with the intention of elucidating both basic and applied methods of research and therapy and including topics of essential interest to reproductive physiologists and immunologists. The Proceedings will be published by the Cambridge University Press in the format of the present symposium.

The topic of this volume was chosen because of its current interest to many scientists and doctors. The conference was held in January 1974, in the Physiological Laboratory, Cambridge, England, by courtesy of Professors R. D. Keynes and C. R. Austin, and finance for the meeting was generously provided by the Cambridge Philosophical Society and by G. D. Searle & Company Ltd. All participants were either British or were visiting Britain.

Further meetings are planned, and the proceedings will be published in the near future. One meeting, to be held at the University College of

North Wales, Bangor, will be concerned with the transfer of immunoglobulins from mother to foetus. A later meeting in Cambridge will be concerned with physiological consequences of immunity to reproductive hormones.

We wish to thank the many people who helped with the meeting and the publication of its papers, especially Sylvia Clutterbuck-Jackson, Cilla Fuller, David Griffin, Audrey Muggleton-Harris, Mr Pine, Joan Pyne, Jean Purdy, Barbara Rankin and Mrs D. Silsbury.

June 1973

R. G. Edwards
M. H. Johnson

PARTICIPANTS

DR M. ADINOLFI — Paediatric Research Unit, Guy's Hospital Medical School, London, SE1 9RT

DR W. R. ALLEN — Animal Research Station, 307, Huntingdon Road, Cambridge

DR A. C. ALLISON — Clinical Research Centre, Watford Road, Harrow, Middlesex, HA1 3UJ

PROFESSOR E. C. AMOROSO, FRS — Institute of Animal Physiology, Babraham, Cambridge

DR K. D. BAGSHAWE — Department of Medical Oncology, Charing Cross Hospital, Fulham

DR W. D. BILLINGTON — Reproductive Immunology Group Department of Pathology, University of Bristol, Bristol BS8 1TD

PROFESSOR W. F. BODMER, FRS — Genetics Laboratory, Department of Biochemistry, University of Oxford, South Parks Road, Oxford OX1 3QU

DR R. BORLAND — Department of Pathology, Tennis Court Road, Cambridge CB2 3EG

PROFESSOR R. R. A. COOMBS, FRS — Immunology Division, Department of Pathology, Tennis Court Road, Cambridge CB2 3EG

PROFESSOR P. CURZEN — Academic Unit, Obstetrics Department, Westminster Medical School, Queen Mary's Hospital, London

DR D. A. L. DAVIES — G. D. Searle & Co., Lane End Road, High Wycombe, Bucks

DR R. G. EDWARDS — Physiology Department, Downing Street, Cambridge CB2 3EG

DR W. PAGE FAULK — MRC Rheumatism Unit, Canadian Red Cross Memorial Hospital, Taplow, Maidenhead

DR R. L. GARDNER — Zoology Department, South Parks Road, Oxford, OX1 3 PS

DR R. B. HEAP — Institute of Animal Physiology, Babraham, Cambridge

MR C. W. S. HOWE* — Physiology Department, Downing Street, Cambridge CB2 3EG

DR D. M. JENKINS — Department of Obstetrics and Gynaecology. The University of Leeds

DR M. H. JOHNSON — Anatomy Department, Downing Street, Cambridge CB2 3EG

DR S. D. LAWLER — Department of Cytogenetics and Immunology, The Royal Marsden Hospital, Fulham Road, London, SW3 6JJ

DR Y. W. LOKE — Department of Pathology, Tennis Court Road, Cambridge, CB2 3EG

DR A. McLAREN — Institute of Animal Genetics, West Mains Road, Edinburgh

DR A. MUNRO — Department of Pathology, Tennis Court Road, Cambridge CB2 3EG

* Present address: Sloan-Kettering Memorial Cancer Center and Cornell University Medical College, New York

INTRODUCTION

By R. G. EDWARDS, C. W. S. HOWE AND M. H. JOHNSON

A classical paradox of immunobiology is highlighted by the survival of an allogeneic foetus in the uterus of a mother who will reject grafts genetically similar to her foetus. The immunological significance of genetic differences between mother and foetus has perhaps been reduced in recent years, for allografts of different tissues will elicit diverse types of rejection phenomena, through factors such as quantitative variations in the alloantigenic determinants on cells and tissues. Despite this qualification, there can be no doubt about the presence of antigenic determinants in foetuses capable of immunising the mother. These antigens are very varied, including transplantation, blood group, carcino-foetal and tissue antigens, and could invoke humoral and cellular responses. Indeed, immunisation of the mother against the antigens of her foetus is known or suspected to result in pathological conditions such as rhesus incompatibility and various other clinical syndromes of pregnancy. Likewise, foetuses have been destroyed or damaged in animals by modifying various immunological factors during gestation.

Nevertheless, most pathological conditions, including rejection, occur very rarely in pregnancy, and several protective factors could be involved. The nature and dosage of some foetal antigens, and their ability to stimulate allograft immunity, are not yet firmly established. This topic was debated by Allison, and will be examined in greater detail in a later volume of this series on the antigenic maturation of the foetus. The immune responsiveness of a mother may be modified during pregnancy by humoral and chemical factors, or by specific tolerising and enhancing factors. These topics are mentioned repeatedly in the papers and discussions. The most important feature of pregnancy central to its survival might be the presence of a tissue which is antigenically neutral and interposed as a selective immunological filter between mother and foetus. Trophoblast cells are obviously the forefront of this filter. The properties of these cells as they develop in the embryo and respond to the endocrine and immunological components of pregnancy were of central importance in the discussions.

Trophoblast differentiates before implantation. Its subsequent diversification and organisation into a variety of distinctive cell types and their spatial and functional interrelationships with other components of the placenta were described by Gardner and Billington. Trophoblast is of foetal origin, and comes into intimate confrontation with maternal tissues;

in some species, cellular fusion occurs between trophoblast and maternal cells. Once formed, trophoblast provides a permanent filter between mother and foetus until the end of pregnancy in most species. Under some circumstances, trophoblastic cells are detached (deported) from the placenta and carried into the maternal circulation, appearing in the lungs and other tissues of the mother. Deportation occurs during normal pregnancies and during malignant changes, and Bagshawe and Lawler discussed the migration of foetal trophoblast cells into the maternal circulation and their function as a potential immunogenic stimulant to the mother.

Most attention was given to the function of the trophoblast as a neutral protective filter. Neutrality could be conferred by the absence of antigen from trophoblast cells, thus simultaneously avoiding stimulation of the mother and susceptibility to her cellular or humoral immunity. On the other hand, antigens expressed on these cells might be masked by protective coats or modified in some manner. Many concepts arise here, including the control of antigenic expression, the spatial topography and turnover of antigens on cell membranes, the modification of antigens by overlying layers or other factors, and the rate of antigen shedding as complexes with antibody. The form of expression of antigens on cells, and the means whereby antigenically-positive cells avoid destruction were considered by Munro. His observations, and other aspects of the control of antigenic differentiation, were related to the trophoblast by several contributors, notably Faulk and Johnson. The notion that some alloantigenic specificities are expressed in the trophoblast emerged during the course of the meeting, although the location and organisation of these antigens and their relation to maternal responsiveness remained controversial. The need for surface protective layers was thus uncertain in the absence of the clear demonstration of antigens on the trophoblast.

The role of the trophoblast as a filter limiting the access of the maternal cells and immunoglobulins to the foetus emerged clearly in the debates, and was considered by Adinolfi; he showed that the transmission of cells from mother to foetus was almost totally prevented by the trophoblast or some other placental tissue. Maternal cells seldom, if ever, gain access to the foetus in numbers sufficient to be effective in causing tissue damage. Some debate even indicated that antigenic specificities present on some part of the trophoblast might be advantageous by capturing antibodies in an allogeneic trap, so reducing or preventing the access of free antibody to the foetus itself.

Another major activity of the placenta is the synthesis and secretion of hormones, and many endocrinologists remain uncertain about the role of some of the placental protein hormones during pregnancy. The endocrine changes during gestation may reduce the immunological responsiveness of the mother to the antigens of her foetus and these concepts were discussed by Borland. Hormones present in the membranes of the trophoblast cell might themselves constitute antigens for the immune lysis of the cell, as shown by results from in-vitro studies and proposed as a source of abortion by Howe.

The relations between trophoblast cells and lymphocytes were also considered since the secretions of the trophoblast could modify lymphocyte traffic or activity. Evidence comes from many sources, using cultures of tissue or direct observations on species such as the horse, made by Allen, in which distinct signs of rejection can be seen in the uterus during pregnancy. This involves the destruction of certain forms of trophoblastic outgrowth by a process closely resembling allograft rejection, although the significance of these observations in relation to the trophoblast itself was not clear.

A meeting devoted to the trophoblast must include a discussion of the physiological consequences arising as a result of allogeneic differences between the mother and her foetus. Such consequences could include foetal wastage, selective implantation, effects on placental size and various other phenomena and these would be determined, at least partially, through the properties of the trophoblast either in allowing transmission of antigens or antibodies, or in itself expressing the antigens responsible for the destruction of the foetus. These points were considered by McLaren, who questioned a great deal of the evidence. Her conclusion, that there was no evidence for several of these effects in pregnancy, was widely accepted by the participants.

If a personal opinion may be ventured, we would suggest that the debate indicated that the properties of the trophoblast as a neutral filter currently offered the most critical single explanation for foetal survival. The extended and exhaustive debates showed clearly the amount of interest in the diverse properties of the trophoblast and we thank the speakers and discussants for their excellent contributions.

ANTIGENIC TOPOGRAPHY AND THE INTERACTION OF ANTIBODIES AND CELLS WITH SURFACE ANTIGENS

BY ALAN MUNRO

Any discussion of this subject has to start from the fluid mosaic model of membrane structure proposed by Singer & Nicholson (1972) which is shown in Fig. 1. The essential feature of this model is a matrix of lipid in which the protein and glycoprotein moieties of the membrane float, and are distributed in a random manner. Once the notion of randomness of these protein units floating in the lipid matrix has been accepted, the whole concept of antigenic topography of a membrane requires re-evaluating.

The fluid mosaic model predicts that surface structures are in a dynamic state and Fig. 2 illustrates the range of results one can obtain. Figure 2*a* shows a cell with a random distribution of a surface protein. In the presence of any reagent which will cause this protein to agglutinate, the protein will be aggregated into small patches on the surface of the cell (Fig. 2*b*) (Pernis,

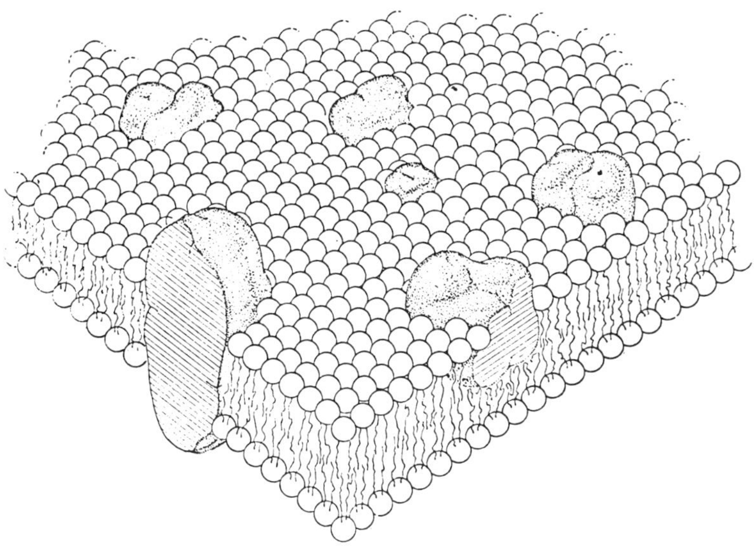

Fig. 1. A schematic three dimensional and cross-sectional view of a fluid mosaic model of a membrane. The solid bodies with striped surfaces represent globular proteins. The matrix consists of a phospholipid bilayer, the small open circles represent the ionic groups of the phospholipid molecules, which make contact with water. The wavy lines represent the fatty acid side chains (from Singer & Nicholson, 1972). Copyright © 1972 by the American Association for the Advancement of Science.

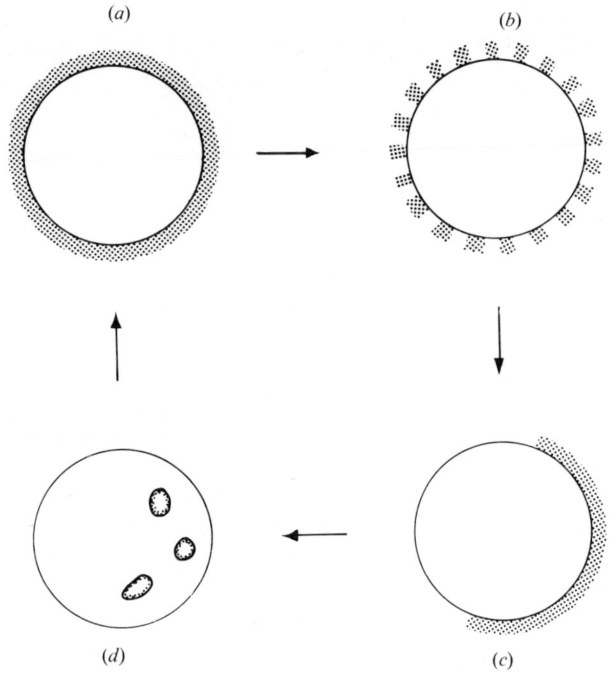

Fig. 2. Diagramatic representation of the possible distribution of a surface structure on a cell. (*a*) The undisturbed random distribution; (*b*) micro-aggregates or 'spots'; (*c*) a cell with a 'cap' of surface structure; (*d*) aggregate surface structure present in pinocytotic vesicles. The diagram also indicates the circle of events that occur after aggregation of a surface structure on a metabolically active cell.

Forni & Amante, 1970; Pinto de Silva, 1972). This will occur at 4 °C but more slowly the lower the temperature. This is not a metabolic phenomenon but a physical reaction. If energy is supplied, these patches of aggregated material are brought together into larger patches and eventually into what is known as a cap at one end of the cell (Fig. 2*c*). This part of the reaction is dependent on energy, requires elevated temperature and is blocked by metabolic inhibitors, (Taylor, Duffus, Raff & de Petris, 1971; Loor, Forni & Pernis, 1972). On further incubation the aggregated surface is internalised into the cell and one can see vesicles inside the cell which contain the aggregated surface structure (Fig. 2*d*) on the inside of the vesicle membrane (de Petris & Raff, 1972; 1973). By this means a cell can be completely stripped of an antigenic determinant. This was first described, although at the time not fully understood, by Boyse, Stockert & Old (1967) for the TL antigens on tumour cells and called antigenic modulation. Many different systems have been shown to behave in this way: for example the immunoglobulin molecules on the surface of mouse lymphocytes (Taylor *et al.*, 1971) and human B lymphocytes (Loor *et al.*, 1972) and cytophilic antibody on basophils and macrophages (Becker *et al.*, 1973; Ferrarini,

Munro & Wilson, 1973); the H-2 and HLA antigens on lymphocytes (Kourilsky *et al.*, 1972; Davis, 1972) and fibroblasts (Edidin & Weiss, 1972). One of the more precise ways of visualising the antigen distribution is by using antisera conjugated to ferritin followed by electron microscopy (Aoki *et al.*, 1969). Figures 3 and 4 show the type of results which can be obtained.

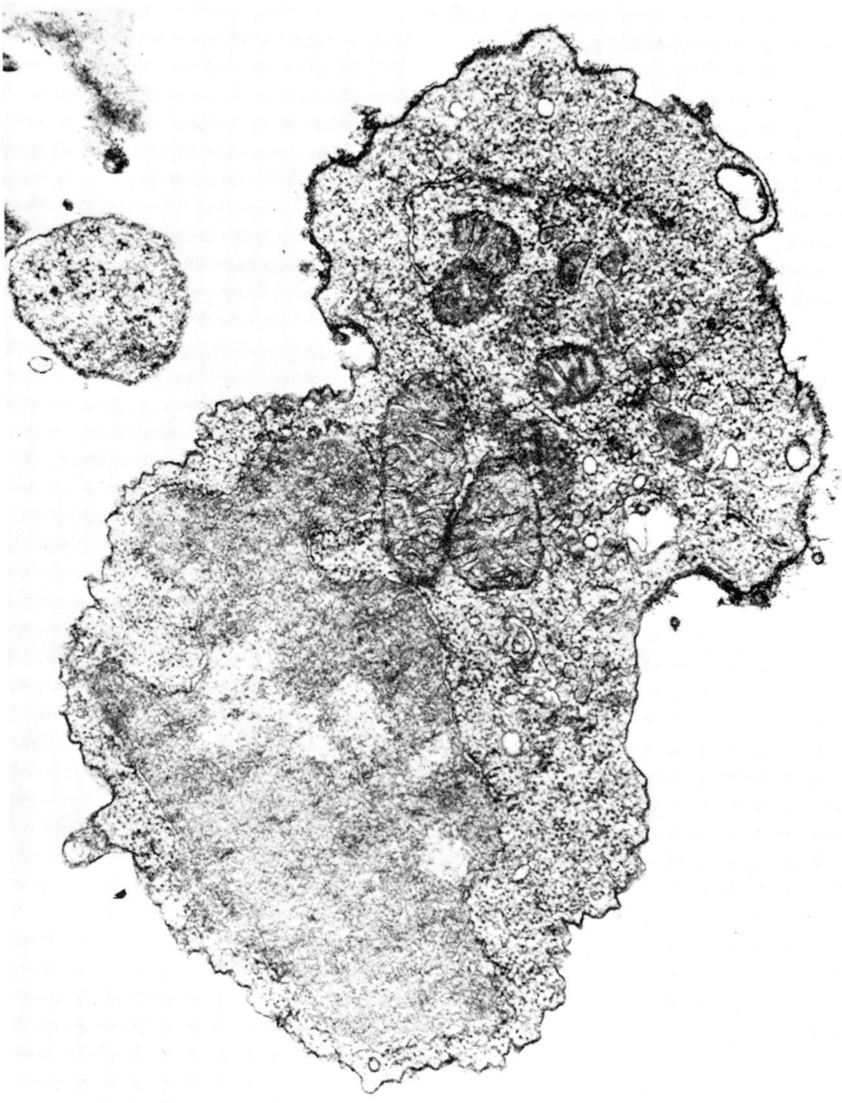

Fig. 3. A mouse lymphocyte labelled at room temperature with ferritin-conjugated antibody to mouse immunoglobulin. The ferritin is localised over the uropod surface while the surface of the cell over the nucleus is unlabelled (from de Petris & Raff, 1972).

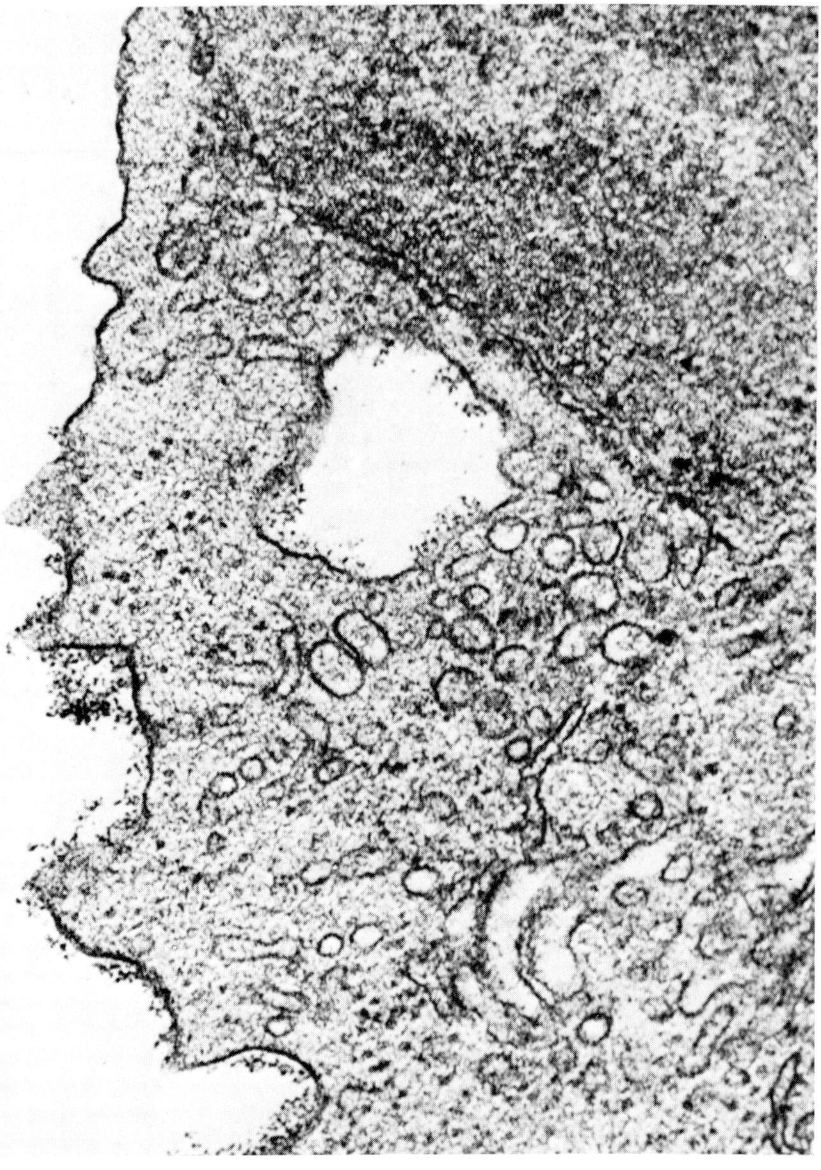

Fig. 4. Detail of a 'cap' of a small lymphocyte labelled at room temperature. Note the pinocytic vacuole partially lined with ferritin-conjugated antibody (from de Petris & Raff, 1972).

One of the difficulties in this field is that in the earlier work it was not appreciated that by using antibody to look at surface structures one was using a divalent reagent, which was causing aggregation on the surface. By using monovalent fragments of antibody such as Fab, it has now been clearly shown that the undisturbed distribution of surface determinants is essentially random (de Petris & Raff, 1974). By now, if a suggestion is made that there is a topographic relationship between different surface protein molecules, the onus is on the proposer to vigorously exclude artificial aggregation on the surface.

Two techniques have been used to search for association of different molecules on the surface of a cell membrane. The first is to look for co-aggregation of different antigenic molecules and has been successfully used to show absence of association of D and K regions of H-2 in mice (Neauport-Sautes, Lilly, Silvestre & Kourilsky, 1973) and Ig antigen receptors on B lymphocytes from the HL-A (Preud'homme *et al.*, 1972) and H-2 antigens (Taylor *et al.*, 1971). As far as I know there has been no example published so far of capping one antigen and finding that another antigen which is known to be a separate molecule, has moved with it (but see Poulik, Bernoco, Bernoco & Ceppellini, 1973). The second method to look for association of surface structures is to study the inhibition of binding of antisera to one determinant by pre-treatment with a second serum. This method has been used by Boyse *et al.* (1968) for H-2, θ and TL on mouse lymphocytes. This work has not been interpreted in light of the fluid mosaic model of Singer & Nicholson but it does suggest that there might be some preferred order amongst different surface structures which is disrupted in capping experiments of the type described above.

So far all the experiments which I have talked about refer to work on cells such as red cells, lymphocytes and fibroblasts. What is the situation when one comes to look at a more complicated arrangement of cells, such as secretory epithelial cells. It is clear that epithelia of this kind are asymmetric. Figure 5 shows an electron micrograph of part of a cell from the intestine of a piglet showing a typical brush border and tight junctions forming the zona adherens. The asymmetry can be seen at the molecular level, for example in the electron microscope one can see glycocalyx on the microvillar membrane. The glycocalyx is absent from the lateral membrane of the cell and the membrane which overlays the basement lamina. This asymmetric distribution of a surface structure does not fit with the fluid mosaic model of membrane and I would like to suggest that one should divide the membrane of such a cell into at least three different zones. One is a zona adherens, i.e. the tight junction, which separates the membrane towards the lumen from the rest of the membrane of the cell. The tight junction may act as a barrier to diffusion in the membrane to prevent the mixing of components in the two different fluid mosaic matrices which are over different parts of the cell. Electron microscopy of rabbit blastocysts show that trophoblast cells have all the features associated with secretory epithelial cells (Fig. 6) in as much as there are microvilli on the outside, tight junctions at the luminal edge of the adjacent cells, and a basal lamina. I would like to suggest that there is a real possibility that the antigenic determinants expressed on the outside of the trophoblast differ from the determinants expressed on the

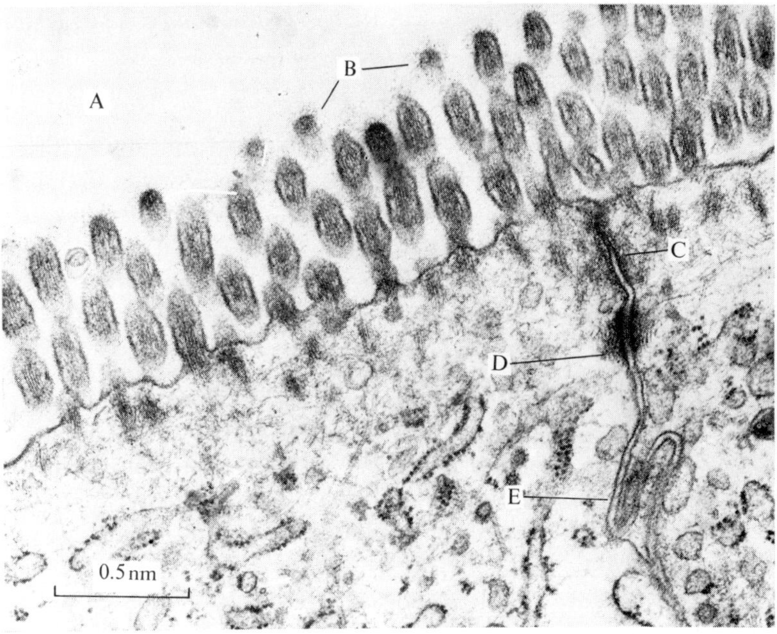

Fig. 5. An electron micrograph of part of a cell from the intestine of a piglet. A, Lumen; B, glycocalyx; C, zonula adherens; D, desmosome; E, intercellular space.

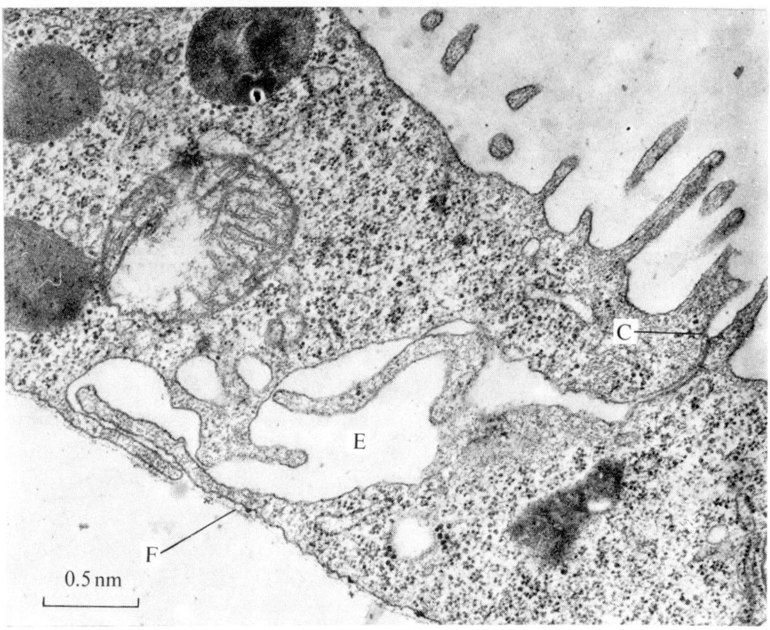

Fig. 6. An electron micrograph of part of a cell from the trophoblast of a rabbit blastocyst. C, zonula adherens; E, intercellular space; F, basement lamina.

inside. Consequently it may not be enough to demonstrate that trophoblast cells have a particular antigen, but it may also be important to show on which surface of the trophoblasts the antigen is expressed.

In general terms there are two types of immunological reaction; those which involve cells and those which involve free antibody and other serum factors. I would like to discuss these two types of reactions for antigens in a fluid mosaic membrane structure. As I have described antibody reacting with cells can lead to spot and cap formation and eventually removal of the antigen from the cell surface but this is not always the case. Certain antisera will not form spots or caps, for example anti-θ serum or some anti-H-2 or anti-HL-A sera (de Petris & Raff, 1974). The requirements for spot formation on a membrane surface are the same as those for lattice formation in solution. For example, if the antigenic molecules possess a single antibody binding site then aggregation will not occur. The effects of aggregation not only accelerate loss of specific surface antigen structures but also should increase the ease of activation via C1q of the complement system, particularly for IgG antibodies. It should be remembered that C1q is itself a polyvalent reagent which reacts with immunoglobulin and in some cases it is possible that it might, itself, cause formation of spots under conditions where antibody alone has failed to do so.

In the case of cellular reactions with surface antigens the fluid mosaic model allows the distribution of receptor and reactant to adjust themselves to give maximum fit. This is true for both direct interaction of surface receptors with antigen and recognition by cells of bound antibody as is seen in antibody-dependent cell mediated lysis. In both cases the close proximity of antigen sites is not essential as the large polyvalent surface of the reacting cell will adjust the distribution of receptors to maximise the binding interaction. This may well explain the extraordinarily high efficiency of antibody when used in antibody-dependent cell mediated lysis.

Finally I should stress that this is a very brief and incomplete review of this subject and that I am indebted to Dr Munn for Figs. 5 and 6 and much helpful discussion.

REFERENCES

AOKI, T., HAMMERLING, U., DEHARVEN, E., BOYSE, E. A. & OLD, L. J. (1969). Antigenic structure of cell surfaces. An immunoferritin study of the occurrence and topography of H-2, θ, and TL alloantigens on mouse cells. *J. exp. Med.*, **130**, 979.

BECKER, K. E., ISHIZAKA, T., METZGER, H., ISHIZAKA, K. & GRIMLEY, P. M. (1973). Surface IgE on human basophils during histamine release. *J. exp. Med.*, **138**, 394.

BOYSE, E. A., OLD, L. J. & STOCKERT, E. (1968). An approach to the mapping of antigens on the cell surface. *Proc. natn. Acad. Sci. USA*, **60**, 886.

BOYSE, E. A., STOCKERT, E., & OLD, L. J. (1967). Modification of the antigenic structure of the cell membrane by thymus-leukemia (TL) antibody. *Proc. natn. Acad. Sci., USA*, **58**, 954.

DE PETRIS, S. & RAFF, M. C. (1972). Distribution of immunoglobulin on the surface of mouse lymphoid cells as determined by immunoferritin electron microscopy. Antibody-induced temperature-dependent redistribution and its implication of membrane structure. *Eur. J. Immunol.*, **2**, 523.

DE PETRIS, S. & RAFF, M. C. (1973). Normal distribution patching and capping of lymphocyte surface immunoglobulin studied by electron microscopy. *Nature New Biol.*, **241**, 257.

DE PETRIS, S. & RAFF, M. C. (1974). Ultrastructural distribution and redistribution of alloantigens and concanavalin A receptors on the surface of mouse lymphocytes. *Eur. J. Immunol.*, **4**, 130.

DAVIS, W. C. (1972). H-2 antigen on cell membranes and explanation for alteration of distribution by indirect labeling techniques. *Science*, **175**, 1006.

EDIDIN, M. & WEISS, A. (1972). Antigen cap formation in cultured fibroblasts: A reflection of membrane fluidity and of cell motility. *Proc. natn. Acad. Sci., USA*, **69**, 2456.

FERRARINI, M., MUNRO, A. & WILSON, ANNE B. (1973). Cytophilic antibody: correlation of its distribution with activation of basophils and macrophages. *Eur. J. Immunol.*, **3**, 364.

KOURILSKY, F. M., SILVESTRE, D., NEAUPORT-SAUTES, C., LOOSFELT, Y. & DAUSSET, J. (1972). Antibody-induced redistribution of HLA antigens at the cell surface. *Eur. J. Immunol.*, **2**, 249.

LOOR, F., FORNI, L. & PERNIS, B. (1972). The dynamic state of the lymphocyte membrane. Factor affecting the distribution and turnover of surface immunoglobulins. *Eur. J. Immunol.*, **2**, 203.

NEAUPORT-SAUTES, C., LILLEY, F., SILVESTRE, D. & KOURILSKI, F. M. (1973). Independence of H-2K and H-2D antigenic determinants on the surface of mouse lymphocytes. *J. exp. Med.*, **137**, 511.

PERNIS, B., FORNI, L. & AMANTE, L. (1970). Immunoglobulin spots on the surface of rabbit lymphocytes. *J. exp. Med.*, **132**, 1001.

PINTO DE SILVA, P. (1972). Translational mobility of the membrane intercalated particles of human erythrocyte ghosts. pH-dependent reversible aggregation. *Cell Biol.*, **53**, 1972.

POULIK, M. D., BERNOCO, M., BERNOCO, P. & CEPPELLINI, D. B. R. (1973). Aggregation of HLA antigens at the lymphocyte surface induced by antiserum to β2-microglobulin. *Science*, **182**, 1352.

PREUD'HOMME, J. L., NEAUPORT-SAUTES, C., PIAT, S., SILVESTRE, D. & KOURILSKY, F. M. (1972). Independence of HLA antigens and immunoglobulin determinants on the surface of human lymphoid cells. *Eur. J. Immunol*, **2**, 297.

SINGER, S. J. & NICHOLSON, G. L. (1972). The fluid mosaic model of the structure of cell membranes. *Science*, **175**, 720.

TAYLOR, R. B., DUFFUS, W. P. H., RAFF, M. C. & DE PETRIS, S. (1971). Redistribution and pinocytosis of lymphocyte surface immunoglobulin molecules induced by anti-immunoglobulin antibody. *Nature New Biol.*, **232**, 223.

DISCUSSION

ALLISON: Some questions have arisen in the last two years about the fluid mosaic model. One problem concerns the tubules which apparently link together some of the proteins in the plasma membrane. Work from various laboratories has shown that sometimes different antigenic constituents such as H-2 are clustered, even when treated with univalent reagents. Low concentrations of colchicine or vinblastine, or lowering of the environmental temperature can change quite markedly the mobility of con A receptors in lymphoid cells (Edelman, Yahara & Wang, 1973). Berlin *et al.* (1974) have studied proteins which mediate the transfer of purines and amino acids across membranes; in normal macrophages these proteins are not internalised when phagocytosis occurs, but they are internalised after colchicine treatment.

The second important point is the occurrence of interactions between the intramembranous proteins themselves, for example in the urinary bladder (Warren & Hicks, 1970), where great plaques of material occur on one pole. These plaques can evidently interact with each other, and under these conditions the fluid mosaic theory does not explain all the observations.

MUNRO: I agree with these comments. I was attempting to start from the simple mosaic theory and show how the concepts had developed.

JOHNSON: Is Tony Allison suggesting that patching and capping are artificial and do not occur *in vivo*, but only *in vitro* because excessive levels of reagent are applied? Edelman showed these phenomena to be dose-dependent, using the con A system.

ALLISON: My point is that the fluid mosaic model, which has become recognised during the past two years is probably not the total explanation of the order of membrane constituents. The basic problem is whether any kind of mosaic structure is predetermined or completely free and has a mobility independent of all the constituents which are floating around. The recent evidence suggests that other factors play a part in organising these units; interactions occur between similar or identical antigenic units themselves so that they are naturally clustered and can move together, and relationships exist with cytoplasmic elements such as microfilaments and microtubules which link these units together in some fashion from the cytoplasmic side of the membrane.

JOHNSON: Does this not imply that specific recognition sites exist on the inside of the membrane equivalent to the specificity of antibody on the outside? Is there any evidence at all, for example, for specific recognition sites on the subsurface microfilaments that recognise, say, H-2D and link all the H-2D moieties in the membrane?

ALLISON: No evidence exists at present. Moieties tend to move together and the work of Boyse and others (1968) may be explained in terms of linked units. The wider use of proper univalent reagents will probably show that two antigenic sites move together.

MUNRO: I have an open mind on this point at the moment. Some structures will associate on the membrane, and the use of monospecific reagents made in coisogenic combinations is critical. Were Boyse, Old & Stockert using such reagents? If not, their results could be explained by the presence of mixtures of antibodies.

DAVIES: Some antisera were, and some were not, made in congenic strains. I had samples of all those antisera from Boyse and there was certainly no indication that they were contaminated.

MUNRO: Yet there is still a doubt that they were using monovalent antisera.

BODMER: I question the complete randomness in the fluid mosaic model, even though this point is slightly irrelevant for the model itself. There are various ways in which non-randomness could arise, including the site of synthesis and the factors influencing the mobility of the cell surface. Many factors could influence the relative positions of molecules, and there is probably no need to emphasise randomness in the extreme sense of the word. Relative mobility would be more acceptable.

FAULK: Some understanding of the presentation of these con A receptors in cell membranes has come about as a result of the work of Yahara & Edelman (1973) mentioned by Allison, and I do not believe these results to be at all artefactual. They reported that incubation of spleen cells with con A at 37 °C resulted in cap formation in only a few of the cells and inhibited cap and patch formation induced by anti-immunoglobulin sera. On the other hand, incubation of cells with con A at 4 °C followed by removal of unbound con A and elevation of the temperature to 37 °C resulted in cap formation in 40% of the cells. They concluded that these effects involve cross-linkage of con A and receptors which occur in two states – mobile and relatively mobile. This suggests to me that aspects of the mosaic structure are organised, and, because of their temperature sensitivity, that microtubules probably participate in this organisation.

MUNRO: What is the molecular heterogeneity of the material in the cell surface which reacts with con A?

ALLISON: There is a lot of heterogeneity in the sense that both glycoprotein and glycolipid are involved.

BODMER: Con A must attach to all sorts of molecules with the appropriate sugars on them, so the results with it would be quite confusing, and especially if it also attaches to glycolipid as well as to glycoprotein.

ALLISON: The redistribution does happen with other membrane proteins as well, such as specific transport sites which are internalised during phagocytosis (Berlin, Oliver, Ukena & Yin 1974).

COOMBS: But why make it so complicated? This result is surely expected with an excess of con A; it would be impossible to produce a lattice. If con A is washed away just the right amount remains to make the links – hence there is no reason to complicate the picture.

ALLISON: That is only part of the story, but other factors are involved as shown by the effects of temperature differential where materials that do not cap will actually do so after exposure to colchicine.

ADINOLFI: Is there any correlation between antigen levels on the cell surface during the cell cycle, or any evidence of a change in distribution?

MUNRO: I do not know of any measurments on the viscosity of cell membranes at different stages of the cell cycle.

EDWARDS: Would the movement of complexes in the membrane relate to the models of cell movement in which changes are found in the tension on the surface of cells? Changes in tension are also well known to occur during the mitotic cycle.

MUNRO: We would really wish to look at the viscosity of the lipid matrix.

COOMBS: Franks (unpublished work) has carried out measurements on synchronised cells, and found no correlation between antigen levels and the cell cycle, although this, of course, is only looking at the frequency of the antigen and not at the rate of spot formation, or rate of aggregation.

DAVIES: Sensitivity to antibody varies at different stages of the cell cycle.

MUNRO: Yes, but this could well be related to the amount of antigenic determinant, and certain antigens are known to vary in amount with the cell cycle.

DAVIES: The Singer model was constructed mainly with red cells in mind. Red cells are a very aberrant type of cell, and very different from a lymphocyte. Is there a symmetrical array of antigens on red cells?

COOMBS: The siting of Rh and ABO antigens on red cells was investigated using ferritin-labelled antibodies, but before the capping was found.

HOWE: Which factors are important for antigen mobility – do identical determinants patch and cap similarly on two different types of cells? Are there any antigens which can be capped on lymphocytes but not on fibroblasts for example?

JOHNSON: Related to this question, are there any cells which never patch?

COOMBS: It must to some extent be a property of the cell because similar antigens exist on many different types of cell. Some are present, for example the A antigen, on most cells of the body, but their reactivity on lymphocytes differs from that on epithelial cells.

ALLISON: But it does cap on all cells. On fibroblasts, H-2 caps at a particular site which is near the nucleus, which is quite different from what occurs on lymphocytes. Capping occurs so quickly – within minutes – that indications of speed are difficult to obtain.

MUNRO: Care is needed to distinguish between capping and spotting. The question was really concerned with aggregation.

ALLISON: Patching happens so quickly that it cannot be measured accurately. Edidin has made measurements of proteins in membranes on muscle cells, for example, and believes they move at about the expected rate for a viscous fluid, so nothing special occurs – the proteins first diffuse within the membrane.

MUNRO: Measurements by Frye & Edidin (1970) of diffusion rates in heterokaryon cells, and by Cone (1972) of the rotation of rhodopsin in the membrane from frogs' eyes, have given values for the viscosity of the membrane which are remarkably similar, from 1–10 and 0.7–6 poise respectively. Obviously, the molecular weight of the determinant being moved will also affect the rate of diffusion, but examining the diffusion equation and the possible range of viscosity and molecular weight should show which has the bigger effect. I would like to know of any work of this type on secretory epithelial cells or on trophoblast surfaces.

DAVIES: What is a molecule in terms of capping phenomena? If entities 'co-cap' they are said to be on the same molecule, but on the other hand pieces of membrane can be extracted in a macromolecular form carrying, for instance, H-2D, H-2K, θ and LY. The problem is to carry out these analyses in relation to gene products. Beta-2 microglobulin may or may not be an integral part of the H-complex. It can be dissociated from purified H-antigens, but its genetics are still unknown.

BILLINGTON: Trophoblast cells are highly secretory, producing steroid and protein hormones, and, so far as I know, most of the work on capping has been done on non-secretory cells.

MUNRO: Cells secreting antibody will form spots and caps quite happily.

FAULK: I would like to make a comment about a non-lymphoid cell. We have examined the movement of collagen in fibroblast membranes using anti-collagen antibody. When these reactions are done at 37 °C, the cells cap, but not at 4 °C. Capping at 37 °C is inhibited by cytochalasin B and less efficiently by vinblastin. These cells are also capped by con A and anti-con A as well as by PHA and anti-PHA, and as well as by anti-H-2-specific antisera (Faulk *et al.*, unpublished).

EDWARDS: Would a similar model work with receptors for gonadotrophins found on many steroid-secreting cells?

FAULK: I do not know the answer about gonadotrophin receptors. There has been no information on the trophoblast cell, as far as I know (see Discussion after Johnson's paper, p. 101).

ALLISON: The concept of antigenic modulation could be interesting in relation to trophoblast antigens. Moderate amounts of maternal antibodies might cause the antigens to disappear by modulation.

REFERENCES

BERLIN, R. D., OLIVER, J. M., UKENA, T. E. & YIN, H. H. (1974). Control of cell surface topography, *Nature, Lond.*, **247**, 45.

BOYSE, E. A., OLD, L. J. & STOCKERT, E. (1968). An approach to the mapping of antigens on the cell surface. *Proc. natn. Acad. Sci., USA*, **60**, 886.

CONE, R. A. (1972) Rotational diffusion of rhodopsin in the visual receptor membrane. *Nature New Biol.*, **236**, 39.

EDELMAN, G. M., YAHARA, I. & WANG, J. L. (1973). Receptor mobility and receptor–cytoplasmic interaction in lymphocytes. *Proc. natn. Acad. Sci. USA*, **70**, 1442.

FRYE, L. D. & EDIDIN, M. (1970). The rapid intermixing of cell surface antigens after formation of mouse–human heterokaryons. *J. Cell Sci.*, **7**, 319.

WARREN, R. C. & HICKS, R. M. (1970). Structure of the subunits in the thick luminal membrane of rat urinary bladder. *Nature, Lond.*, **227**, 280.

YAHARA, I. & EDELMAN, G. M. (1973). The effect of concanavalin A on the mobility of lymphocyte surface receptors. *Exptl. Cell Res.*, **81**, 143.

ANTIGENS SHARED BY TUMOUR CELLS AND FOETAL OR GONADAL CELLS

BY ANTHONY ALLISON

The concept that antigens are shared by foetal and malignant cells apparently dates from the observations in 1906 of G. Schöne, a pupil of Ehrlich, that mice which had been injected with foetal tissue became resistant to transplants of tumours that grew in and killed control mice or mice injected with adult tissues. The distinguished Polish immunologist, L. Hirszfeld (1929), discussed the presence on tumours of antigens cross-reacting with those of embryos. The modern period of research on this subject began with the discovery of what was subsequently termed α-foetoprotein in mice bearing liver tumours by the Russian scientist Abelev in 1963 and of the carcinoembryonic antigen by the Canadians Gold and Freedman in 1965.

During the past decade much other information has accumulated about the antigens shared by tumour cells and embryonic cells. Some reports have also appeared about antigens common to gonadal cells and tumour cells. These results are of theoretical interest and some are also of practical value in the diagnosis and treatment of human cancers.

α-foetoprotein (AFP)

Abelev *et al.* (1963) raised antibodies against the serum proteins of newborn mice and absorbed them with sera of adult mice. Antibodies in the absorbed sera reacted with a component of newborn mouse serum and also with sera of adult mice carrying transplantable hepatomas but not other tumours. The foetal serum protein defined by the absorbed antiserum had the electrophoretic mobility of an α-globulin. Tatarinov (1965) made similar observations with human sera. He found that antisera reacting with a component of human foetal serum also gave precipitation with the sera of two adults bearing primary carcinoma of the liver. Later studies by Gitlin & Boesman (1967) and others showed the existence in early mammalian foetal sera of a major component distinct from fetuin and present also in the sera of animals carrying hepatomas or teratomas. The protein was termed α-foetoprotein.

Various methods have been used to detect AFP. Double immunodiffusion, which has been widely used, can detect about 1 μg/ml whereas 0.25 μg/ml can be detected by radioimmunoassay (Ruoslahti & Seppälä 1971). With the use of sensitive methods of assay some differences formerly thought to be qualitative have proved to be quantitative. Thus, small amounts of AFP persist in normal adult serum.

In human foetuses the maximum concentrations of AFP are about 3–4 mg/ml serum, attained about the thirteenth week of intra-uterine development, falling thereafter to low levels in umbilical cord serum. In pregnant women levels of AFP increase to 100–500 mg/ml during the third trimester (Seppälä & Ruoslahti, 1972), dropping shortly before term to normal.

AFP is formed mainly by the rodent and human foetal liver, but also by the yolk sac and gastro-intestinal tract (Gitlin, 1971). Other foetal tissues, including kidney, lung and spleen, do not synthesise detectable amounts of AFP. Fluorescein-labelled antibodies show that by the sixth week of foetal life most liver cells are synthesising AFP, but as the levels in serum fall the proportion of hepatocytes involved in the synthesis falls, until mainly those around the central vein are involved (Engelhardt *et al*., 1969). Less than 20% of cells in hepatomas synthesise AFP (Nishioka *et al*., 1972). In regenerating liver synthesis of AFP is increased in a minority of cells around blood vessels (Engelhardt *et al*., 1969).

Human AFP is a protein of molecular weight 64 000 daltons (Nishi, 1970) with an amino acid composition like that of albumin and about 4% carbohydrate (Ruoslahti, Seppälä, Phiko & Vuopio, 1971). Whether AFP has any physiological role is uncertain; like albumin AFP can bind oestrogens, but AFP does not bind testosterone, which is bound to albumin (Uriel, de Nechaud & Dupiers, 1972).

AFP appears also in liver diseases associated with inflammation and/or regeneration, especially in infants (Abelev *et al*., 1971). In experimental animals acute poisoning with carbon tetrachloride leads to a rise in AFP levels two to three days later.

Using the immunodiffusion assay, the test for AFP is specific for hepatocellular cancer and teratocarcinoma; patients with seminoma have AFP levels in the normal range. Tumours of other origins, including pancreatic, gastric and prostatic carcinomas, have raised levels of AFP when they metastasise to the liver. The incidence of positive AFP tests in patients with primary hepatocellular carcinoma is much higher in patients between 10 and 30 years (many of whom are indigenous Africans) than in subjects over 40 years. Even with the most sensitive methods now available, some 5–10% of patients with hepatocellular carcinoma do not show detectable increases in AFP. A high percentage of monkeys treated with *N*-nitrosodiethylamine developed hepatomas and most tumours produced AFP (Hull *et al*., 1969).

The question arises whether maternal antibodies against AFP are ever formed and what their effects might be. Maternal antibodies have not yet been reported, but they might be difficult to detect in the presence of an excess of antigen. Slade (1973) showed that injections of sheep antibodies against rabbit AFP into pregnant rabbits result in rapid death of the majority of foetuses, within 26 hours. The yolk sac was most obviously damaged. Since this is the primary site of AFP synthesis in the rabbit, as in other species investigated, and the yolk sac also plays an important part in the transmission of maternal antibodies to the foetus, it is not surprising that it constitutes a major target for antibodies against AFP. Possibly the antibodies react with AFP associated with the plasma membranes of yolk sac epithelial cells. Antibodies against yolk sac are known to produce congenital malformations and foetal mortality in pregnant rats (Leung, Brent & Koszalka, 1972) and this reaction is directed mainly against the visceral yolk sac. Conceivably some congenital malformations and abortions may be due to maternal immunisation against AFP or some other yolk-sac product.

Carcinoembryonic antigen (CEA)

Gold & Freedman (1965) immunised rabbits with pooled saline extracts of human colonic adenocarcinomas and absorbed the sera obtained with saline extracts of normal mucosa taken at operation from an adjacent segment of the colon. After further absorption with human plasma, fibrin and killed gut bacteria, the antisera gave a single line of precipitation in double diffusion tests. The antigen so defined was found in primary tumours of the alimentary tract, from the oesophagus to the rectum, including the pancreas, and also in neoplasms metastasing from these sites (Gold & Freedman, 1965). Negative reactions were reported in other tumours and adult mucous membrane, but the antigen was present in the alimentary tract, pancreas and liver of human foetuses aged between two to six months. The antigen common to tumours and the foetal alimentary tract was termed carcinoembryonic antigen (CEA).

The presence of an antigen common to tumours and embryonic gastrointestinal tract has been repeatedly confirmed. CEA has also been found in some colonic polyps, as well as some inflammatory lesions of the gastrointestinal and colonic regions (Burtin, Martin, Sabine & von Kleist, 1972). The development of radioimmunoassay (Thomson, Krupey, Freedman & Gold, 1969) has greatly increased the sensitivity of detection of CEA, which has been found in the sera of normal adults and also in their urine (Hall *et al.*, 1972).

CEA is a water-soluble glycoprotein, consisting mainly of carbohydrate residues, including galactose, mannose, glucose, fucose, glucosamine and a variable amount of sialic acid (Krupey *et al.*, 1968). The presence of mannose and absence of *N*-acetylgalactosamine, as well as the amino acid composition, distinguish CEA from H blood group substance (Krupey, Gold & Freedman, 1968; Laurence & Neville, 1972). CEA may well be heterogeneous (Laurence & Neville, 1972).

Immunofluorescence studies show CEA to be localised in the glycocalyx of the plasma cell membranes of colonic tumour cells, in particular that part of the membrane bordering on the neoplastic acini (von Kleist & Burtin, 1969). Loss of normal membrane antigens often accompanies the appearance of CEA (Burtin, von Kleist & Sabine, 1971). The carbohydrate in the glycocalyx of cells is relatively rapidly turned over, with a half-life of about one day (Kraemer, 1966) and if this is also true of CEA it would explain its presence in the plasma. The carbohydrate composition of CEA suggests that it may be related to the core unit described by Spiro (1970) as a feature of several glycoproteins linked to asparagine. The absence of *N*-acetylgalactosamine suggests a lack of relationship with serine-linked glycoproteins which include the blood-group glycoproteins.

The mechanism underlying the formation of CEA is not yet understood. The synthesis of glycoproteins and glycolipids depends on the sequential action of a series of glycosyltransferases. According to Hakomori & Murakami (1968), tumour cells often fail to complete the synthesis of their glycolipid antigens, and the same may be true of glycoproteins. Alternatively, normal antigens could be modified by additional glycosyltransferase activity, adding sugars which cover those normally present, or specific enzymes may cleave residues from the normal products.

Simmons & Perlmann (1973) have suggested that CEA is an incomplete blood-group substance that lacks the type 1 chain in most of the material studied. Since A-, B-, H- and Le^a-specific sequences in the type 2 chain were also missing, they supposed that gastrointestinal adenocarcinomas may have defects localised at the ABO, H and Le loci. In view of the biochemical findings further evidence in support of this view is required.

The clinical significance of tests for CEA has been reviewed by Laurence & Neville (1972). Patients with high levels of CEA ($> 40\ \eta$g/ml) nearly always have malignant disease; however this detects only less than 20 % of patients with mammary, colonic and bronchial carcinomas. Determination of CEA levels may be useful in assessing whether tumours have been successfully treated, and also in the diagnosis of bladder tumours.

Mouse tissue cell antigens which may be the analogue of CEA were described by Boyle, Davies & Haughton (1963). Rabbit antisera against mouse ascites tumour cells were absorbed with normal mouse serum and used to detect cell-associated antigens by immunodiffusion. One of the antigens was detected in tumour and embryonic cells but not in several cell types from adult mice investigated.

α2H-foetoprotein (α2HF)

Much less information is available about this protein, which is present in foetal liver and in serum up to the second month of postnatal life

(Buffé *et al.*, 1970). It is a 17 S macroglobulin containing iron, which is detectable by radioimmunodiffusion in the serum of a proportion of children and a lower proportion of adults with tumours of diverse origins, including nephroblastoma, leukemia, lymphoma and brain tumours. Raised levels of α2HF are rare in children without malignant disease but are seen in occasional adults with inflammatory diseases and even apparently healthy persons. The concentration of this component in foetal serum is about 1.5 μg/ml.

βS-foetoprotein

Takahashi, Yashi, Anzai & Wada (1967) described another component of human foetal serum, liver and other tissues which disappears from the body fluids five to seven months after birth. It appears to be a glycoprotein of relatively high molecular weight (> 200 000) present in the cytoplasm of foetal hepatocytes. Raised levels of βS-foetoprotein are found by double immunodiffusion in the sera of a minority of patients with hepatocarcinoma, cholangiocarcinoma, gastric carcinoma, leukaemia and lymphoma (Wada *et al.*, 1971).

Foetal sulphoglycoprotein antigen (FSA)

Häkkinen, Jarvi & Gironroos (1968) reported that three sulphoglycoproteins appear in the gastrointestinal tract of human foetuses about seven weeks old. One is a foetal sulphoglycoprotein antigen (FSA) which is found in the superficial epithelial cells of the stomach in foetuses and babies during the first nine months of postnatal life. Häkkinen & Viikari (1969) found FSA in the gastric juice of 96% of patients with gastric carcinomas. Immunofluorescence showed FSA in tumour cells and the adjacent mucosal epithelial cells. It was found in 14% of patients with peptic ulceration and in some patients with cancers of other sites.

Carcino – foetal glial antigen

Trouillas (1971) combined a study of the antigenicity of brain tumours (gliomas) with immunotherapy of tumour-bearing patients. The patients were inoculated with saline extracts of their own tumour together with Freund's complete adjuvant. The patients became hypersensitive to challenge with tumour extracts and produced precipitating and cytotoxic antibodies reacting with tumour extracts or cells. Using these precipitins a common antigen in the astrocytoma–glioblastoma group of tumours was demonstrated. The α-lipoprotein antigen was specific to these tumours but was found also in foetal brain between the second and eight to ninth month of gestation.

Common antigens in foetal and tumour cells in experimental animals

Several approaches have been taken to the problem of demonstrating antigens common to foetal tissues and tumours. Stonehill & Bendich (1970) used extracts of whole mouse embryos (9–19 day gestation) to immunise rabbits. After absorption with adult tissues the sera still reacted in double immunodiffusion tests with embryonic tissues, mouse tumours and adult skin. The 72 tumours used had arisen spontaneously or been induced by leukemogenic viruses, chemicals or X-radiation.

Klavins, Mesa-Tejada & Weiss (1971) reported that absorbed antisera raised against human 6–7-week old embryos, cross reacted with extracts of human lung, breast, colonic, hepatic, renal, bronchial and skin carcinomas. The only normal adult tissue showing a reaction was epidermis, which contains a layer of rapidly dividing cells. The foetal and tumour antigens showed reactions of identity and had similar gel filtration properties, suggesting a molecular weight of the order of 66 000–68 000 daltons.

Pearson & Freeman (1968) immunised adult hamsters and rabbits with hamster tumour cells transformed by polyoma virus. The tumours had an antigen present also in normal hamster embryonic cells but not in adult hamster or embryonic mouse cells. Spleen cells from mature hamsters immunised with normal hamster embryonic cells were cytotoxic for the polyoma tumour cells. Ting (1968) inoculated mice with syngeneic embryonic tissues but failed to induce resistance against transplantation of syngeneic polyomavirus-induced tumour cells. Blair (1970) found that virus-induced mammary tumours developed significantly earlier in female mice receiving implants of embryonic tissue than in those implanted with mammary gland tissue or injected with saline. Conceivably some type of enhancement due to an antigen common to the tumours and embryonic tissues was involved.

Coggin, Ambrose & Anderson (1970) reported that hamster and mouse foetal cells contain an antigen cross-reactive with SV40-induced tumour cells. When 10-day, but not 14-day, irradiated foetal cells were used to immunise adult hamsters, sera were obtained which prevented the multiplication of SV40 tumour cells in diffusion chambers. The immunised adults showed also a slightly increased resistance to transplants of tumour cells. Males responded to immunisation with foetal cells better than females, but the latter developed some cytotoxic antibodies during pregnancy.

Reactions of cells and sera from pregnant animals with tumour cells

Brawn (1970) reported that lymph node cells from pregnant BALB/c mice multiparous by BALB/c males inhibited colony formation *in vitro* by several lines of primary syngeneic sarcomas induced by 3-methylcholanthrene in female mice. Lymph node cells from virgin BALB/c females of comparable age had no such effect. Colony formation by normal female BALB/c or female BALB/cxC3H fibroblastic cells was not inhibited. These results were taken as evidence for immunisation of the mother during the course of syngeneic pregnancies by embryonic antigens shared with syngeneic tumour cells. These results have recently been confirmed by the Hellströms and their colleagues, who have shown that the *in vitro* cytotoxicity is due, at least in part, to cells sensitive to anti-θ serum, presumably sensitised T lymphocytes (Hellström, 1974 personal communication).

Duff & Rapp (1970) found that sera from pregnant hamsters contain antibodies which react with an antigen on the surface of hamster cells transformed by SV40, as shown by immunofluorescence. They concluded that the surface antigen on SV40-transformed hamster cells may be immunologically related to an antigen found during embryonic development.

Baldwin and his colleagues (Baldwin, Glaves & Vose, 1974; Baldwin, Glaves, Pimm & Vose, 1972) have shown the presence in transplantable 3-methylcholanthrene-induced sarcomas and 4-dimethylaminoazobenzene(DAB)-induced hepatomas in rats of antigens which react with the sera of pregnant syngeneic rats. It was possible to differentiate between these antigens and the tumour-rejection antigens by their specificities. Moreover, tumours which were incapable of eliciting tumour-rejection responses in syngeneic hosts (mammary carcinomas induced by 2-acetylaminofluorene or spontaneously arising mammary carcinomas or sarcomas) had surface antigens as demonstrable by cytotoxicity tests using sera from pregnant rats in the presence of complement.

Effect of anti-tumour immunity on pregnancy in the mouse

Parmiani & Della Porta (1973) immunised female BALB/c mice by injections of sarcoma cells and excision of the developing tumours. Anti-tumour immunity was checked by demonstrating resistance to inocula of sarcoma cells that could induce tumours in unimmunised mice. Other BALB/c females were injected with mitomycin-treated minces of 14-day old syngeneic embryos, containing 4×10^6 viable cells, or comparable injections of syngeneic adult cells. Anti-tumour immunity significantly reduced the rate of pregnancy from 96% in control mice to 62% during the first pregnancy and from 92% to 50% during the second. Stronger inhibition was observed in the females immunised with embryonic tissues, in which the successful matings were as low as 39% in the first pregnancy.

However, this inhibition was transient and had almost disappeared in the second pregnancy (89% rate). After both immunisations there was a reduction of litter size at the borderline of statistical significance. Immunisation with syngeneic or allogeneic adult tissues had no demonstrable effect on subsequent pregnancies. These results were taken as indirect evidence that embryonic antigens exist on chemically-induced tumour cells. Lowering the rate of pregnancy and decreasing the litter size by an anti-embryo vaccine was attributed to an immune rejection of the conceptus even when only differentiation antigens are involved. This is an interesting idea. However, it must be pointed out that non-specific effects can reduce pregnancy rates, and the recovery by the mothers between the first and second matings of virtually normal litter rates is inconsistent with the view that they are immunised against embryonic cells in such a way as to jeopardise pregnancy. Immunisation would be expected to have a longer-lasting effect. The reductions of litter size were only marginal and of doubtful significance. It would therefore be interesting to repeat the work of Parmiani & Della Porta to see whether comparable results can be obtained in other strains of mice and with other tumours.

An antigen common to mouse eggs and tumour cells

Moskalewski & Koprowski (1972) described an egg antigen which is detected in mouse eggs by their susceptibility to lysis with guinea-pig antisera. The latter are obtained by immunisation of guinea-pigs with unfertilised mouse eggs from which the zona pellucida had been removed by digestion with pronase. The egg antigen is present in small oocytes, mature eggs and in post-implantation embryos throughout the late blastula stage. The presence of the zona pellucida greatly decreases sensitivity to the antibody. No antigen could be detected in post-implantation embryos, somatic cells or spermatozoa. Baranska, Koldovsky & Koprowski (1970) reported that guinea-pig antisera against the egg antigen were cytotoxic for SV40-transformed 3T3-cells but not for the corresponding untransformed cells. The reaction was not due to H-2 antigens, as shown by suitable absorption. Further studies were reported by Koprowski, Sawicki & Koldovsky (1971). Antisera against virus-transformed mouse cells were cytotoxic for mouse eggs and this reactivity could not be removed by absorption with syngeneic adult mouse cells. Mouse cells transformed by polyoma virus did not share the antigen common to eggs and SV40-virus-transformed cells.

Antigens common to teratocarcinomas, sperm cells and morulae

Artzt, Bennett & Jacob (1974) have described an antigen common to mouse primitive teratocarcinoma (PTC) cells, sperm and morulae. This antigen was shown by serological methods to be specified by a wild-type T locus gene, the mutant form of which is lethal in homozygotes at the morula stage. The cultured PTC cells elicit, after hyperimmunisation of syngeneic male (-129/Su-CP) mice, the formation of specific antibodies cytotoxic to various PTC lines, but not to a variety of differentiated cells which originated from the same tumour. These anti-PTC sera failed to react with all other mouse cells tested except mouse morula cells and male germ cells, especially mature sperm. The reaction with morula cells increased from barely perceptible at the one-cell stage to a maximum at the eight-cell stage, suggesting the progressive expression of a surface antigen or antigens.

A relationship to antigens determined by alleles of the T locus was sought because a series of surface antigens genetically determined by these alleles has been detected on sperm cells. Moreover, mutations at the T locus, which determine surface antigens of sperm, are known to interfere with early embryonic development in specific ways attributable to cell surface defects. The specific anti-PTC serum was found to be absorbed completely by sperm from animals with the appropriate T locus wild allele $+\,t^{12}$. Homozygotes for this allele show blocked development at the morula stage.

Besides being of interest in the context of a relationship between tumour-cell and embryonic-antigens, these results support the concept that genetically controlled differentiation antigens have essential functional roles during embryogenesis.

Carcinoplacental alkaline phosphatase

In experimental animals bearing tumours, isoenzymes characteristic of foetal or neonatal organs can appear, and there are relative increases in the concentrations of components present in the foetus. Examples are lactic dehydrogenase and aldolase (Leese, 1969). One such change, of special interest because it is reflected in the body fluids, is placental alkaline phosphatase. This is present in maternal serum during the third trimester of pregnancy, but does not occur in foetal tissues and serum. It is never found in normal males. Placental alkaline phosphatase is heat stable and inhibited by L-phenylalanine (Fishman *et al.*, 1968), which is not true of the other forms of the enzyme. Placental alkaline phosphatase was detected in the serum of a patient with bronchial carcinoma named Regan (whose name is sometimes used to designate the isoenzyme) and it has been found in the sera of a minority of patients with a wide variety of cancers. The isoenzyme is synthesised in the tumours and is indistinguishable biochemically and immunologically from the placental enzyme (Nathanson & Fishman, 1971). It is a useful guide to the progression or regression of tumours following treatment.

Ectopic hormone production

Many tumours produce hormones such as insulin and gonadotrophins (Castleman, Scully & McNeely, 1972) and human placental lactogen (Weintraub & Rosen, 1971). Such ectopic hormone production is most frequently found in carcinoma of the bronchus, but human placental lactogen has also been found in hepatic tumours, which can produce AFP as well. It is clear that the hormones are synthesised in the tumours themselves and some appear to be identical in chemical composition to the natural hormones.

Explanations for common antigens in tumour and embryonic cells

Two explanations have been offered for the presence of antigens common to tumour cells and embryonic cells. One depends on the concept, expressed by Conheim nearly a century ago, that tumour cells do not arise from fully differentiated cells but from embryonic 'cell nests' present in most tissues. Formulated in contemporary terms, this view would distinguish between fully differentiated cells which do not divide, such as neurons, muscle cells, blood leukocytes or plasma cells, and precursor cells which retain the capacity to divide. Only the latter could undergo malignant transformation, and it is postulated that they have certain antigens common to embryonic cells. In fact, various stages of differentiation can be observed in malignant cells. Thus, minimum-deviation hepatomas retain many of the biochemical functions characteristic of normal livers, plasma cell tumours can secrete apparently normal immunoglobulins and rhabdomyosarcomas have regularly ordered actomyosin filaments. But even in these cases the cells which are actually dividing may be relatively undifferentiated, with some of their progeny undergoing differentiation along particular lines.

The second explanation for the presence of embryonic antigens in tumour cells is that genes which are repressed in normal adult cells become de-repressed in the tumour cells. The synthesis of insulin and other ectopic hormones in tumours is quoted as an example of such de-repression. However, the de-repression does not appear to be random: large amounts of AFP are synthesised by tumours of the liver and teratocarcinomas, the carcinofoetal glial antigen is common in the central nervous system, not elsewhere, and teratocarcinomas have antigens present in spermatozoa and morulae. It is exceptional to find certain gene products, such as haemoglobin or myoglobin, synthesised by tumour cells. Thus the de-repression has some organ specificity. Perhaps as organs differentiate there is serial repression of blocks of genes, eventually leaving those appropriate for a particular organ, say foetal liver. In a final differentiation step a further block of genes, including that specifying AFP, is largely repressed, only a trace of function remaining in the adult organ (perhaps in a very small number of dividing hepatocytes). When the cells become malignant, the last stage of repression is most readily reversed. In the absence of evidence it is inappropriate to discuss various models by which this might occur. One possibility is that the last stage of repression involves control of transcription rather than translation. Evidence for transcriptional control in the hepatocyte and hepatoma cells has been presented by Tomkins and his colleages (Thompson, Tomkins & Curran, 1966). Transcriptional control is easily reversed (for example, by appropriate hormonal signals) and such reversal could readily occur in tumour cells. In addition to the common, organ-specific onco-foetal antigens, individual tumours might have other de-repressed functions, such as ectopic synthesis of polypeptide hormones. Indeed, the more closely tumour products are studied by sensitive methods such as radioimmunoassay, the more commonly they are found. The frequency with which polypeptide hormones and onco-foetal antigens have been reported in tumours may reflect the efficiency with which they have been sought rather than their proponderance over other products. Teratocarcinomas are interesting because they show biochemical as well as structural multipotentiality, producing in relatively large amounts AFP as well as antigens common to spermatozoa and morulae, and doubtless many other products.

Immunological implications of onco-foetal antigens

The main message for reproductive physiology that emerges from the study of onco-foetal antigens is that pregnancy is regularly associated with immunisation against certain foetal products. The sera and lymphocytes of pregnant animals, especially multiparous animals, consistently show a variety of reactions with tumour cells, and this is presumably due to immunisation against antigens common to tumour and foetal cells. The mechanism by which the immunisation occurs is not yet understood, but it is clear that the mothers cannot be tolerant to all foetal antigens, to many of which they were themselves exposed in embryonic life. Whether the mother normally retains tolerance to some foetal antigens, for example AFP, perhaps because of the presence of small amounts of this material in the blood of the adult, is also unknown. The presence of small amounts of maternal antibody in the presence of a relatively large amount of antigen, presumably mainly of foetal origin, would be difficult to detect. By analogy with other systems (Allison, 1974), it may be supposed that tolerance to AFP, as to other proteins circulating in low dose, may involve unresponsiveness of T lymphocytes to the antigen, while B lymphocytes for the antigen are present in normal numbers. The latter can be induced to respond immunologically and make autoantibody against AFP as a result of immunisation with a cross-reacting AFP from another mammalian species (Abelev, 1974). Whether animals with autoantibodies to AFP are resistant to the induction of liver tumours is at present under investigation. In view of the finding of Slade (1973) that sheep antisera to rabbit AFP have devastating effects on pregnancy, it would be interesting to know whether animals with autoantibody to AFP can maintain normal pregnancy. Indeed, any information on whether maternal immunisation against foetal products affects pregnancy would be interesting. This is why the studies of Parmiani & Della Porta (1973) on pregnancies in mice immunised against tumour foetal cells, would be worth repeating to determine their reproducibility and to eliminate trivial causes for the observed effects.

The use of an antibody against a teratocarcinoma to study the presence of products specified by genes at the T locus on morulae (Artzt *et al.*, 1974) and their relationship to mortality at the morula stage, illustrates the application of onco-foetal antigens in fundamental studies of reproductive physiology and differentiation. The onco-foetal antigens provide convenient tools which have only just come to be exploited in the laboratory. In the clinic they have already become established as useful indications of the presence and type of cancer and the response to treatment (Laurence & Neville, 1972).

REFERENCES

ABELEV, G. I. (1971). Evaluation of the use of highly sensitive methods for determination of the α-foetoprotein in the diagnosis of hepatocellular cancer and teratoblastoma. *Byull eksp. Biol. Med.*, **71**, 75.

ABELEV, G. I. (1974). Alpha-foetoprotein: biological and clinical aspects. *Bull. Cancer*, **12**, 1.

ABELEV, G. I., PEROVA, S. D., KHRAMKOVA, N. I. POSTINIKOVA, Z. A. & IRLIN, I. S. (1963). Production of embryonal α-globulin by transplantable mouse hepatomas. *Transplantation*, **1**, 174.

ALLISON, A. C. (1974). Interactions of T and B lymphocytes in self tolerance and autoimmunity. In *Immunological Tolerance* (ed. D. Katz and B. Benacerraf). Academic Press: New York (in press).

ARTZT, K., BENNETT, D. & JACOB, F. (1974). Primitive teratocarcinoma cells express a differentiation antigen specified by a gene at the T locus in the mouse. *Proc. natn. Acad. Sci., USA*, **71**, 811.

BALDWIN, R. W., GLAVES, D., PIMM, M. V. & VOSE, B. M. (1972). Tumour specific and embryonic antigen expression on chemically induced rat tumours. *Ann Inst. Pasteur*, **122**, 715.

BALDWIN, R. W., GLAVES, D. & VOSE, B. M. (1974). Differentiation between the embryonic and tumour specific antigens on chemically induced rat tumours. *Br. J. Cancer*, **29**, 1.

BARANSKA, W., KOLDOVSKY, P. & KOPROWSKI, H. (1970). Antigenic study of unfertilized mouse eggs: cross-reactivity with SV40-induced antigens. *Proc. natn. Acad. Sci., USA*, **67**, 193.

BLAIR, P. B. (1970). Search for cross-reacting antigenicity between mammary tumour virus-induced mammary tumours and embryonic antigens: effect of immunization on the development of spontaneous mammary tumours. *Cancer Res.*, **30**, 1199.

BOYLE, W., DAVIES, D. A. L. & HAUGHTON, G. (1963). Mouse tissue cell antigens. Some properties of cell-bound components. *Immunology*, **6**, 499.

BRAWN, R. J. (1970). Possible association of embryonal antigen(s) with several primary 3-methylcholanthrene-induced murine sarcomas. *Int. J. Cancer*, **6**, 245.

BUFFÉ, D., RIMBAULT, C., LEMERLE, J., SCHWEISGUTH, O. & BURTIN, P. (1970). Presence d'une ferroproteine d'origine tissulaire L'a$_2$H, dans le serum des enfants porteurs de tumeurs. *Int. J. Cancer*, **5**, 85.

BURTIN, P., von KLEIST, S. & SABINE C. (1971). Loss of normal colonic membrane antigen in human cancers of the colon. *Cancer Res.*, **31**, 1038.

BURTIN, P., MARTIN, E., SABINE, M. C. & VON KLEIST, S. (1972). Immunological study of polyps of the colon. *J. natn. Cancer Inst.*, **48**, 25.

CASTLEMAN, B., SCULLY, R. E., & MCNEELY, B. W. (1972). Case records of Massachusetts general hospital. Case 13–1972. *New Eng. J. Med.*, **286**, 713.

COGGIN, J. H., AMBROSE, K. R. & ANDERSON, N. G. (1970). Fetal antigen capable of inducing immunity against SV40 hamster tumour cells. *J. Immunol.* **105**, 524.

DUFF, R. & RAPP, F. (1970). Reaction of serum from pregnant hamsters with surface of cells transformed by SV40. *J. Immunol.*, **105**, 521.

ENGELHARDT, N. V., SHIPOVA, L. YA., GUSEV, A. I., YAZOVA, A. K. & TER-GRIGOROVA, E. N. (1969). Observations on the α_F-globulin in sections of human embryo liver and the liver of newborn mice by the fluorescent antibody method. *Byull. eksp. Biol. Med.*, **68**, 62.

FISHMAN, W. H., INGLIS, N. R., GREEN, S., ANTISS, C. L., GOSH, N. K., REIF, A. E., RUSTIGAN, R., KRANT, M. J. & STOLBACH, L. L. (1968). Immunology and biochemistry of the Regan isoenzyme of alkaline phosphatase in human cancer. *Nature, Lond.*, **219**, 697.

GITLIN, D. (1971). Sites of alpha-foetoprotein synthesis. *New England J. Med*, **285**, 1436.

GITLIN, D. & BOESMAN, M. (1967). Fetus specific proteins in several mammals and their relation to human α-fetoprotein. *Comp. Biochem. Physiol.*, **21**, 327.

GOLD, P. & FREEDMAN, S. O. (1965). Demonstration of tumour-specific antigens in human colonic carcinoma by immunological tolerance and absorption technique. *J. exp. Med.*, **121**, 439.

HÄKKINEN, I., JARVI, O. & GIRONROOS, J. (1968). Sulphoglycoprotein antigens in the human alimentary canal and gastric cancer. An immunohistochemical study. *Int. J. Cancer*, **3**, 572.

HÄKKINEN, I. & VIIKARI, S. (1969). Occurrences of fetal sulphoglycoprotein antigens in gastric juice of patients with gastric disease. *Ann. Surg.*, **169**, 277.

HAKOMORI, S. & MURAKAMI, W. T. (1968). Glycolipids of hamster fibroblasts and derived malignant transformed cell lines. *Proc. natn. Acad. Sci., USA*, **59**, 254.

HALL, R. R., LAURENCE, D. J. R., DARCY, D., STEVENS, U., JAMES, R., ROBERTS, S. & NEVILLE, A. M. (1972). Carcinoembryonic antigen in the urine of patients with urothelial carcinoma. *Brit. Med. J.*, **3**, 609.

HIRSZFELD, L. (1929). Untersuchungen über die serologischen, eigenschaften der gewebe: über serologische eigenschaten der neubildungen. *Z. Immun-Forsch, exp. Ther.*, **64**, 81.

HULL, E., CARBONE, P. GITLIN, D., O'GARA, R. & KELLEY, M. (1969). Alpha-fetoprotein in monkeys with hepatoma. *J. natn. Cancer Inst.*, **42**, 1035.

KLAVINS, J. V., MESA-TEJADA, R. & WEISS, M. (1971). Human carcinoma antigens cross-reacting with anti-embryonic antibodies. *Nature New Biol.*, **234**, 153.

KOPROWSKI, H., SAWICKI, W. & KOLDOVSKY, P. (1971). Immunologic cross-reactivity between antigen of unfertilized mouse eggs and mouse cells transformed by simian virus 40. *J. natn. Cancer Inst.*, **46**, 1317.

KRAEMER, P. M. (1966). Regeneration of sialic acid on the surface of Chinese hamster cells in culture. 1. General characteristics of the replacement process. *J. cell comp. Physiol.*, **68**, 85.

KRUPEY, J., GOLD., P. & FREEDMAN, S. O. (1968). Physiochemical studies of the carcinoembryonic antigen of the human digestive system. *J. exp. Med.*, **128**, 387.

LAURENCE, D. J. R. & NEVILLE, A. M. (1972). Foetal antigens and their role in diagnosis and clinical management of human neoplasms: A review. *Br. J. Cancer*, **26**, 335.

LEESE, C. L. (1969). Enzymes and Isoenzymes. *Homoeostatic Regulators* (eds. G. E. W. Wolstenholme & J. Knight), Ciba Foundation Symposium, p. 144.

LEUNG, C. C., BRENT, R. L. & KOSZALKA, T. D. (1972). The production of congenital malformations using tissue antisera. XI Antisera to acellular visceral yolk sac. *Teratology*, **6**, 317.

MOSKALEWSKI, S. & KOPROWSKI, H. (1972). Presence of egg antigen in mammalian oocytes and preimplantation embryos. *Nature, Lond.*, **238**, 168.

NATHANSON, L. & FISHMAN, W. H. (1971). New observations on the Regan isoenzyme of alkaline phophatase in cancer patients. *Cancer, N.Y.*, **27**, 1388.

NISHI, S. (1970). Isolation and characterisation of α-human foetal α-globulin from the sera of fetus and a hepatoma patient. *Cancer Res.*, **30**, 2507.

NISHIOKA, M., IBATA, T., OKITA, K., HARADA, T. & FUJITA, T. (1972). Localisation of α-fetoprotein hepatoma tissues by immunofluorescence. *Cancer Res.*, **32**, 162.

PARMIANI, G. & DELLA PORTA, G. (1973). Effects of antitumour immunity on pregnancy in the mouse. *Nature New Biol.*, **241**, 26.

PEARSON, G. & FREEMAN, G. (1968). Evidence suggesting a relationship between polyoma virus-induced transplantation antigen and normal embryonic antigen. *Cancer Res.*, **28**, 1665.

RUOSLAHTI, E. & SEPPÄLÄ, M. (1971). Studies of carcino-fetal proteins. III. Development of a radioimmunoassay for α-fetoprotein. Demonstration of α-fetoproteins in serum of healthy human adults. *Int. J. Cancer*, **8**, 374.

RUOSLAHTI, E., SEPPALA, M., PHIKO, J. & VUOPIO, P. (1971). Studies of carcino-fetal proteins. II. Biochemical comparison of α-fetoprotein from human fetuses and patients with hepatocellular cancer. *Int. J. Cancer*, **8**, 283.

SCHÖNE, G. (1906). Untersuchungen über karzino mimmunität bei Mänsen. *München Med. Wchnshr.*, **53**, 2517.

SEPPÄLÄ, M. & RUOSLAHTI, E. (1972). Radioimmunoassay of maternal serum alphafetoprotein during pregnancy and delivery. *Am. J. Obstet. Gynec.*, **112**, 208.

SIMMONS, D. A. & PERLMANN, P. (1973). Carcinoembryonic antigen and blood group substances. *Cancer Res.*, **33**, 313.

SLADE, B. (1973). Antibodies to α-foetoprotein cause foetal mortality in rabbits. *Nature, Lond.*, **246**, 493.

SPIRO, R. G. (1970). Glycoproteins. *Ann. Rev. Biochem.*, **39**, 599.

STONEHILL, E. H. & BENDICH, A. (1970). Retrogenetic expression: the reappearance of embryonal antigens in cancer cells. *Nature. Lond.*, **288**, 370.

TAKAHASHI, A., YASHI, A., ANZAI, T. & WADA, T. (1967). Presence of a unique serum protein in sera obtained from patients with neoplastic diseases and in embryonic and neonatal sera. *Clin Chim. Acta.*, **17**, 5.

TATARINOV, Y. S. (1965). The content of an embryospecific α-globulin in the serum of human embryos, of new-born babies and of adults in cases of primary carcinoma of the liver. *Vopr. med. Khim.*, **11, 20.**

THOMSON, D. M. P., KRUPEY, J., FREEDMAN, S. O. & GOLD. P. (1969). The radioimmunoassay of circulating carcinoembryonic antigens of the human digestive system. *Proc. natn. Acad. Sci., USA*, **64**, 161.

THOMPSON, E. B., TOMKINS, G. M. & CURRAN, J. F. (1966). Induction of tyrosine α-ketoglutarate transaminase by steroid hormones in a newly established tissue culture cell line. *Proc. natn. Acad. Sci. USA*, **56**, 296.

TING, R. C., (1968). Failure to induce transplantation resistance against polyoma tumour cells with syngeneic embryonic tissues. *Nature, Lond.*, **217**, 858.

TROUILLAS, P. (1971). Carcinofetal antigen in glial tumours. *Lancet*, **ii**, 552.

URIEL, J., DE NECHAUD, B. & DUPIERS, M. (1972). Oestrogen-binding properties of rat, horse and human fetospecific proteins by immunoautoradiographic methods. *Biochem. biophys. Res. Comm.*, **46**, 1175.

VON KLEIST, S. & BURTIN, P. (1969). Localization cellulaire d'un antigene embryonnaire de tumeurs coliques humaines. *Int. J. Cancer*, **4**, 874.

WADA, T., ANZAI, T., YACHI, A., TAKAHASHI, A. & SAKAMOTO, S. (1971). Incidence of three different fetal proteins in sera of patients with primary hepatoma. In *Protides of the Biological Fluids* (ed. H. Peeters), p. 221. Oxford: Pergamon Press.

WEINTRAUB, B. D. & ROSEN, S. W. (1971). Ectopic production of human chorionic somatomammotrophin by non-trophoblastic cancers. *J. clin. Endocr. Metab.*, **32**, 94.

DISCUSSION

MUNRO: Did you say that the antigen present with brain tumours was, unlike CEA (carcinoembryonic antigen) and alpha-foetoprotein, not detectable in serum?

ALLISON: With alpha-foetoprotein, a high proportion is secreted into the serum; with carcinoembryonic antigen, the concentration in the tumours is much higher than in serum although it is detectable in serum. I do not know how carefully the carcinofoetal glial antigen has been looked for in serum: radioimmuno-assays might be useful clinically.

JENKINS: What happens to alpha-foetoprotein levels in women on the pill? While the blood levels in the foetus and amniotic fluid are falling, are not maternal levels actually rising?

ADINOLFI: The maternal levels of alpha-foetoprotein start to increase after about 12–14 weeks of gestation, reaching values ranging between 50 and 500 ng/ml (Seppälä & Ruoslahti, 1973). Increased levels have not been detected in women on the pill (Seppälä, 1973).

ALLISON: But the maternal levels are always lower than those in the foetus.

JENKINS: You are implying that all the alpha-foetoprotein in the mother has come from the baby. This may not be the case: it could be produced in the mother as an effect of pregnancy hormones.

ADINOLFI: Abnormally high levels of alpha-foetoprotein have been detected in mothers with foetuses with anencephaly or spina bifida and in the presence of foetal or placental distress. In these cases the maternal alpha-foetoprotein is derived from the foetal circulation. Genetic variants are needed to establish whether in normal pregnancies the protein present is of foetal or maternal origin.

MUNRO: Are there any genetic polymorphisms in alpha-foetoprotein, e.g. electrophoretic polymorphism?

ADINOLFI: A microheterogeneity has been described; the two forms of alpha-foetoprotein have been observed in foetal sera and in sera of some patients with hepatoma (Alpert, Drysdale & Isselbacher, 1973). At present there is no evidence of a genetic polymorphism of this protein.

FAULK: I question the argument that rising levels of antigens in maternal circulation at the time of delivery and their decline soon afterwards indicates foetal production of these components. Several plasma proteins such as lipoproteins, ceruloplasmin and other globulins rise in maternal blood as gestation progresses, and many of these do not cross the placenta (Scheinberg, Cook & Murphy, 1954; Putnam, 1960; Brambell, 1970).

BAGSHAWE: A point arises here about substances secreted by the placenta or foetus into the maternal circulation and their immunological relationship to substances normally present in the normal non-pregnant subject. A fairly general pattern emerges with the qualification about the specificity of radioimmunoassays for specific proteins. Luteinising hormone (LH) is always present in the body fluids of the normal non-pregnant subject. Chorionic gonadotrophin has a very close immunological similarity with human LH and although the beta subunits of the two hormones exhibit some differences in their amino acid sequences there are close chemical similarities (Bahl, Carlsen, Bellisario & Swaminathan, 1972; Saxena & Rathnam, 1971). A similar situation exists with placental lactogen, which is closely related to growth hormone (Kaplan & Grumbach, 1964). Likewise, placental alkaline phosphatase can be distinguished antigenically from other forms of alkaline phosphatase, but there are also cross-reactions (Seppälä & Ehnholm, 1969). The normal background of polypeptide secretions from various tissues may maintain tolerance, so that immune reactions do not occur when the maternal organism is confronted with these foetal products.

ALLISON: The question of identity or non-identity is important from the point of view of immunological tolerance. According to the model of Weigle and ourselves, tolerance to body constituents which circulate in small amounts, such as hormones, is due to selective failure of T lymphocytes to respond. Lymphocytes of the B lineage, able to bind antigen and respond immunologically when autoantigens are suitably presented to them, are demonstrable. If a product similar but not identical to that normally present is formed, it can be immunogenic and even, by a hapten-carrier effect, stimulate the formation of autoantibodies.

BAGSHAWE: I am not aware of any report of antibodies occurring spontaneously to alpha-foetoprotein.

ADINOLFI: Since maternal sera contain alpha-foetoprotein, the antibody could be found by dissociation of the immune complexes. Antibodies against this protein have occasionally been detected in patients with hepatitis (Florin-Christensen, Roux & Arana, 1974).

FAULK: Have you or anyone else tried to stimulate maternal lymphocytes with alpha-foetoprotein?

ADINOLFI: The protein has been considered to have an inhibitory effect on the lymphocyte proliferation induced by PHA, con A, PKW, anti-IgG and in the mixed lymphocyte culture (Belanger *et al.*, 1974). However, these are still preliminary studies and it is too early to reach definitive conclusions.

JOHNSON: I am puzzled by the very high levels of CEA and alpha-foetoprotein; the functions of these substances seem to be unknown. Is there any evidence to indicate that these molecules attach to the surface of cells to act, possibly as a masking layer, and the surfeit free in body fluids is to cope with antibody? Only when antibody levels are extremely high as, for example, in the studies of Slade (1973), or when the alpha-foetoprotein is relatively deficient, does the immunological destruction of cells which have been coated with this erstwhile protective layer occur. Does this seem feasible and is there any evidence for it?

ALLISON: The interesting point in the work of Slade is that antibody can actually damage the embryo. It specifically damages the yolk sac and the liver, which suggests that the alpha-foetoprotein is present in the membrane of these cells at least. Whether the protein can be picked up by other cell types is unknown, but it might occur. The alpha-foetoprotein may be present in excess as a protection against the possibility of maternal immunisation, and the same may apply for other antigens. However, it seems likely that they have other functions.

BAGSHAWE: If this were a physiological mechanism it would, of course, carry the danger of immune complex nephritis which could be quite a substantial hazard.

ALLISON: This hazard would only apply with a lot of antibody, but not if only a small amount of antibody were formed by the mother.

JOHNSON: Tung (1974) found that the deposition of immune complexes in mice rises markedly with multiparity.

In connection with the cross-reaction between CEA and the ABH system, should Szulman's results (1964) be re-interpreted in view of the wide and prolific distribution of ABH substance in human foetuses in the first trimester, which then declines in distribution and intensity. His experiments were done about ten years ago and he may have detected the distribution of CEA. How well has the histological distribution of CEA been mapped out in the human embryo?

ALLISON: The distribution of the two antigens in the gut by immunofluorescence is similar. Szulman's work should be repeated, using anti-CEA serum, to see how well they correspond.

DAVIES: Mouse CEA was found very many years ago but was named by Gold only much later. It was evidently produced by the liver and attached to tumour cells by a specific receptor, identified as cytolipin-H, which has the sphingosine fatty acid lactose configuration. CEA is indeed a regeneration antigen. Where the cells are proliferating in the foetus, in tumour growth, in regenerating liver or in recovery after burns, this substance appears (Boyle, Davies & Haughton, 1963). It is highly unlikely to be a precursor of blood group substance. When a range of other rodents were immunised with mouse CEA and the sera tested by precipitation, spurs unique to each species were associated with the common precipitation line, showing the presence of several specificities in the different sera (Haughton, 1962). These do not occur with blood group substance in any kind of degraded form. A misinterpretation arose because slight ABO activity was detected in CEA, possibly due to a little galactose.

ALLISON: It is not as simple as that. The blood group antigens have been degraded and a considerable increase found in cross-reactivity.

LAWLER: Tumour cells are known to lose the expression of ABO antigens which might perhaps be correlated with the appearance of CEA. Moreover, all the chemistry of the ABH groups has been analysed from what is secreted, and the ABO system is expressed on cells in a different way. The Lewis system is basically a system of secretion and absorption which could be different from the ABO system on the cell membrane.

DAVIES: In some tumour cells, there is a tendency to go from A or B to H in culture.

BODMER: Of course, A disappears from fibroblasts in culture (Franks, 1968).

LAWLER: But it persists if the medium is supplemented with the right material.

BODMER: More data are needed on this point. In our laboratory Fellous (unpublished) has shown that the combination of the A_1 and Lewis b specificities may be present as judged by micro-complement fixation methods on fibroblasts. We also find that the A antigen may be re-expressed when fibroblasts are fused with HeLa cells. Thus the mechanism of suppression is not at all clear. It could be due to the activity of an enzyme, to lack of activity of an enzyme or to degradation. Alternatively, a sialic acid residue may be added which hides the antigen. The whole question of these sorts of polysaccharide specificities is complex, because they are found on different types of molecules and change in their expression with growth conditions. In the classic case of the Lewis specificity on red cells, which is picked up from the serum, a glycolipid seems to be absorbed.

DAVIES: The two forms of ABO are the secreted mucopolysaccharide and the glycolipid on the cell surface. Both forms evidently occur in red cells, for example, and both may occur on lymphocytes. With regard to ABO on CEA, CEA is at least not a glycolipid.

BAGSHAWE: One point bears on this, concerning trophoblast and trophoblastic tumours. The CEA level in serum is normally very low, but when cell destruction is induced by cytotoxic drugs it increases very sharply. If cytotoxic drugs are given ineffectively and do not cause extensive cell destruction then no such increase occurs. When cell destruction occurs, CEA is evidently liberated in quite large amounts, apparently from trophoblastic tumours (Bagshawe, 1974).

HOWE: In a recent series of 900 patients with non-trophoblastic tumours, just under 10% of them had evidence of HCG or its subunits in the serum, and the incidence was higher in patients with tumours of the gastro-intestinal tract. During the treatment of choriocarcinoma with cytotoxic drugs leading to increase in CEA, is there also an increase in levels of HCG?

BAGSHAWE: A transient increase in HCG levels is found in serum when cytotoxic drugs are given to patient with choriocarcinoma (Bagshawe, 1969). Ectopic HCG production by non-trophoblastic tumours is obviously interesting. Of a wide range of tumours examined by Braunstein, Vaitukaitis, Carbone & Ross (1973), 8% produced either HCG or its β subunit. Our own figures are lower, but in a much smaller series. This situation is not unique to HCG. Weintraub & Rosen (1971) find that about 8% of a wide range of tumours produce placental lactogen. Fishman and his colleagues (Fishman, Inglis, Stolbach & Krant, 1968) find quite a large proportion of patients with a wide range of tumours, particularly ovarian carcinomas, express placental alkaline phosphatase. There are good and sensitive test systems for these antigenic products of cells and when methods for detecting other proteins of equal sensitivity are developed, a much wider variety may also be detected.

HOWE: Why is this peculiar association found between the gastro-intestinal tract with both the foetal antigens and HCG?

BAGSHAWE: I doubt there is a very close association. Certainly a lot of lung tumours produce chorionic gonadotrophin, but without examining the statistics closely and also the type of population being studied, it would be very premature to place much importance on the association with the gastro-intestinal tract (Braunstein, Vaitukaitis, Carbone & Ross, 1973).

CURZEN: The transient rise in levels of CEA in patients with trophoblastic neoplasms treated with cytotoxic drugs may be due to the destruction of tumour cells, but could alternatively be a regenerating antigen produced by normal gut cells which tend to be damaged by cytotoxic drugs.

BAGSHAWE: I doubt it, because when cytotoxic drugs are given, gut cells can be damaged without harming the choriocarcinoma, and no rise in CEA is found.

ADINOLFI: Recently, Popp (1972) has detected Hb variants similar to embryonic Hb in aged mice. It may be of interest to test old individuals for the presence of CEA.

ALLISON: The whole question of surveillance comes into this story. Most of us accept the notion of surveillance against tumours, with the possibility of immune mechanisms being involved. The analogy with allotype suppression which can be mediated by T cell lymphocytes and manipulated, with the possibility of re-emergence in old people, is interesting. Autoantibodies are, of course, common in ageing people.

BILLINGTON: Phase-specific antigens are related to specific ontogenetic stages. Some early Russian work (see Volkova & Maysky, 1969) claimed to show that, following ablation of certain maternal organs during pregnancy, 'serum factors' appeared which were believed to regulate the differentiation of the corresponding organ in the foetus, and suggestions were also made that there was a necessary balance between maternal phase-specific antibodies and the embryonic antigens for normal development.

A further point concerns work by Parmiani & Della Porta (1973) showing that in animals immunised against syngeneic embryonic tissue, reproductive performance was reduced, with lower rates of pregnancy and decreased litter size. The implication again was of some embryo-specific antigen.

ALLISON: The mice were actually immunised with tumours or with normal embryonic tissue and their reproductive performance was reduced.

FAULK: There is also an alternative line of evidence, for the adult animals immunised with adult antigen display the same type of foetal wastage. This has been reviewed by Brent (1966).

HOWE: Della Porta did not find this, for his experiments emphasised the importance of foetal-specific antigens after doing the experiments with adult tissue.

BILLINGTON: I was not wholly convinced by the result, since they observed only a marginal decrease in the reproductive performance and were assessed simply on the basis of reduced pregnancy rate, litter size and foetal resorption. This is a perilous thing to rely on because the reproductive performance of any mouse is difficult to measure. There was no attempt to evaluate the effects on any other basis, nor to establish the precise time and site of action.

ALLISON: Many results (see Coggin & Anderson, 1974) are not really very dramatic. They are there, they appear to be reproducible, but they are not very large effects.

JENKINS: In this connection Dr Allison mentioned the work of Breyere & Burhoe (1963) on tolerance or unresponsiveness to paternal-strain antigen as a result of multiparity. The possibility was raised that in multiparous females there might be a higher incidence of colonic tumours because of a similar unresponsiveness to carcinoembryonic antigen carried on the tumour cells. Allison was not very impressed by that possibility and rightly. We have not found an increase in colonic tumours in multiparous human females compared with nulliparous human females.

EDWARDS: Would anyone like to take up the point about cross-reactions on eggs and embryos?

ALLISON: One possibility is that egg antigens may be a leukaemogenic virus antigen, which is activated in the course of transformation by SV.40. These latent viruses, which pass through the egg, may be manifested on the egg surface. In connection with the phase-specific antigens we discussed earlier, it would be of interest to know whether these antigens were merely, in fact, related to mitotic cells and whether there is indeed any organ specificity.

EDWARDS: Unfertilised eggs are a synchronised cell population in meiosis.

COOMBS: Would a spindle protein be considered as a phase-specific antigen?

BODMER: Yes! Any substance which is secreted and only made at a certain phase in the cell cycle.

EDWARDS: Spindle proteins are hardly phase-specific proteins because they are conserved in the cytoplasm during interphase.

ALLISON: Yes, it is largely microtubular, and there is good evidence that this is shunted about from cytoplasmic microtubules to spindle microtubules.

COOMBS: Why couldn't it be shown at other times?

ALLISON: It depends on the sensitivity of detection. Colchicine binding, for example, shows no difference with cell cycle. Certain proteins, and presumably also antigens, can be expressed after virus infection, e.g. thymidine kinase. This enzyme is determined by host cell genes, as appropriate genetic markers show, and it would appear as a specific kind of antigen if manifested on the cell surface. Could we call these things 'cycle-specific' because, as far as I know, there is no specific term.

FAULK: There is evidence in the literature that con A receptors are dependent on cell cycle (Smets, 1973).

DAVIES: Some sort of capping or aggregation might result in a greater sensitivity because the molecules are all grouped, rather than genuine changes in activity of the receptors.

ALLISON: With radioactive antibody, binding differs enormously at different stages of the cell cycle.

REFERENCES

ALPERT, E., DRYSDALE, J. W. & ISSELBACHER, K. J. (1973). Isoelectric focusing of human α-fetoprotein: an aid in purification and characterization of microheterogeneity. *Ann. N.Y. Acad. Sci.*, **209**, 387.

BAGSHAWE, K. D. (1969). *Choriocarcinoma*. Edward Arnold: London.

BAGSHAWE, K. D. (1974). Tumour-associated antigens. *Br. med. Bull.*, **30**, 68.

BAHL, O. P., CARLSEN, R. B., BELLISARIO, R. & SWAMINATHAN, N. (1972). Human chorionic gonadotrophin amino-acid sequence of alpha and beta subunits. *Biochem. biophys. Res. Comm.*, **48**, 416.

BELANGER, L., WAITHE, W., DAGUILLARD, F., LAROCHELLE, J. & DEFOUR, D. (1974). A study of the influence of AFP on the 'in vitro' lymphocyte response. In *Alpha-feto-protein* (ed. R. Masseyeff), p. 423, Proc. of the INSERM Symposium, Nice, France.

BOYLE, W., DAVIES, D. A. L. & HAUGHTON, G. (1963). Mouse tissue-cell antigens: some properties of cell bound components. *Immunology*, **6**, 499.

BRAMBELL, F. W. R. (1970). *The Transmission of Passive Immunity from Mother to Young*. p. 234. North-Holland: Amsterdam.

BRAUNSTEIN, G. D., VAITUKAITIS, J. L., CARBONE, P. P. & ROSS, G. T. (1973). Ectopic production of human chorionic gonadotrophin by neoplasms. *Ann. Internal Med.*, **78**, 39.

BRENT, R. L. (1966). Immunologic aspects of developmental biology. *Adv. Teratol.*, **1**, 82.

BREYERE, E. J. & BURHOE, S. O., (1963). The nature of the 'partial' tolerance induced by parity. *J. natn. Cancer Inst.*, **31**, 179.

COGGIN, J. H. & ANDERSON, N. G. (1974). Cancer differentiation and embryonic antigens: some central problems. *Adv. Cancer Res.*, **19**, 105.

FISHMAN, W. H., INGLIS, N. I., STOLBACH, L. L. & KRANT, M. J. (1968). A serum alkaline phosphatase isoenzyme of human neoplastic cell origin. *Cancer Res.*, **28**, 150.

FLORIN-CHRISTENSEN, A., ROUX, M. E. B. & ARANA, R. M. (1974). Cryoglobulins in acute and chronic liver diseases. *Clin. exp. Immunol.*, **16**, 599.

FRANKS, D. (1968). Antigens as markers on cultured mammalian cells. *Biol. Rev.*, **43**, 17.

HAUGHTON, G. (1962). Some cell-bound, species specific antigens of mouse tumour cells. *Ann. N.Y. Acad. Sci.*, **101**, 131.

KAPLAN, S. L. & GRUMBACH, M. M. (1964). Studies of human and simian placental hormone with growth hormone-like and prolactin-like activities. *J. clin. Endocr. Metab.*, **24**, 80.

PARMIANI, G. & DELLA PORTA, G. (1973). Effects of anti-tumour immunity on pregnancy in the mouse. *Nature New Biol.*, **241**, 26.

POPP, R. A. (1972). Variant hemoglobins in aged animals. In *Embryonic and fetal antigens in cancer*, vol. **2**. p. 67.

PUTNAM, F. W. (ed.) (1960). *The Plasma Proteins*, vol. II, pp. 37, 69, and 323. Academic Press: New York.

SAXENA, B. B. & RATHNAM, P. (1971). Chemical and immunological characteristics of the subunits of human pituitary FSH and LH. *Excerpta Medica series* no. **241**, Proc. 2nd International Symposium, p. 122.

SCHEINBERG, H., COOK, C. D. & MURPHY, J. A. (1954). The concentration of copper and ceruloplasmin in maternal and infant plasma at delivery. *J. clin. Invest.*, **33**, 963.

SEPPÄLÄ, M. (1973). Alpha fetoprotein levels in women taking oral contraceptives. *Int. J. Fertil.*, **18**, 206.

SEPPÄLÄ, M. & EHNHOLM, C. (1969). Immunological relationship of human placental and intestinal alkaline phosphatase. *Scand. J. Clin. Lab. Invest.*, **23**, *Suppl.* **108**, 49.

SEPPÄLÄ, M. & RUOSLAHTI, E. (1973). Alpha fetoprotein in maternal serum: A new marker for detection of fetal distress and intrauterine death. *Am. J. Obstet. Gynec.*, **115**, 48.

SLADE, B. (1973). Antibodies to α-foetoprotein cause foetal mortality in rabbits. *Nature, Lond.*, **246**, 493.

SMETS, L. A. (1973). Agglutination with con-A dependent on cell cycle. *Nature New Biol.*, **245**, 113.

SZULMAN, A. E. (1964). The histological distribution of the blood group substances in man as disclosed by immune fluerescence. III The A, B and H antigens in embryos and foetuses from 18 mm in length. *J. exp. Med.*, **119**, 503.

TUNG, K. S. K. (1974). Immune complex in the renal glomerulus during normal pregnancy. A study in the guinea-pig and the mouse. *J. Immunol.*, **112**, 186.

VOLKOVA, L. S. & MAYSKY, I. N. (1969). Immunological interaction between mother and embryo. In *Immunology and Reproduction* (ed. R. G. Edwards), p. 211. I.P.P.F.: London.

WEINTRAUB, B. D. & ROSEN, S. W. (1971). Ectopic production of human chorionic sommatomammotrophin by nontrophoblastic cancers. *J. clin. Endocr. Metab.*, **32**, 94.

ORIGINS AND PROPERTIES OF TROPHOBLAST

By RICHARD GARDNER

The term trophoblast is applied to an almost bewildering variety of cell types found at the periphery of the mammalian conceptus. However, these cells have at least one common feature; they all seem to be mitotic descendants of cells present in the outer wall or trophectoderm of the blastocyst. Trophoblast cells are of particular interest to the immunobiologist because they constitute the immediate cellular boundary between maternal and embryonic tissues in both the implanting blastocyst and chorio-allantoic placenta of an established foetal homograft. The aim of this paper is to discuss the pre-implantation origins, properties, and early postimplantation development of trophoblast cells as revealed by recent experiments. I will be mainly concerned with the mouse since most relevant experiments have been carried out on embryos of this species. The extent to which present findings may apply to other mammals will be considered later.

A brief description of the organisation of the mouse blastocyst and fate of its different parts is necessary in order to provide a basis for interpreting the experimental data. Unfortunately, knowledge of the fate map of the blastocyst has depended, until recently, on examination of serial histological sections of successive stages rather than direct analysis of cell lineages (Snell & Stevens, 1966). Nevertheless, recent experiments employing transplantation of genetically marked cells and tissues has confirmed that such purely descriptive studies are largely correct (Gardner, 1971, 1974; Gardner & Johnson, 1973; Gardner, Papaioannou & Barton, 1973).

At $3\frac{1}{2}$ days post-coitum the mouse embryo is a hollow, fluid-filled spherical blastocyst composed of approximately 60 cells (Fig. 1). Three-quarters of the cells form the surface of the blastocyst, the trophectoderm, and are flattened, polygonal cells held together by tight junctions. The remaining cells are completely enclosed and attached to the inner surface of part of the trophectoderm as a discrete disc appropriately called the inner cell mass (ICM). The ICM gives rise to the embryo proper, and to certain extra-embryonic structures (Fig. 2). The blastocyst lies free in the uterine lumen at this stage, and is separated from maternal tissues by a mucopolysaccharide coat, the zona pellucida. The zona is lost late on day 4 of development, and attachment between the trophectoderm and uterine epithelium is thereafter rapidly established. Regional differentiation with-

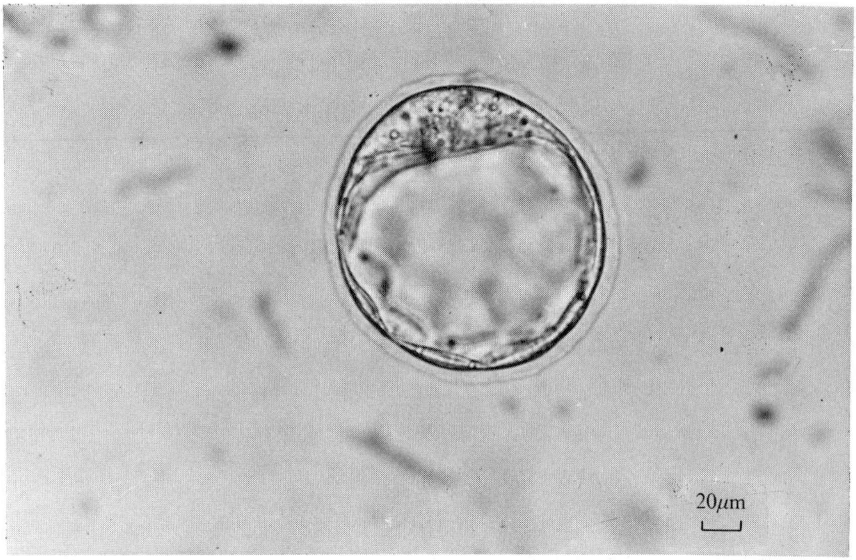

Fig. 1. A $3\frac{1}{2}$-day post-coitum mouse blastocyst still surrounded by the zona pellucida. The inner cell mass is very closely applied to the inner surface of part of the trophoblast wall (polar trophoblast) while the remaining trophoblast cells (mural trophoblast) surround the blastocoelic cavity.

in the trophectoderm is then soon evident. The mural trophoblast cells surrounding the blastocoele enlarge, while the polar trophoblast cells overlying the ICM remain unchanged (Fig. 3). This transformation of mural trophoblast into the so-called primary giant cells begins at the abembryonic pole of the blastocyst at about midnight on day 4, and has reached the edge of the ICM by midday on day 5 (Dickson, 1966). DNA levels greater than four times the haploid value can be detected in some cells of blastocysts at this stage, even after culture *in vitro* (Barlow, Owen & Graham, 1972; Barlow & Sherman, 1972). Hence giant cell transforma-

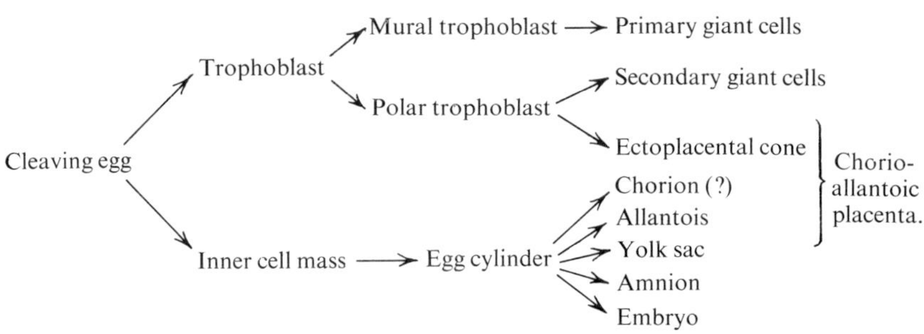

Fig. 2. Fate map of trophoblast and ICM of the $3\frac{1}{2}$-day mouse blastocyst according to the histological study of Snell & Stevens (1966).

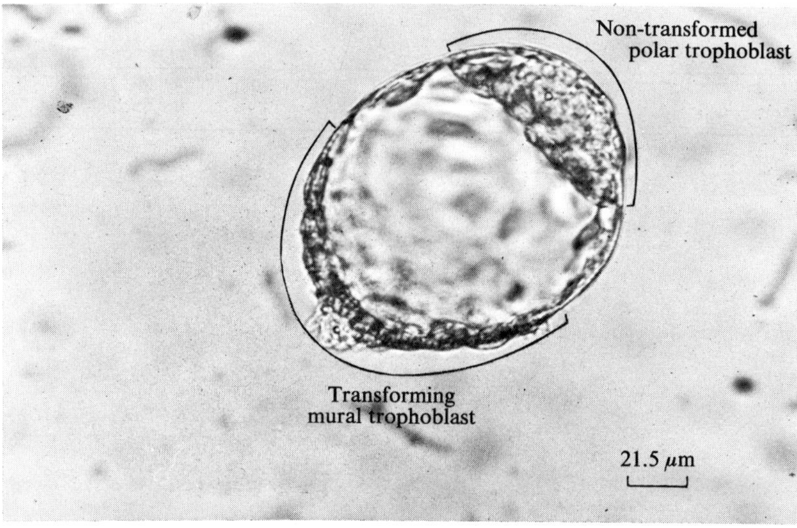

Fig. 3. An approximately 4½-day post-coitum blastocyst. The zona pellucida has disappeared, and transformation of the mural trophoblast into primary giant cells is progressing from the abembryonic pole of the blastocyst towards the edge of the inner cell mass.

tion can occur independently of the maternal environment. The polar trophoblast cells remain diploid, and give rise to a conical mass of cells, the ectoplacental cone, through sustained mitotic activity (Fig. 4).

As development proceeds large numbers of secondary giant cells, morphologically indistinguishable from primary ones, form in the periphery of the cone, and spread laterally and ventrally so as to surround the entire conceptus eventually. By about 8½ days the chorion becomes flattened against the base of the cone, and is itself penetrated by the upgrowing tip of the allantois (Fig. 5). In due course embryonic blood vessels grow along the allantoic stalk and the ectoplacental cone, together with the chorion and allantois, becomes the definitive chorio-allantoic placenta. Three layers of trophoblast cells are interposed between maternal and foetal blood in the mature placenta (Kirby & Bradbury, 1965). These cells presumably originate from cone cells that did not transform earlier into giant cells, and possibly also from the chorion.

Both the origin and mode of formation of trophoblastic giant cells has been controversial until recently. Some authors favoured a maternal origin for these cells, others an embryonic origin. Yet others argued that giant cells developed as a result of fusion between cells of maternal and embryonic origin (Galassi, 1967; Avery & Hunt, 1969), or by fusion of embryonic trophoblast cells (Saccoman, Morgan & Wells, 1967). These issues have been resolved by recent experiments. The fact that typical

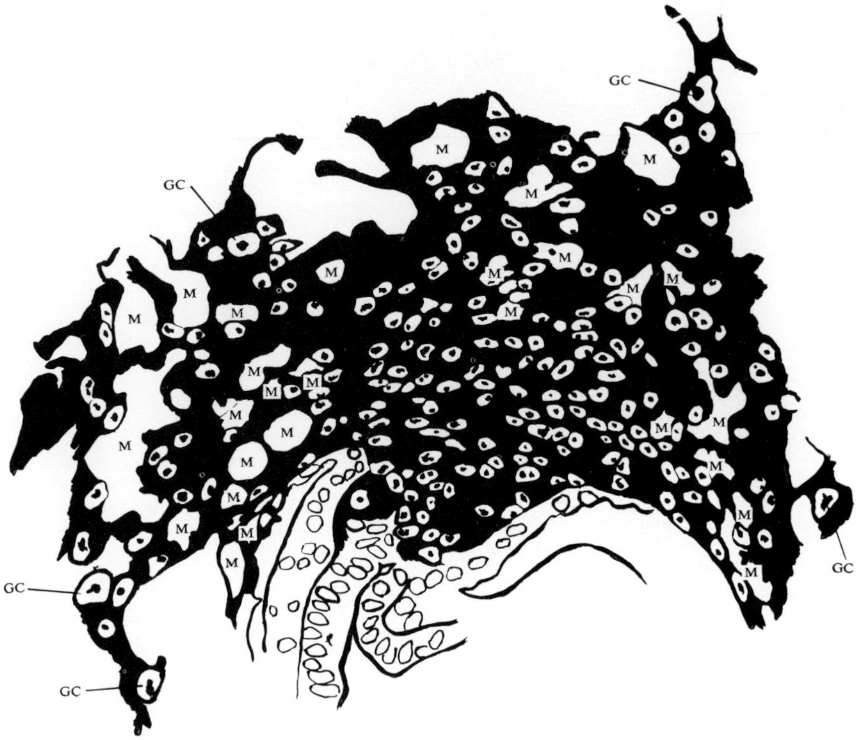

Fig. 4. A camera lucida drawing of a central section of the ecto-placental cone from a $7\frac{1}{2}$-day post-coitum embryo. Cell boundaries are virtually impossible to resolve by light microscopy due to the very close packing of the trophoblast cells. Hence the full extent of trophoblast cytoplasm is indicated in black and nuclei as enclosed white areas containing black dots (nucleoli). The cone consists of a solid central core of trophoblast cells and a more loosely arranged peripheral area broken up by maternal sinusoids (M). Note that the peripheral trophoblast nuclei are generally larger than the more central ones, and a few in the extreme periphery have attained dimensions typical of giant cells (GC).

giant cells develop following transplantation of blastocysts to ectopic sites (Fawcett, Wislocki & Waldo, 1947), and also in outgrowths of pre-implantation stages *in vitro* (Gwatkin, 1966), argues against a maternal origin. Further evidence against a maternal origin has been obtained by transferring blastocysts homozygous for one isozyme of glucose phosphate isomerase (GPI) to pseudopregnant mice homozygous for the other isozyme. Electrophoretic analysis of the giant cells of implanted embryos revealed little or no maternal contribution (Gardner, Papaioannou & Barton, unpublished results). This type of experiment also failed to show any of the hybrid enzyme which is expected if maternal cells regularly fuse with trophoblast cells (Chapman, Ansell & McLaren, 1972; Gearhart & Mintz, 1972; Gardner *et al.*, 1973). Likewise, giant trophoblast cells developing *in vivo* or *in vitro* from chimaeric blastocysts obtained by aggregating pairs of cleaving eggs of opposite GPI genotypes showed no

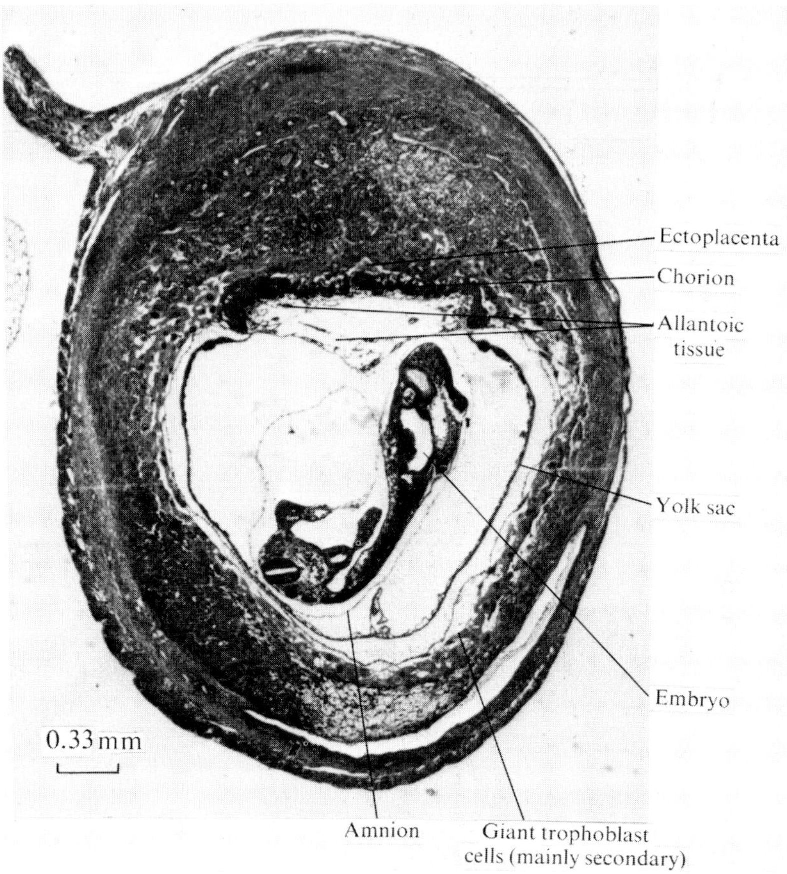

Fig. 5. Transverse section of a uterine horn containing a 9-day conceptus to show the various trophoblast derivatives and their relationship with other extra-embryonic tissues.

hybrid enzyme (Chapman *et al.*, 1972; Gearhart & Mintz, 1972). In contrast, multinucleate skeletal muscle fibres of such chimaeras regularly show hybrid enzyme, as expected from their mode of formation *in vitro* by fusion of uninucleate myoblasts (Mintz & Baker, 1967; Chapman *et al.*, 1972; Konigsberg, 1963).

Lack of evidence of cell fusion has led to the conclusion that giant cells attain their very high DNA content (up to 1024 times the haploid value) by endoreduplication. Density gradient centrifugation and DNA hybridisation studies suggest that the DNA accumulates by endoreduplication of the entire genome rather than by amplification of part of it (Sherman, McLaren & Walker, 1972). Whether these cells are polytene or polyploid has yet to be decided (Zybina, 1961, 1963; Jollie, 1964).

Having outlined the early development of trophoblast cells in the mouse embryo, I should now like to consider the embryonic origins and properties of these cells, in relation to the following questions.

1. What is the mechanism of differentiation of the trophectoderm cells of the blastocyst?
2. What properties serve to distinguish trophoblast cells from other cells of the blastocyst?
3. Are trophoblast cells already determined – i.e. committed to particular developmental pathways at the blastocyst stage?
4. What factors are responsible for the regional differentiation of mural versus polar trophoblast cells?
5. What are the functions of early trophoblast cells?

Mechanism of differentiation of trophectoderm cells

Parthenogenetically activated eggs and fertilised eggs cultured *in vitro* from the one-cell stage can develop into morphologically normal blastocysts capable of implanting in the uterus (Tarkowski, Witkowska & Nowicka, 1970; Graham, 1972; Whitten & Biggers, 1968). Thus, neither the fertilising sperm nor the maternal environment during cleavage are decisive factors in trophoblast differentiation. The central issue has been whether differentiation depends on the segregation of cytoplasmic determinants present in the undivided egg, or on the interrelations of blastomeres during cleavage. The former hypothesis received support from cytochemical work done by Dalcq and his colleagues (Dalcq, 1957), and the latter hypothesis from experimental manipulations made possible when cleavage stages could be maintained *in vitro*. Cytochemical analysis showed two distinct cytoplasmic regions in the egg which appeared to be segregated into the trophoblast and ICM during cleavage. No evidence was provided of a causal relationship between the type of cytoplasm inherited by a blastomere and its developmental fate, and recent reappraisal indicates that the observed cytoplasmic differences may have been an artifact (Solter, Damjanov & Skreb, 1973). The fact that regional differences are not visible in the living egg precludes direct investigation of this hypothesis, though most experimental data conflict with it.

Experimental approaches have included analysis of the development of isolated blastomeres in the rat, mouse and rabbit. Both blastomeres separated from two-cell rat eggs could develop to the egg-cylinder stage after transfer (Nicholas & Hall, 1942). Similarly, one out of every two blastomeres yielded young at term in both the mouse and rabbit (Tarkowski, 1959; Seidel, 1952, 1960). A small percentage of single blastomeres from eight-cell rabbit eggs can also develop to term (Moore, Adams & Rowson, 1968).

However, more informative are studies in which all sister blastomeres isolated from two-, four-, and eight-cell stages have been cultured through to the blastocyst stage (Mulnard, 1965; Tarkowski & Wroblewska, 1967). A variety of developmental forms are obtained in this way, some of which elude satisfactory classification. Quite frequently, all blastomeres develop into trophoblastic vesicles, and there is invariably a conspicuous absence of the complementary structures developing from sister-blastomeres that Dalcq's hypothesis predicts.

Similarly, there is no orderly deployment of cells in blastocysts obtained by aggregating pairs of [^{3}H]thymidine-labelled eggs with unlabelled eggs (Mintz, 1965). Larger numbers of eggs aggregated later in cleavage can also form morphologically normal giant blastocysts (Mintz, 1965).

The foregoing experiments have led to suggestions that blastomeres are labile in early cleavage (Mintz, 1965), and that it is the position that cells eventually occupy that determines their fate (Tarkowski & Wroblewska, 1967). More specifically, Tarkowski & Wroblewska suggest that all blastomeres are destined to become trophoblasts unless they become enclosed within the cleaving cell mass, in which case they become determined to form ICM cells. This hypothesis predicts that by altering the position of a blastomere from outside to inside, or vice versa, it should be possible to alter its fate. Two groups of workers have attempted to test this prediction.

Wilson and his colleagues (Wilson, Bolton & Cuttler, 1972) developed a technique for injecting drops of silicone oil either into the peripheral cytoplasm of blastomeres, or centrally near the interface between cells. All peripheral drops were later found in trophectoderm cells, while seven out of 29 central drops were found in the ICM. The results indicate that cleavage takes place with little or no reorganisation of cortical cytoplasm. However, if pairs of cleaving eggs injected with peripheral drops were aggregated, eight out of 77 drops were noted subsequently in the ICM, compared with none out of 113 in peripherally injected controls.

Hillman and her colleagues (Hillman, Sherman & Graham, 1972) placed blastomeres disaggregated from four- and eight-cell eggs labelled with [^{3}H]thymidine outside unlabelled four- to sixteen-cell embryos. More than 90% of the labelled cells were later found in the outside layer of the blastocyst. When the labelled blastomeres were sandwiched between the blastomeres of unlabelled two- and four-cell stages, only 60% of the labelled cells were found later in the trophectoderm of the blastocyst. Individual, intact, labelled embryos were also completely surrounded by 14 unlabelled embryos to envelop them more efficiently. Labelled cells were found exclusively in the ICMs of some of the resulting giant blastocysts.

However, all the above experiments may be criticised because the results were analysed solely in terms of the position of cells in the blastocyst. Demonstrating that cells alter their position following experimental manipulation does not prove that they have altered their developmental fate. This criticism is strengthened by the 'false' blastocysts encountered in the blastomere isolation experiments of Tarkowski &

Wroblewska (1967). These structures consisted of a single layer of cells, but some adjacent cells were more rounded than others, giving the impression of an ICM. Such cells might well have been ICM cells forced to occupy a peripheral position as a result of the very small size of the embryo.

Hence it is necessary not only to demonstrate that blastomeres interiorised or exteriorised by manipulation during cleavage are subsequently found mainly in the ICM or trophoblast layers of the blastocyst, but that they actually contribute cells to the differentiated tissues to which ICM and trophectoderm give rise. The study of Hillman *et al.* (1972) largely meets this requirement, since these authors carried out similar rearrangements to those employing [^{3}H]thymidine-labelled embryos, using GPI or pigmentation markers, and analysed the results in foetal stages or postnatally. Blastomeres placed on the outside of cleavage stages contributed largely to trophoblast and yolk sac, whereas those placed inside contributed to the foetus and other membranes.

Thus, it seems probable that in order to differentiate into a trophoblast rather than an ICM cell, a blastomere must attain a peripheral location during cleavage. The way in which cell position controls determination is still obscure. If mouse eggs are cultured in medium containing an appropriate concentration of [^{3}H]thymidine from before the 16-cell stage, trophoblastic vesicles devoid of ICM cells are obtained (Snow, 1973). If, however, culture is begun later, an ICM develops, though it often consists of fewer cells than normal. It may, therefore, be between the eight- and 16-cell stage that decisive determinative events occur.

Properties of trophectoderm versus ICM

It is when the blastocoele forms between the 16- and 32-cell stage that segregation of cells into trophectoderm and ICM is first unequivocal. Even then, it is only the presence of tight junctions that enables trophoblasts to be distinguished ultrastructurally from ICM cells (Enders & Schlafke, 1965). However, the properties of the two living tissues are quite distinct.

Isolation of pure ICM and mural trophoblast tissue from $3\frac{1}{2}$-day blastocysts can be achieved microsurgically. The properties of the two tissues can thus be compared both in culture, and after transfer to the uteri of recipient mice (Gardner, 1971, 1972). Three obvious differences were found. Cultured fragments of mural trophoblast reformed into vesicles resembling small blastocysts lacking ICMs. They did not aggregate with similar vesicles or fragments despite culture for up to 14 hours. Failure to aggregate may be due to the restraint imposed on cell movement by the tight junctions. The frequency of implantation of transferred trophoblastic vesicles was the same as that of intact blastocysts transplanted to the opposite uterine horns of the same recipients.

Isolated ICMs did not form vesicles in culture, but pairs or groups readily aggregated to form spherical masses of cells which failed to implant *in utero*.

A fourth difference between the two tissues emerged recently during abortive attempts to label individual trophoblast cells by intracellular injection of melanin granules. The melanin granules were obtained from the eyes of CBA mice. Trophoblast cells of zona-free blastocysts incubated in drops of medium containing sedimented granules rapidly immobilised many granules on their exposed surfaces, as judged by the cessation of Brownian movement. Examination of stained 1 μm-sections of fixed blastocysts embedded in araldite demonstrated that the trophoblast cells had phagocytosed many granules after two to three hours. Intracellular granules were a little larger than those on the surface, but persisted in the living cells for up to 36 hours.

Phagocytosis by cleavage stages, isolated ICMs, ICMs following injection of granules into the blastocoele, and by $4\frac{1}{2}$-day implanted blastocysts, was then studied. The zona pellucida was first removed mechanically from $3\frac{1}{2}$-day blastocysts and earlier embryos, because pronase treatment seemed to inhibit phagocytosis. Preliminary results may be summarised as follows.

Cleavage stages do not phagocytose melanin granules, though later ones immobilise some granules on their outer surfaces (see Fig. 6). Coincident with blastulation, trophoblast cells show marked immobilisation and uptake of granules, while ICM cells do not. Distinct regional differences in behaviour towards the granules is shown by the trophectoderm of the $4\frac{1}{2}$-day implanting blastocyst. Polar cells behave like all trophoblast cells of the $3\frac{1}{2}$-day blastocyst, with numerous surface granules and slightly larger intracellular ones. Mural trophoblast cells, however, have more intracellular than surface granules, and those in the cytoplasm are very large and pale, indicating heightened intracellular digestion.

Phagocytic activity is a very prominent feature of trophoblasts in the implanted embryo (Owers, 1971; Billington, 1971), particularly the peripheral giant cells which may be packed with maternal red cells and other cellular debris. Evidently, the capacity to take up particles of at least 1 μm diameter is a property peculiar to trophoblast cells from the time they first make their appearance some hours before the zona pellucida is lost.

ICM and trophoblast thus differ strikingly in properties (Table 1) despite lack of ultrastructural differences between their constituent cells.

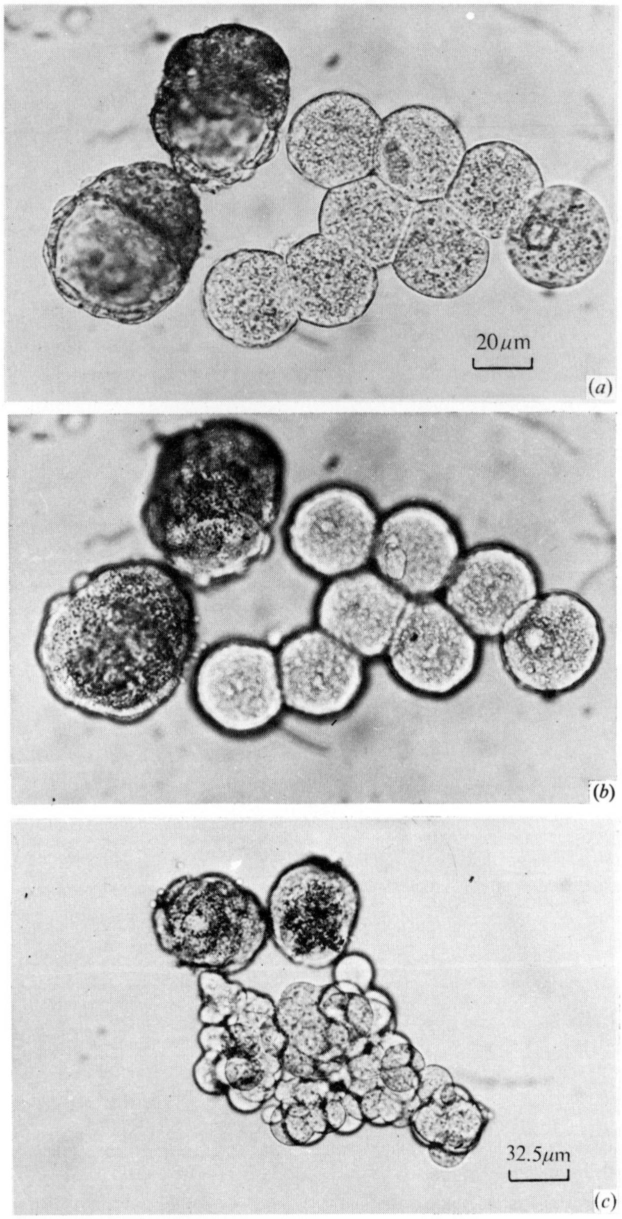

Fig. 6. Zona-free cleavage stages and $3\frac{1}{2}$-day blastocysts following co-culture for 2 hours in ocular melanin granules. (*a*) Four 2-cell stages that have aggregated during incubation, and two blastocysts cultured with them. Note the complete absence of melanin granules on the surface of the blastomeres. (*b*) The same embryos shown in (*a*) but with the focus altered to show the numerous melanin granules adhering to the surface of the blastocysts. (*c*) A group of $2\frac{1}{2}$-day post-coitum morulae lacking adherent melanin granules following culture with two blastocysts that are heavily labelled.

TABLE 1 *Differences in properties between ICM and trophoblast tissue isolated from $3\frac{1}{2}$-day post-coitum mouse blastocysts*

	Property investigated			
Tissue	Aggregation of tissue fragments	Formation of blastocoelic fluid	Initiation of implantation *in utero*	Phagocytosis of melanin granules
ICM	+	−	−	−
Trophoblast	−	+	+	+

Determination of trophoblast cells

Vesicles of mural trophoblast typically give rise only to a limited number of giant cells following implantation, in accordance with their presumptive fate (Gardner, 1972, and Fig. 2). ICM derivatives were present in only two out of 68 implantation chambers, and almost certainly represent instances in which some ICM cells were included in the isolated trophoblast. Therefore, vesicles of mural trophoblast cannot regulate for the loss of ICM tissue.

Trophoblast cells obtained by disaggregating trophoblastic fragments also failed to colonise ICM derivatives of host embryos after injection into the blastocoele of blastocysts carrying distinct genetic makers (Gardner, 1971, 1974). Transferred ICM cells regularly colonised host embryos, even if a single cell was injected (Gardner, 1971; Gardner & Lyon, 1971). However, this result is not conclusive because in no case was a contribution of the transferred trophoblast cells to trophoblastic derivatives of the host embryo detected. This could be either because the injected trophoblast cells transformed into giant cells and hence did not increase in number to an extent that could be detected by the available markers, or because the presence of tight junctions renders them prone to lethal damage during disaggregation.

The problem of trophoblast determination has also been tackled by analysing the GPI of presumptive trophoblast and ICM derivatives of implanted embryos that had developed from blastocysts reconstituted by inserting ICMs of one genotype into trophoblastic vesicles of the opposite genotype (Gardner *et al.*, 1973). Altogether, nineteen implanted embryos were analysed, and with a single exception, there was no significant ICM contribution to the giant cells and ectoplacental trophoblast. Also, no rat cells were detected by immune fluorescence in trophoblast of implanted embryos obtained after injecting rat ICMs into mouse blastocysts (Gardner & Johnson, 1973). These results provide very strong support for the idea expressed earlier, that all trophoblast cells of the implanted embryo originate from the trophectoderm of the blastocyst.

However, in the reconstituted blastocyst experiments, a significant contribution of trophectoderm cells to presumptive ICM derivatives was observed in most instances in which the ICM had continued to develop (Gardner *et al.*, 1973). Since contamination of isolated trophoblast with ICM cells is rare, this means that either some trophoblast cells are undetermined in the $3\frac{1}{2}$-day blastocyst, or that the trophectoderm normally produces certain extra-embryonic structures that are assumed to be derived from the ICM. A likely candidate for trophoblastic derivation is the extra-embryonic ectoderm which is shown clearly to be in continuity with the polar trophoblast in the definitive study of early mouse development (Snell & Stevens, 1966, Fig. 12.7). The extra-embryonic ectoderm contributes to the chorion, which we attempted to isolate with the embryonic fraction rather than the trophoblast in our study (Gardner *et al.*, 1973). This possibility can only be decided by reconstituting blastocysts from rat ICM and mouse trophoblast, and analysing the spatial distribution of rat and mouse cells in later development by immune fluorescence (Gardner & Johnson, 1973).

This is clearly an important issue to resolve regarding the determination of trophoblast, and also because division of the conceptus at the base of the ectoplacental cone has been assumed to separate trophoblast derived parts from those of ICM origin, in investigations of the antigenicity of trophoblast and embryo (e.g. Simmons & Russell, 1966; Kirby, 1971; Simmons, 1971) (see also Johnson, p. 93).

Although the trophectoderm cells are probably determined in the $3\frac{1}{2}$-day blastocyst, conclusive evidence has yet to be obtained.

Regional differentiation of the trophectoderm

Data presented earlier demonstrates that by $4\frac{1}{2}$ days the trophectoderm of the implanting blastocyst is differentiated into two regions. The mural trophoblast cells have ceased dividing, show enhanced digestion of phagocytosed granules, and are committed to forming primary giant cells. Polar trophoblast cells overlying the ICM remain diploid and their continued division produces the ectoplacental cone. Vesicles of mural trophoblast isolated at $3\frac{1}{2}$ days develop into giant cells without proliferating in recipient uteri. However, if such vesicles are recombined with isolated ICMs prior to transfer, trophoblast proliferation continues, and morphologically normal conceptuses with ectoplacental cones and innumerable giant cells are obtained. Furthermore, isozymal markers show that the proliferated trophoblast has developed from the mural trophoblast of the reconstituted embryo (Gardner *et al.*, 1973). Therefore, at least some of the mural trophoblast cells are not committed to form giant cells at $3\frac{1}{2}$ days. These experiments, together with the observation that the junction between giant and diploid cells invariably coincides with the edge of the ICM, clearly implicate the ICM in this regional differentiation of the early trophoblast cells.

Barlow & Sherman (1972) suggest that formation of the blastocoele provides the stimulus for giant transformation of mural trophoblast, and that the ICM serves to isolate polar trophoblast from this stimulus. However, the blastocoele obviously cannot be involved in the production of the large number of identical secondary giant cells that appear in the periphery of the cone a little later in development.

The fact that primary and secondary giant cells are morphologically identical, though differing both in their time and site of origin, is more readily accounted for by the following hypothesis (Gardner *et al.*, 1973). Trophoblast cells lose the capacity to divide at the later blastocyst stage, unless they are in intimate contact with the ICM. Hence all mural trophoblast cells cease dividing and most, if not all, transform into primary giant cells. Mitotic activity of the polar trophoblast is maintained as a result of an inductive interaction with the underlying ICM, and leads to formation of the ectoplacental cone. However, as progeny of the dividing cells are pushed by growth towards the periphery of the cone they are no longer influenced by the ICM or its derivatives. They therefore stop dividing, though many continue to replicate their DNA and thus transform into secondary giant cells. Continued division of cells adjacent to the ICM in the basal regions of the ectoplacenta could thus generate vast numbers of giant cells, and also maintain a reservoir of diploid cells destined to form the later trophoblastic components of the chorio-allantoic placenta.

If this interpretation of the data is correct, regional differentiation of the trophoblast in the implanting blastocyst is determined by the position in which the inner cell mass develops. Rat and mouse ICMs developed independently in three conceptuses obtained after transferring rat ICMs into mouse blastocysts. Immune fluorescence established that the rat embryo had stimulated overlying mouse mural trophoblast to form a second ectoplacental cone in all cases (Gardner & Johnson, 1973, and unpublished observations). Apart from the fact that it is local, and apparently not species-specific, there is at present no information about the nature of the interaction between the ICM and overlying trophoblast.

Continued proliferation of trophoblast is governed by continued development of the ICM on the above model. This indeed appears to be the case from examination of embryos that have developed abnormally *in utero* and in ectopic sites (Gardner, 1972, 1974; Johnson, 1972). Such growth control may exist throughout pregnancy, since sites of trophoblast proliferation seem to be confined to the placenta, and more specifically to those areas where trophoblast cells are in intimate relationship with ICM derivatives such as the allantois.

This leaves unanswered the question of why early trophoblast cells should soon lose their intrinsic capacity for division while retaining the ability to replicate their DNA. Intense phagocytic activity appears to be a major function of peripheral trophoblast cells in the implanted embryo (see later). Continued replication of DNA following cessation of cell division enables a considerable increase in cell size, and hence in cytoplasmic volume relative to surface area. These changes could enhance the efficiency of phagocytosis if intracellular digestion rather than uptake of foreign material is rate limiting, as is indeed suggested by experiments with melanin granules.

Functions of early trophoblast cells

Secretion of steroid and gonadotrophic hormones, nutrition of the embryo, invasion of maternal tissue, phagocytosis, and many other functions are attributed to cells of trophoblastic origin. I intend to discuss possible functions of trophoblast cells in the pre- and early post-implantation period. Trophectoderm cells are responsible for the secretion of blastocoelic fluid (Daniel, 1963; Gardner, 1972). This is evidently a regulated fluid which may provide an optimal nutritive environment for cells of the ICM (Lutwak-Mann, 1971). In species with central implantation such as the rabbit, the trophoblastic pump (Gamow & Daniel, 1970) may also maintain the position of the blastocyst by expansion until the trophoblast attaches to the uterine wall. It is clear from the trophoblast isolation experiments discussed earlier that it is the trophectoderm rather than ICM that initiates implantation. It is probably mural rather than polar trophoblast cells that first make intimate contact with the uterine epithelium (Eaton & Green, 1963).

It is questionable whether trophoblast cells actively invade maternal tissues and destroy living cells during implantation in the mouse. Injection of inert mineral oil into the uteri of pseudopregnant mice results in local epithelial breakdown, decidualisation, and later death of decidual cells that begins at the centre of the swelling and spreads towards the periphery. An implantation chamber is thus formed in the absence of an embryo, and its size at any stage corresponds approximately with that which an embryo of the equivalent post-implantation stage would occupy (Finn, 1971). The most obvious difference between these deciduomata and decidua containing embryos is that the former contain many dead cells and much debris which the latter do not. Implantation chambers produced by trophoblastic vesicles resemble those evoked by oil in this respect (Gardner, 1972). Peripheral giant cells are very numerous in normal implanted embryos and show evidence of very intense phagocytic activity. Hence the presence of debris is probably due to the fact that giant cells are few in chambers formed by trophoblastic vesicles, and absent altogether in those produced by oil. The nature of the stimulus is not known, but it seems that once implantation is initiated epithelial cells die spontaneously, and the decidual chamber forms and enlarges by spontaneous pycnosis of the decidual cells (Finn, 1971). An important role of the giant cells may thus be the removal of the resulting cellular debris rather than the destruction of viable cells. The giant cells might also play a role in nutrition of the embryo, by making the breakdown products of phagocytosis available to it.

CONCLUSIONS

The remarkable diversity in form of the blastocyst, mode of implantation and placentation in mammals, makes untenable the notion that detailed findings in the mouse will apply universally. Nonetheless, one might reasonably argue that such fundamental problems as the mechanism of trophoblast differentiation and control of its subsequent proliferation, might have a common solution in different mammals. However, in the absence of detailed experimental data comparable to that available for the mouse, it is difficult to assess the situation in other species. In some, the ICM develops externally and this clearly poses problems for the positional theory of trophoblast differentiation (Hamilton & Mossman, 1972).

Again, there are only descriptive studies to use when attempting to decide whether control of trophoblast proliferation by the ICM applies to other species. Proliferating trophoblast is underlaid by mesoblast in the human embryo, but it is still uncertain whether the latter is of trophoblastic or ICM origin (Boyd & Hamilton, 1970). The rabbit blastocyst attaches to the uterine epithelium by means of knobs of trophoblast cells which proliferate at sites quite remote from the ICM. However, these knobs apparently form after the endoderm has spread from the ICM round the inner surface of the trophoblast (Steer, 1970).

A species which apparently does not conform to the mouse model is the guinea pig. Here a cone of proliferating trophoblast forms diametrically opposite the ICM before any spreading of ICM cells. Furthermore, guinea pig blastocysts cultured *in vitro* beyond the time of implantation show very active trophoblast proliferation for several days before the cytotrophoblast gradually transforms into syncytial trophoblast (Blandau, 1971). The ICM is said to cease development and soon die under these conditions. This is entirely different from the behaviour of mouse trophoblast. The ICM of mouse blastocysts also survive poorly in simple culture conditions, but there is little trophoblast proliferation. Trophoblastic vesicles do not proliferate at all (see McLaren in Discussion after this paper). Nevertheless, careful experiments will have to be done before one accepts that the guinea pig really is different. The local accumulation of trophoblast cells at the abembryonic pole of the blastocyst could be due to spreading of cells from a proliferative centre over the ICM. A thick zona is still present at this stage, and could thus cause the local accumulation by preventing enlargement of the blastocyst. Also, growth of blastocysts tends to become rapidly disorganised in simple culture systems, and it is therefore very difficult to rule out the possibility that cells of ICM origin are playing a vital role during the period of tropho-proliferation. Culture of vesicles of guinea-pig trophoblast is the only decisive experiment. Hence it is perhaps unwise to speculate further until experimental data are available for other species.

SUMMARY

The early development of trophoblast in the mouse embryo is outlined briefly, as also is the mode of formation of trophoblastic giant cells. This serves to highlight certain questions concerning the determination, differentiation, properties and possible functions of early trophoblast cells that recent experimental investigations have sought to answer.

Isolation and rearrangement of blastomeres support the idea that cell fate is determined by position during cleavage rather than by segregation of preformed cytoplasmic factors. Cells occupying a peripheral position during cleavage tend to become trophoblast cells, while inside-cells tend to contribute progeny to the inner cell mass (ICM) and its derivatives. The properties of trophoblast and ICM are quite distinct in the $3\frac{1}{2}$-day post-coitum blastocyst. Furthermore cells of these two tissues appear to be committed to mutually exclusive pathways of development.

The majority of trophoblast cells transform into non-dividing giant cells at the blastocyst stage, while those overlying the ICM continue to divide, though many of their progeny also ultimately form giant cells. Various observations and experiments support the hypothesis that continued mitotic activity of trophoblast depends on a local inductive interaction with the ICM. The possible significance of giant cell formation is discussed as also are the functions of early trophoblast cells. The extent to which experimental findings in the mouse can be extrapolated to other species is also considered briefly.

I wish to thank Dr C. F. Graham, Dr M. H. Johnson, Miss S. J. Kelly, and Miss J. Rossant for valuable discussion. The World Health Organization, Ford Foundation, and Medical Research Council supported the author's experimental work discussed in this paper.

REFERENCES

AVERY, G. B. & HUNT, C. V. (1969), The differentiation of trophoblast giant cells in the mouse studied in kidney capsule grafts. *Transplant Proc.*, **1**, 61.

BARLOW, P., OWEN, D. A. J. & GRAHAM, C. (1972). DNA synthesis in the preimplantation mouse embryo. *J. Embryol. exp. Morph.*, **27**, 431.

BARLOW, P. & SHERMAN, M. I. (1972). The biochemistry of differentiation of mouse trophoblast: studies on polyploidy. *J. Embryol. exp. Morph.*, **27**, 447.

BILLINGTON, W. D. (1971). Biology of the trophoblast. In *Advances in Reproductive Physiology*, **5** (ed. M. W. H. Bishop) p. 27. Logos Press: London.

BLANDAU, R. J. (1971). Culture of guinea pig blastocyst. In *The Biology of the Blastocyst* (ed. R. J. Blandau) p. 59. University of Chicago Press: Chicago.

BOYD, J. D. & HAMILTON, W. J. (1970). *The Human Placenta*, Heffer: Cambridge.

CHAPMAN, V. E., ANSELL, J. D. & MCLAREN, A. (1972). Trophoblast giant cell differentiation in the mouse: expression of glucose phosphate isomerase (GPI-1) electrophoretic variants in transferred and chimeric embryos. *Devl. Biol.*, **26**, 153.

DALCQ, A. M. (1957). *Introduction to General Embryology*. Oxford University Press: Oxford.

DANIEL, J. C. Jr. (1963). Some kinetics of blastocyst formation as studied by the process of reconstitution. *J. exp. Zool.*, **154**, 231.

DICKSON, A. D. (1966). The form of the mouse blastocyst. *J. Anat.*, **100**, 335.

EATON, G. J. & GREEN, M. N. (1963). Giant cell differentiation and lethality of homozygous yellow mouse embryos. *Genetica*, **34**, 155.

ENDERS, A. C. & SCHLAFKE, S. J. (1965). The fine structure of the blastocyst: some comparative studies. In *Preimplantation stages of Pregnancy* (eds. G. E. W. Wolstenholme & M. O'Connor), p. 29. Churchill: London.

FAWCETT, D. W., WISLOCKI, G. B. & WALDO, C. M. (1947). The development of mouse ova in the anterior chamber of the eye and in the abdominal cavity. *Am. J. Anat.*, **81**, 413.

FINN, C. A. (1971). The biology of decidual cells. In *Advances in Reproductive Physiology*, **5** (ed. M. W. H. Bishop), p. 1. Logos Press: London.

GALASSI, L. (1967). Reutilisation of maternal nuclear material by embryonic and trophoblast cells in the rat for the synthesis of deoxyribonucleic acid. *J. Histochem. Cytochem.*, **15**, 573.

GAMOW, E. & DANIEL, J. C. Jr. (1970). Fluid transport in the rabbit blastocyst. *Wilhelm Roux Arch. Entw Mech. Org.*, **164**, 261.

GARDNER, R. L. (1971). Manipulations on the blastocyst. In *Advances in the Biosciences*, no. **6**: *Schering Symposium on Intrinsic and Extrinsic Factors in Early Mammalian Development* (ed. G. Raspé), p. 279. Pergamon: Oxford.

GARDNER, R. L. (1972). An investigation of inner cell mass and trophoblast tissues following their isolation from the mouse blastocyst. *J. Embryol. exp. Morph.*, **28**, 279.

GARDNER, R. L. (1974). Microsurgical approaches to the study of early mammalian development. In the *Seventh Harold C. Mack Symposium: Birth Defects and Fetal Development., Endocrine and Metabolic Factors*, (in press).

GARDNER, R. L. & JOHNSON, M. H. (1973). Investigation of early mammalian developments using interspecific chimaeras between rat and mouse. *Nature New Biol.*, **246**, 86.

GARDNER, R. L. & LYON, M. F. (1971). X chromosome inactivation studied by injection of a single cell into the mouse blastocyst. *Nature, Lond.*, **231**, 385.

GARDNER, R. L., PAPAIOANNOU, V. E. & BARTON, S. C. (1973). Origin of the ectoplacental cone and secondary giant cells in mouse blastocysts reconstituted from isolated trophoblast and inner cell mass. *J. Embryol. exp. Morph.*, **30**, 561.

GEARHART, J. D. & MINTZ, B. (1972). Glucosephosphate isomerase subunit-reassociation tests for maternal–fetal and fetal–fetal cell fusion in the mouse placenta. *Devl. Biol.*, **29**, 55.

GRAHAM, C. F. (1972). Genetic manipulation of mouse embryos. In *Advances in the Biosciences*, **8**, *Workshop on Mechanisms and Prospects of Genetic Exchange* (ed. G. Raspé), p. 263. Pergamon: Oxford.

GWATKIN, R. B. L. (1966). Defined media and development of mammalian eggs *in vitro*. *Ann. N.Y. Acad. Sci.*, **139**, 79.

HAMILTON, W. J. & MOSSMAN, H. W. (1972). *Hamilton, Boyd and Mossman's Human Embryology*. Heffer: Cambridge.

HILLMAN, N., SHERMAN, M. I. & GRAHAM, C. (1972). The effect of spatial arrangement on cell determination during mouse development. *J. Embryol. exp. Morph.*, **28**, 263.

JOHNSON, M. H. (1972). Relationship between inner cell mass derivatives and trophoblast proliferation in ectopic pregnancy. *J. Embryol. exp. Morph.*, **28**, 306 (appendix).

JOLLIE, W. P. (1964). Radioautographic observations on variations in deoxyribonucleic acid synthesis in rat placenta with increasing gestational age. *Am. J. Anat.*, **114**, 161.

KIRBY, D. R. S. (1971). The transplantation of mouse eggs and trophoblast to extrauterine sites. *In Methods in Mammalian Embryology* (ed. J. C. Daniel, Jr) p. 146. W. H. Freeman: San Francisco.

KIRBY, D. R. S. & BRADBURY, S. (1965). The hemo-chorial mouse placenta. *Anat. Rec.*, **152**, 279.

KONIGSBERG, I. R. (1963). Clonal analysis of myogenesis. *Science*, **140**, 1273.

LUTWAK-MANN, C. (1971). The rabbit blastocyst and its environment: physiological and biochemical aspects. In *The Biology of the Blastocyst* (ed. R. J. Blandau), p. 243. University of Chicago Press: Chicago.

MINTZ, B. (1965). Experimental genetic mosaicism in the mouse. *In Preimplantation stages of Pregnancy* (eds. G. E. W. Wolstenholme & M. O'Connor), p. 194. Churchill: London.

MINTZ, B. & BAKER, W. W. (1967). Normal mammalian muscle differentiation and gene control of isocitrate dehydrogenase synthesis. *Proc. natn. Acad. Sci., USA*, **58**, 592.

MOORE, N. W., ADAMS, C. E. & ROWSON, L. E. A. (1968). Developmental potential of single blastomeres of the rabbit egg. *J. Reprod. Fert.* **17**, 527.

MULNARD, J. G. (1965). Studies of regulation of mouse ova *in vitro*. *In Preimplantation stages of Pregnancy* (ed. G. E. W. Wolstenholme & M. O'Connor), p. 123. Churchill: London.

NICHOLAS, J. S. & HALL, B. V. (1942). Experiments on developing rats. II. The development of isolated blastomeres and fused eggs. *J. exp. Zool.*, **90**, 441.

OWERS, N. O. (1971). Ingestive properties of guinea pig trophoblast in tissue culture: a possible lysosomal mechanism. *In The Biology of the Blastocyst* (ed. R. J. Blandau), p. 225. University of Chicago Press: Chicago.

SACCOMAN, F. M., MORGAN, C. F. & WELLS, L. J. (1967). Radioautographic studies of DNA synthesis in the developing extraembryonic membranes of the mouse. *Anat. Rec.*, **158**, 197.

SEIDEL, F. (1952). Die entwicklungspotenzen einer isolierten blastomere des zweizellenstadiums im saugetierei. *Naturwissenschaften*, **39**, 355.

SEIDEL, F. (1960). Die entwicklungsfahigkeiten isolierter furchungszellen aus dem ei des kaninchens, *Oryctolagus cuniculus*. *Wilhelm Roux Arch. EntwMech. Org.*, **152**, 43.

SHERMAN, M. I., MCLAREN, A. & WALKER, P. M. B. (1972). Mechanism of accumulation of DNA in giant cells of mouse trophoblast. *Nature New Biol.*, **238**, 175.

SIMMONS, R. L. (1971). Viviparity, histocompatibility, and fetal survival. *In Advances in the Biosciences*, **6**, *Schering Symposium on Intrinsic and Extrinsic Factors in Early Mammalian Development* (ed. G. Raspe), p. 405. Pergamon: Oxford.

SIMMONS, R. L. & RUSSELL, P. S. (1966). The histocompatibility antigens of fertilized mouse eggs and trophoblast. *Ann. N. Y. Acad. Sci.*, **129**, 35.

SNELL, G. D. & STEVENS, L. C. (1966). Early embryology. In *Biology of the Laboratory Mouse*, 2nd edn., (ed. E. L. Green), p. 205. McGraw-Hill: New York.

SNOW, M. H. L. (1973). The differential effect of (^{3}H)-thymidine upon two populations of cells in pre-implantation mouse embryos. *In The Cell Cycle in Development and Differentiation: British Society for Developmental Biology Symposium*, (eds. M. Balls & F. S. Billett) p. 311. Cambridge University Press: London.

SOLTER, D., DAMJANOV, I. & SKREB, N. (1973). Distribution of hydrolytic enzymes in early rat and mouse embryos – a reappraisal. *Z. Anat. Entwickl. Gesch.*, **139**, 119.

STEER, H. W. (1970). The ultrastructure of the extraembryonic region of the preimplanted rabbit blastocyst before trophoblastic knob formation. *J. Anat.*, **106**, 263.

TARKOWSKI, A. K. (1959). Experiments on the development of isolated blastomeres of mouse eggs. *Nature, Lond.*, **184**, 1286.

TARKOWSKI, A. K., WITKOWSKA, A. & NOWICKA, J. (1970). Experimental parthogenesis in the mouse. *Nature, Lond.*, **226**, 162.

TARKOWSKI, A. K. & WROBLEWSKA, J. (1967). Development of blastomeres of mouse eggs isolated at the four- and eight-cell stage. *J. Embryol. exp. Morph.*, **18**, 155.

WHITTEN, W. K. & BIGGERS, J. D. (1968). Complete development *in vitro* of the pre-implantation stages of the mouse in a simple chemically-defined medium. *J. Reprod. Fert.*, **17**, 399.

WILSON, I. B., BOLTON, E. & CUTTLER, R. H. (1972). Preimplantation differentiation in the mouse egg as revealed by microinjection of vital markers. *J. Embryol. exp. Morph.*, **27**, 467.

ZYBINA, E. V. (1961). Endomitosis and polyteny of trophoblast giant cells. *Dokl. Akad. Nauk. SSSR.* **140**, 1177 (in Russian). (English transl: *Dokl. Biol. Sci.*, 1962, **140**, 797.)

ZYBINA, E. V. (1963). Cytophotometric determination of DNA content in nuclei of trophoblast giant cells. *Dokl. Akad. Nauk, SSSR*, **153**, 1428 (in Russian). (English transl: *Dokl. Biol. Sci.*, 1964, **153**, 1512.)

DISCUSSION

BILLINGTON: Does the trophectoderm have an intrinsic property of developing into giant cells, a property inhibited by the inner cell mass? If this is true, the trophoblastic vesicles should display a greater transformation into giant cells when replaced in the uterus than occurs in blastocysts. This possibility could be tested by inserting several inner cell masses into the vesicles, returning them to the uterus and looking for a reduced giant cell transformation.

GARDNER: Trophoblast cells lose their intrinsic capacity to divide but do not all inevitably form giant cells. Trophoblast vesicles contain, on average, about 30 cells on transfer, and after implantation many of them fail to transform, or die during transformation so that a final number of a dozen mature giant cells are found. We do not know how many mural cells form proper giant cells in the intact blastocyst. Thirty at most are found in trophoblastic vesicles and a minority of giant cell clusters containing more than a dozen cells are found in ectopic transplants (Johnson, 1972). The inner cell mass had presumably died or been damaged after transfer. Trophoblast vesicles are very small and difficulties arise over the insertion of one inner cell mass, let alone several. We have done an essentially similar experiment, in which two inner cell masses were placed into a blastocyst, and the ectoplacenta appeared to be larger although cell counts have not been done.

BILLINGTON: What about the giant cell transformation in that situation? Your theory would predict that it would be reduced.

GARDNER: When several inner cell masses are placed into a blastocyst the rate of implantation is reduced, and morphological data implies that implantation involves an intimate attachment between the mural trophoblast and the uterine epithelium. The mural trophoblast cells might have been reduced to such an extent that they could not evoke a decidual reaction.

BILLINGTON: One of the functions of the primary giant cells is supposedly to anchor the blastocyst whilst its development proceeds.

MCLAREN: Our own experiments on the control by the inner cell mass of trophoblast proliferation *in vitro*, fully confirm Gardner's findings. Trophoblast vesicles were derived by Snow's technique (Snow, 1973*a,b*). Cleaving mouse eggs are cultured in tritiated thymidine, which differentially destroys or prevents the development of the inner cell mass. The trophoblastic vesicles lacking an inner cell mass were cultured and formed a single-cell layer on a glass or plastic surface. Ansell & Snow (personal communication) counted the number of cells in the outgrowths from both trophoblastic vesicles and intact blastocysts and classified the giant cells in them. With intact blastocysts, the number of cells increased considerably in the course of outgrowth – in other words, cells were proliferating, but by no means all of them became giant cells, for a population of small proliferating cells was seen. The number of giant cells in the outgrowth was greater than the initial number of trophoblast cells. On the other hand, in trophoblast vesicles, which started with fewer trophoblast cells, there was no increase in cell number; the number was maintained at the initial level, but all of the cells present underwent the giant cell transformation. We do not know if the polar trophoblast is present in our trophoblast vesicles; it is absent in Gardner's work, because the microsurgical technique removes it. Should it be present in ours, then the polar trophoblast is capable of undergoing the giant cell transformation, but one proviso is that the polar trophoblast may be inhibited by the tritiated thymidine just as the inner cell mass is. The rate of outgrowth is actually faster from the vesicles than from intact blastocysts, as though the inner cell mass was in some way delaying the outgrowth of trophoblast cells *in vitro*.

GARDNER: We have been frustrated because the polar trophoblast is so difficult to isolate. We would like to remove just the inside cells, leaving the whole outside layer intact, because the polar trophoblast might be able to proliferate quite independently of the inner cell mass, whereas the mural trophoblast needs the inner cell mass for proliferation. This suggestion is unlikely, and the only evidence has arisen from injections of melanin granules locally under the zona pellucida. The granules are phagocytosed by only one trophoblast cell, hence a cell at any point in the blastocyst can be marked. This can be done over the inner cell mass region, with the zona pellucida remaining, although the polar trophoblast occasionally herniates out through the microsurgical slit to form a secondary blastocoelic cavity. After herniation, the inner cell mass becomes rearranged and attached to a different part of the trophoblast, and the granules are found later in the middle of a giant cell. This evidence argues against any intrinsic difference between polar and mural trophoblast at this early stage.

MCLAREN: One nice experiment would be to inject three or four extra inner cell masses into a blastocyst, allow outgrowths *in vitro* and then count the giant cells. This would answer Billington's question as to whether extra inner cell mass material would inhibit giant cell transformation.

GARDNER: One problem with in-vitro culture is to distinguish primary from secondary giant cells. Although injection of extra inner cell masses would be expected to reduce primary giant cell number, it should also increase the number of secondary giant cells. We feel that this distinction is more readily made *in utero* where the normal shape of the embryo is conserved. A second problem, concerns the nature of growth *in vitro*, for normal development rather than just outgrowth may occur (Hsu 1972), so that slight changes in culture conditions could favour the inner cell mass, and complicate quantitative studies.

McLaren: Providing control blastocysts were present in the same cultures, this problem could be overcome.

Allison: The development of endocytic capacity is of functional importance in relation both to nutrition and implantation. The relation between DNA synthesis and this form of differentiation could be very interesting. In haemopoietic cells, macrophages and polymorphs originate from a single precursor, a non-phagocytic cell, and DNA synthesis and differentiation appear to be mutually exclusive. Whilst they are synthesising DNA and dividing the cells have not yet acquired the capacity to undertake phagocytosis. A similar situation may arise in outgrowths from blastocysts. Small doses of thymidine would mark the cell synthesising DNA and at the same time the phagocytic label, e.g. latex, bacteria or some other particle, would indicate phagocytosis.

Gardner: While trophoblast cells are active phagocytically, they are still synthesising very large amounts of DNA (Nagl, 1972), an increase probably occurring over a period of several days. The two activities may be going on rapidly at the same time.

Allison: Cell division may be absent at this time, however. Is the block to division in G-2?

Gardner: I think so.

Allison: Do trophoblast cells actually have a whole lysosomal system as judged histochemically or by other tests? They might be able to digest materials, and actually use amino acids so that proteins, for example, in the culture media are degraded.

Gardner: We do not know about the early embryo. Mural trophoblast cells in the blastocyst take up melanin granules which then swell enormously, become very pale and persist for only a short time; this may indicate enhanced breakdown.

Curzen: Acid phosphatase is certainly present in the early cytotrophoblast of human placentae at approximately eight to ten weeks, but no evidence is available on the very early stages (Curzen, 1967).

Johnson: Once a cell has lost the chance to become trophoblast, does it ever have a second chance. Can the inner cell mass give rise to trophoblast subsequently? I believe that there are data from work on teratomas showing that ectopic transplants of blastocysts yield primarily trophoblastic teratomas, but transplants of egg cylinder, which include extra-embryonic tissue, produce yolk sac endoderm, but no trophoblast. However, if the extra-embryonic portion is excluded from the egg cylinder, neither yolk sac endoderm nor trophoblast will differentiate, but other sorts of tissue will do so. This argues in favour of a serial loss of totipotency.

GARDNER: The only evidence that I know on the failure of inner cell mass cells to become trophoblast cells is provided by Janet Rossant (personal communication). She has taken morulae or mid-cleavage stages carrying one type of glucose phosphate isomerase (GPI) and put them in contact with isolated inner cell masses carrying a different type of GPI. Twenty-four hours later a normal blastocyst developed with an inner cell mass about twice the size of normal. These 'embryos' would implant, and when the derivatives of the inner cell mass and trophoblast respectively were analysed in foetuses, the donor inner cell mass was found to have colonised only the embryo and certain extra-embryonic membranes. It never contributed to the trophoblast. The cells were placed on the outside of the morula, and seemed to enter it immediately without dividing much at the time. These cells contributed only to the later embryo and not to the trophoblast.

BODMER: Is there any possibility of doing biochemistry on the inner cell mass or the blastocoelic fluid? If fluid could be accumulated and then added *in vitro* to embryos, it may stimulate cell proliferation.

GARDNER: The mouse blastocyst is at most only 100 μm in diameter, and collecting fluid from it would be quite a formidable task. In rabbits, analyses of blastocoelic fluid have been carried out (Lutwak-Mann, 1971). I do not know of any studies on the effect of this fluid on cell proliferation.

REFERENCES

CURZEN P. (1967). Histochemical investigation of the concept of placental insufficiency. *J. Obstet. Gynec. Br. Commonw.*, **74**, 385.

HSU, Y-C. (1972). Differentiation *in vitro* of mouse embryos beyond the implantation stage. *Nature, Lond.*, **239**, 200.

JOHNSON, M. H. (1972). Relationship between inner cell mass derivatives and trophoblast proliferation in ectopic pregnancy. *J. Embryol. exp. Morph.*, **28**, 306 (appendix).

LUTWAK-MANN, C. (1971). The rabbit blastocyst and its environment: physiological and biochemical aspects. In *The Biology of the Blastocyst* (ed. R. J. Blandau), p. 243. University of Chicago Press: Chicago.

NAGL, W. (1972). Giant sex chromatin in endopolyploid trophoblast nuclei of the rat. *Experientia*, **28**, 217.

SNOW, M. H. L. (1973*a*). The differential effect of [^{3}H]-thymidine upon two populations of cells in pre-implantation mouse embryos. In *The Cell Cycle in Development and Differentiation* (eds. M. Balls & F. S. Billett), p. 311. Brit. Soc. Dev. Biol. Symp., Cambridge University Press: London.

SNOW, M. H. L. (1973*b*). Abnormal development of pre-implantation mouse embryos grown *in vitro* with [^{3}H] thymidine, *J. Embryol. exp. Morph.*, **29**, 601.

ORGANIZATION, ULTRASTRUCTURE AND HISTOCHEMISTRY OF THE PLACENTA: IMMUNOLOGICAL CONSIDERATIONS

By DAVID BILLINGTON

Studies on the immunobiology of trophoblast have been extremely restricted in their scope. Not only have very few species been considered, but the various forms of trophoblast, existing at different phases of pregnancy and comprising major components of the mature placenta, have received little attention. Current views in this field have been derived almost solely from information on the trophectoderm layer of the blastocyst, the early proliferating trophoblast and organised placenta of the mouse, and the early villous trophoblast and full-term placenta of man. In addition it should be noted that the life span of trophoblast is enormously variable, since the gestation period of a Eutherian mammal can range from 16 days in the hamster to 22 months in the elephant. This contrasts markedly with the relative uniformity of the immunological system in mammals.

It is not possible in this brief review to consider all the different types of trophoblast in the many mammalian species. It should be recognized, however, that there are two basic types of placenta; the chorio-vitelline placenta, where a vascularized yolk-sac fuses with the trophoblast or the chorion, and the chorio-allantoic placenta, where the chorion is vascularized by allantoic vessels (Amoroso, 1952). The former type occurs as a transient structure during the development of a number of mammals and also as the definitive placenta of most marsupials. The latter type is characteristic of all the true mammals, and a few marsupials. It is thus the chorio-allantoic placenta that is of most concern in the present context. This shows a great diversity of form, and classification depends partly upon gross morphology but largely, and more importantly from the functional point of view, on the degree of invasiveness of the trophoblast into the maternal tissues (Grosser, 1909). In the pig, the trophoblast wall of the blastocyst does not develop invasive properties and simply lies in intimate apposition to the uterine epithelium. This is therefore referred to as an epithelio-chorial placenta. In other ungulate species there may be a slight invasion of the uterine wall, with the trophoblast in contact with connective tissue, and forming a syndesmochorial placenta. When the trophoblast erodes through to the endothelium of the maternal blood vessels in the uterus this establishes an endothelio-chorial placenta, as in many carnivores. The final degree of normal trophoblast invasiveness is seen in the haemochorial placenta, such as that in man, other primates, and the rodents. Here the uterine capillaries are breached and the trophoblast is bathed in maternal blood. Electron microscopical studies have now shown that this trophoblast may be in the form of either a single layer (man, guinea-pig), two layers (rabbit) or three layers (rat, mouse). In the single and double-layer forms the outer trophoblast appears to be syncytial, whilst that of the three-layer forms is cellular, covering two layers of syncytium (Enders, 1965).

In all mammalian species trophoblast is the only foetal element in direct and continuous contact with the maternal tissues of the placenta and thus capable of acting as a physical barrier between the two organisms at this level. It can also be seen that trophoblast may be in contact with different maternal tissues in the different placental types. The potential susceptibility of the trophoblast to maternal immune attack may thus to some extent be influenced by its position.

Since murine and human trophoblast have been most extensively investigated, placental morphogenesis will be outlined only in these species. The choice of mouse and man for detailed study is clearly justified for many reasons, but it must be emphasized that they may not give more than a limited insight into mechanisms of foetal survival in mammals with other placental forms. The evolution of the viviparous state and the immunological system has occurred side by side, and probably interdependently, and all types of existing maternal-foetal association must have been biologically successful in their development.

Mouse placental development

The trophectoderm in the abembryonic and lateral areas of the blastocyst gives rise to primary trophoblast giant cells. Their function is not finally established, but probably includes an anchoring mechanism during the early stages of implantation. The trophectoderm overlying the inner cell mass at the embryonic pole of the blastocyst undergoes rapid proliferation to form the so-called ectoplacental cone, composed of highly invasive trophoblast cells. From its superficial regions arise the secondary trophoblast giant cells, and these migrate outwards to lie at the margins of the developing placenta and downwards to surround the early embryo. These cells become highly polyploid, by a process now known to be endomitosis (Chapman, Ansell & McLaren, 1972), but their function is obscure. The proliferative centre of the ectoplacental cone produces tissue that differentiates in some as yet unknown way to form the main foetal components of the placenta, the labyrinthine and spongio-trophoblast. The composition of the mature placenta is illustrated in diagrammatic form in Fig. 1. All three types of trophoblast are in contact with maternal tissues (Fig. 2). The giant cells and the spongio-trophoblast abut uterine decidual

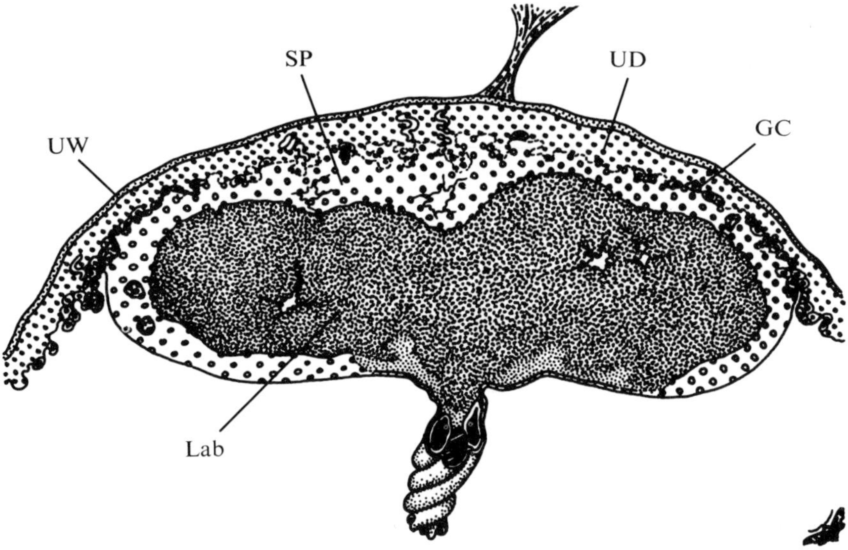

Fig. 1. Diagrammatic representation of a mid-line section of the definitive mouse placenta.

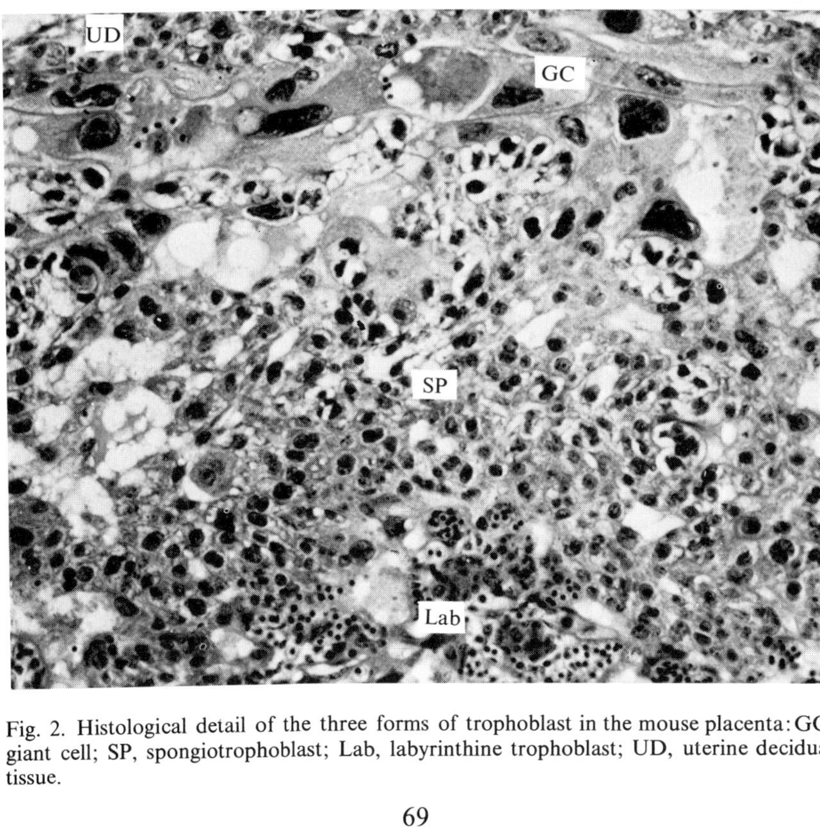

Fig. 2. Histological detail of the three forms of trophoblast in the mouse placenta: GC, giant cell; SP, spongiotrophoblast; Lab, labyrinthine trophoblast; UD, uterine decidual tissue.

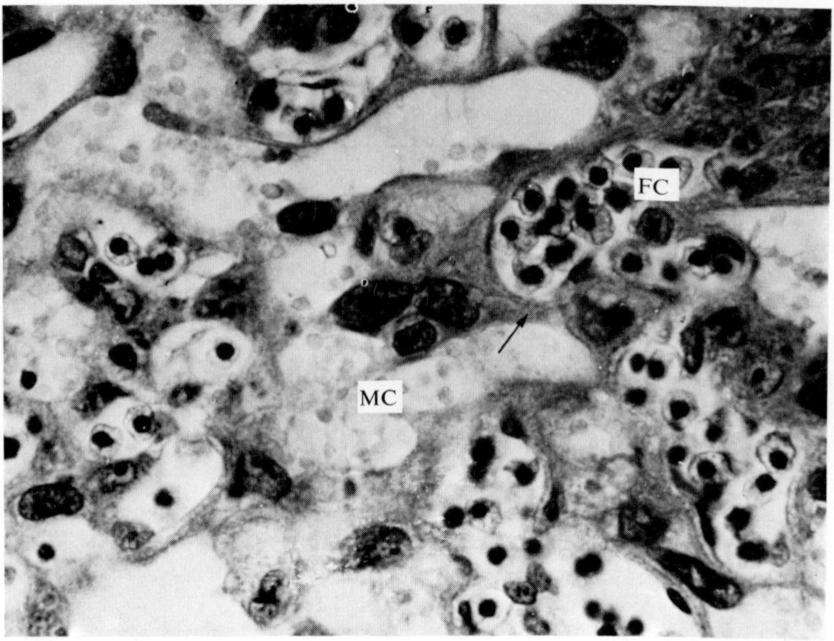

Fig. 3. Placental labyrinth of the mouse. The nucleated foetal blood cells (FC) are separated from the smaller maternal blood cells (MC) by a trophoblast membrane (arrow) that can be seen to be trilaminar under the electron microscope. The large dense nuclei are trophoblastic.

tissues with close cellular contact. The labyrinthine trophoblast is actually trilaminar, and the outer cellular layer is bathed by the maternal blood in the placental sinuses (Fig. 3).

Human placental development

Soon after implantation, the trophectoderm of the human blastocyst gives rise to two basic forms of trophoblast, an inner cellular cytotrophoblast and an outer syncytiotrophoblast (Fig. 4). The syncytial form is derived by differentiation, and breakdown of cell membranes, of the cellular layer, and undergoes quantitative increase throughout pregnancy. At an early stage columns of cytotrophoblast push through the syncytium, forming the villous primordia, and spread out at the ends to form a 'shell' around the embryo. The characteristic mature placental villi develop ultimately with an outer covering of syncytium and an underlying layer of cytotrophoblast which in the later stages of pregnancy becomes discontinuous. It is therefore the villous syncytium that is bathed in maternal blood, and the trophoblast of the basal plate that is in close contact with the uterine decidual tissue in the organized placenta.

The fibrinoid demonstrable by standard histochemical techniques must now be clearly distinguished from the membrane-bound mucoprotein on the trophoblast surface. As pointed out recently by Wynn (1971), the presence of a thick fibrinoid layer on the surface of labyrinthine or villous trophoblast in the haemochorial placenta would almost certainly be incompatible with normal placental transfer function. The trophoblast mucoprotein can be detected by optical or electron histochemistry, and it is possible to correlate the sites of histochemical reactivity for mucoprotein with the ultrastructural morphology of the trophoblast. One particularly useful method involves the binding of colloidal iron onto the cell surface and its visualization in the electron microscope (Fig. 6). Many, if not all, cell types in fact possess a mucoprotein coat, or 'glycocalyx' as it is often referred to (Rambourg, Neutra & Le Blond, 1966). The important point to establish, therefore, is whether the trophoblast mucin differs in any way that could account for the immunological privilege of this tissue. The most marked feature so far identified, at least on human trophoblast, is an extreme degree of sulphation. This has been confirmed by the use of a number of direct histochemical tests (Bradbury *et al.*, 1970), and supported by biochemical analysis of placental extracts (Calatroni & Di Ferrante, 1969). It is suggested that the sulphated mucoprotein could offer protection to the trophoblast either by masking antigenic determinants on the cell surface or by producing an electrostatic barrier with a high negative charge to repel maternal lymphocytes, as originally proposed by Currie & Bagshawe (1967).

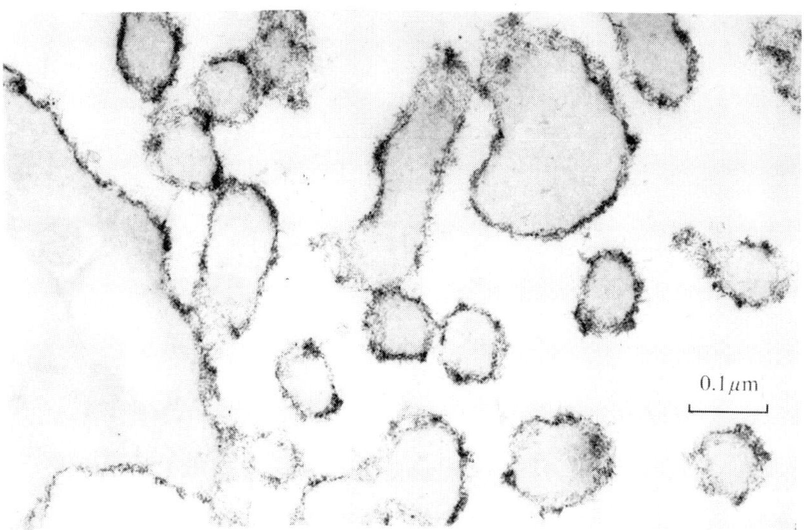

Fig. 6. Electron micrograph of human placental villi showing colloidal iron binding on the syncytiotrophoblast surface. The marker represents 0.1 μm (Bradbury *et al.*, 1970).

Earlier evidence indicated that the trophoblast mucoprotein contained significant quantities of sialic acid, and that this had an important role in trophoblast immunobiology (Kirby *et al.*, 1964; Currie & Bagshawe, 1967). This was supported by experimental studies on mouse ectoplacental cone trophoblast showing that the enzyme neuraminidase, which cleaves terminal *N*-acetyl neuraminic acid from cell surface sialomucoproteins, rendered the trophoblast immunogenic (Currie, van Doorninck & Bagshawe, 1968). However, this finding has not been confirmed by other workers (Simmons, Lipschultz, Rios & Ray, 1971), and needs further investigation. Using a different approach, we have also failed to expose antigenic sites after neuraminidase treatment of early mouse trophoblast (Jenkinson & Billington, 1974*a*). In-vitro outgrowths of ectoplacental cone, either untreated or pre-incubated with the enzyme, were not susceptible to lysis by immune lymphoid cells. It would appear that this is not due to a lack of neuraminidase-sensitive material on the trophoblast since similar cultures were examined by electron microscopy and colloidal-iron binding shown to be reduced following enzyme treatment (Fig. 7) (R. F. Searle, unpublished results). The present conclusion must be that the mouse trophoblast mucoprotein is sialic (actually, *N*-acetyl neuraminic) acid containing, but that it has not yet been proven to be involved in the 'masking' of surface antigens, in the manner reported for certain tumour cells (Sanford, 1967; Currie & Bagshawe, 1968).

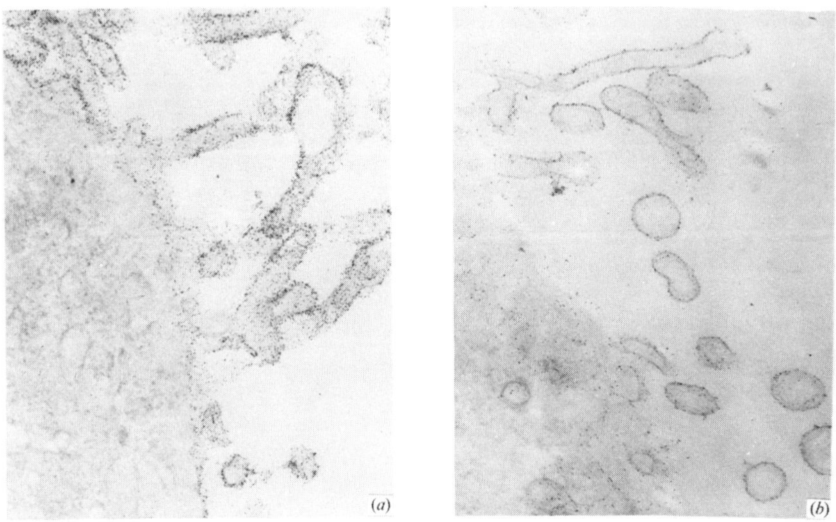

Fig. 7. Electron micrographs of mouse ectoplacental cone trophoblast outgrowths *in vitro* (courtesy of Dr R. F. Searle).
(*a*) Colloidal iron staining on the surface of control outgrowths
(*b*) Similar preparation following neuraminidase treatment. Note the marked reduction in colloidal iron binding.

Although the term 'sialomucin' is also frequently used for the surface coat of human trophoblast it cannot be said that this is as yet entirely justified. There is little direct evidence for the presence of sialic acid, and our own histochemical investigations have failed to detect any significant levels. Incubation of sections in neuraminidase had no appreciable effect on the intensity of staining reactions on human villous trophoblast, nor did a direct test for sialic acid prove positive (Bradbury *et al.*, 1970). It may be that although the surface mucins of mouse and human trophoblast afford immunological protection there is some difference in the precise mechanisms involved. But since there is still no conclusive proof of the existence of histocompatibility antigens on the surface of any of the different forms of trophoblast, this is merely speculation.

Antigens of the foetal membranes

There appears to be a wide belief that the placenta is the only region of importance as far as the problem of the survival of the embryo as an intra-uterine allograft is concerned. However, although the precise relationships vary from species to species, there is always extensive contact between either the foetal chorionic membrane (e.g. Fig. 5) or the yolk sac (e.g. Fig. 8) and the uterine tissues away from the placental site. Over the non-placental portion of its surface the conceptus can also be regarded as potentially vulnerable to maternal immune attack. This therefore raises the question of the antigenic status of the foetal membranes.

We have recently investigated the antigens of the mouse yolk sac at different stages of development by means of a cell-mediated microcytotoxicity system. Cultures of cells from the entire yolk sac, and pure endodermal cells from the outer layer, were prepared by trypsinization of dissected or intact membranes and incubated with spleen cells from highly immune allogeneic donors. Specific target cell destruction occurred with all material obtained from days 10 to 16 of pregnancy (Jenkinson & Billington, 1974*b*). The mouse yolk sac, which is in direct contact with the uterine wall after the breakdown of Reichert's membrane at about day 14, therefore possesses antigens on its outer (endodermal) surface. From a study on the grafting of yolk sac fragments to prepared sites on the thorax of pre-sensitized recipient mice, Avery & Hunt (1967) also concluded that the membrane as a whole was antigenic. However, they considered that the yolk sac might still function as a barrier by virtue of an outer layer of lower antigenicity than that of its other cellular elements. This cannot now be accepted, and the lack of susceptibility of the yolk sac to immune attack *in vivo* must depend upon some other factor(s). The lack of vascularization of the membrane by uterine vessels is undoubtedly of significance, and in addition it is possible that maternal lymphocytes are

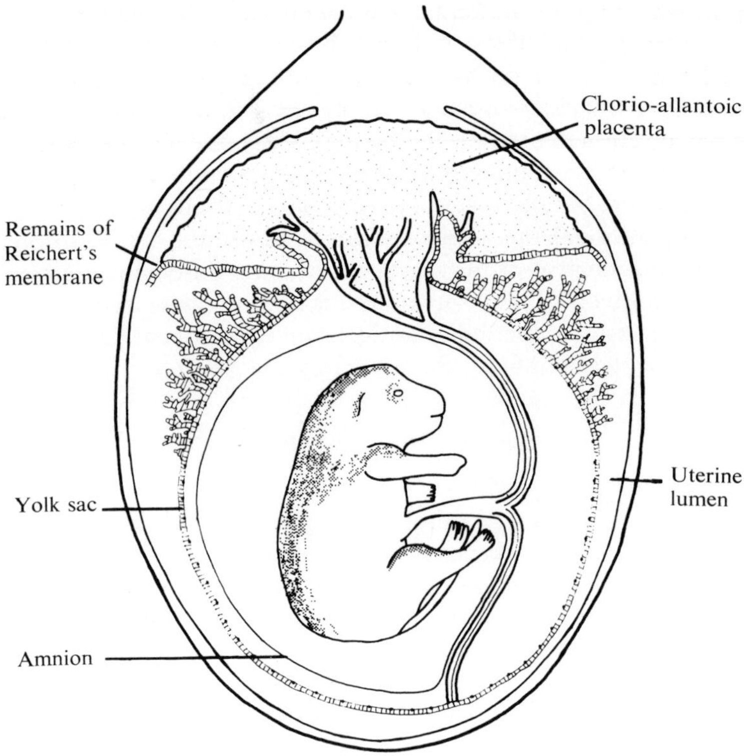

Fig. 8. Following the breakdown of Reichert's membrane at the beginning of the second half of pregnancy the yolk sac is in direct contact with the uterine wall. The situation in the rat and mouse is almost identical.

unable to gain access across the regenerated uterine epithelium. The latter suggestion would not suffice in relation to the human chorion since this membrane is in direct contact with uterine decidual tissue throughout pregnancy. The human embryo has an interstitial type of implantation and does not therefore develop in the lumen of the uterus.

Finally, it is of interest that in experimental situations the yolk sac can be susceptible to humoral antibody. Heterologous antisera to kidney, placenta and yolk sac preparations interfere with yolk sac function following injection into pregnant rats, and secondarily induce malformations of the foetus (Brent, 1971).

SUMMARY

The organization of the placenta differs widely among mammalian species, and the trophoblast elements exhibit different biological forms during placental morphogenesis and in the mature organ. This must be taken into account in the assessment of current information on trophoblast immunobiology. Histochemical studies indicate that deposition of fibrinoid material is a common occurrence at the main deciduo–trophoblastic interface in the haemochorial placenta. It is not yet clear whether this is of primary immunological significance. The fibrinoid must now be distinguished from the membrane-bound mucoprotein coating the free surface of the trophoblast in the mature placenta. The chemical nature of the mucoprotein may provide a clue to the possible role of this substance in the protection of the trophoblast against maternal immune reactions. Investigations on foetal membranes of the mouse indicate that these are antigenic and of importance in the immunological interrelationships between mother and foetus.

REFERENCES

AMOROSO, E. C. (1952). Placentation. In *Marshall's Physiology of Reproduction* (ed. A. S. Parkes), Vol. II, 3rd edn, p. 127. Longmans, Green: London.

AVERY, G. B. & HUNT, C. V. (1967). The foetal membranes as a barrier to transplantation immunity. *Transplantation*, **5**, 444.

BARDAWIL, W. A. & TOY, B. L. (1959). The natural history of choriocarcinoma: problems of immunity and spontaneous regression. *Ann. N.Y. Acad. Sci.*, **80**, 197.

BRADBURY, S., BILLINGTON, W. D. & KIRBY, D. R. S. (1965). A histochemical and electron microscopical study of the fibrinoid of the mouse placenta. *J. R. Microsc. Soc.*, **84**, 199.

BRADBURY, S., BILLINGTON, W. D., KIRBY, D. R. S. & WILLIAMS, E. A. (1969). Surface mucin of human trophoblast. *Am. J. Obstet. Gynec.*, **104**, 416.

BRADBURY, S., BILLINGTON, W. D., KIRBY, D. R. S. & WILLIAMS, E. A. (1970). Histochemical characterization of the surface mucoprotein of normal and abnormal human trophoblast. *Histochem. J.*, **2**, 263.

BRENT, R. L. (1971). The effect of immune reaction on fetal development. *Adv. in the Biosciences*, **6**, 421.

CALATRONI, A. & DI FERRANTE, N. (1969). The glycosaminoglycans of human term placenta. *Carbohyd. Res.*, **10**, 535.

CHAPMAN, V. M., ANSELL, J. D. & MCLAREN, A. (1972). Trophoblast giant cell differentiation in the mouse: expression of glucose phosphate isomerase (GPI-1) electrophoretic variants in transferred and chimeric embryos. *Devl. Biol.*, **29**, 48.

CURRIE, G. A. & BAGSHAWE, K. D. (1967). The masking of antigens on trophoblast and cancer cells. *Lancet*, **i**, 708.

CURRIE, G. A. & BAGSHAWE, K. D. (1968). The effect of neuraminidase on the immunogenicity of the Landschutz ascites tumour: site and mode of action. *Br. J. Cancer*, **22**, 588.

CURRIE, G. A., VAN DOORNINCK, W. & BAGSHAWE, K. D. (1968). Effect of neuraminidase on the immunogenicity of early mouse trophoblast. *Nature, Lond.*, **219**, 191.

ENDERS, A. C. (1965). A comparative study of the fine structure of the trophoblast in several haemochorial placentas. *Am. J. Anat.*, **116**, 29.

GROSSER, O. (1909). *Vergleichende Anatomie und Entwicklungsgeschichte der Eihäute und der Placenta*, Wilhelm Braumüller: Wein und Leipzig.

HAMILTON, W. J. & BOYD, J. D. (1970). Development of the human placenta. In *Scientific Foundations of Obstetrics and Gynaecology* (eds E. E. Philipp, J. Barnes & M. Newton), p. 185. Heinemann: London.

JENKINSON, E. J. & BILLINGTON, W. D. (1974*a*). Differential susceptibility of mouse trophoblast and embryonic tissue to immune cell lysis. *Transplantation*, **18**, 286.

JENKINSON, E. J. & BILLINGTON, W. D. (1974*b*). Studies on the immunobiology of mouse foetal membranes: the effect of cell-mediated immunity on yolk sac cells *in vitro*. *J. Reprod. Fertil.*, (in press).

KIRBY, D. R. S., BILLINGTON, W. D., BRADBURY, S. & GOLDSTEIN, D. J. (1964). Antigen barrier of the mouse placenta. *Nature, Lond.*, **204**, 548.

RAMBOURG, A., NEUTRA, M. & LE BLOND, C. P. (1966). Presence of a 'cell coat' rich in carbohydrate at the surface of cells in the rat. *Anat. Rec.*, **154**, 41.

SANFORD, B. H. (1967). An alteration in tumour histocompatibility induced by neuraminidase. *Transplantation*, **5**, 1273.

SIMMONS, R. L., LIPSCHULTZ, M. L., RIOS, A. & RAY, P. K. (1971). Failure of neuraminidase to unmask histocompatibility antigens on trophoblast. *Nature New Biol.*, **231**, 111.

WYNN, R. M. (1967). Comparative electron microscopy of the placental junctional zone. *Obstet. Gynec.*, **29**, 644.

WYNN, R. M. (1971). Immunological implications of comparative placental ultrastructure. In *The Biology of the Blastocyst* (ed. R. J. Blandau), p. 495. University of Chicago Press: Chicago and London.

DISCUSSION

EDWARDS: Recent electron microscopical studies of the interface between trophoblast and decidual cells have shown no evidence of fibrinoid layers in the human placenta or in the placentas of several other species (Tekelioglu-Uysal, Edwards & Kisnisci, 1975; Martinek, 1970, 1971). The areas of fibrinoid deposition were discrete and not continuous. Any idea of fibrinoid or other barriers of that sort must be abandoned in the face of this evidence.

BILLINGTON: The situation varies in different placental forms. The main point is that the histochemically-demonstrable deposits, which have been called 'fibrinoid', are quite separate from the surface mucoproteins that may have some function in relation to the antigenic properties of the trophoblast. The comparative studies of Wynn (1967*a*) on various placental forms have shown thick deposits of fibrinoid in many haemochorial placentae. However, Wynn was only examining the fully differentiated placenta and such deposits may not occur in the early stages of development. The very heavy deposits may result from some form of interplay between the invasive trophoblast and the maternal tissue, since the deposition is much more prevalent in those placentae where the trophoblast is highly invasive. The fibrinoid probably results from the degeneration products of the interaction between these two cell types, and although conclusions about immunological function cannot be drawn, the material does separate certain foetal and maternal tissues in many species.

CURZEN: A difficulty of semantics arises here. The word 'fibrinoid' has been used in the past and it must be clearly defined. The layer of Nitabuch in the basal plate was mentioned; histochemically this is periodic acid/Schiff-reactive material, which is not found around the trophoblast of the chorionic villi in the human placenta. The surface layer in the mouse trophoblast, and indeed in the human trophoblast round the chorionic villi, was described as being alcianophilic and colloidal iron positive. Has a layer analogous to that in the basal plate of the human been found in the mouse placenta using periodic acid/Schiff staining?

BILLINGTON: There can be no question of fibrinoid material, as placentologists describe it, being present around the human chorionic villus, because fibrinoid material in that site would preclude the normal physiological functions ascribed to trophoblast. It can only be a glycocalyx sort of material in that situation. In the past there has been a semantic difficulty about the use of the term 'fibrinoid', and I now avoid it.

CURZEN: Can you see any difference between alcianophilic layers on trophoblast as distinct from other epithelial type tissues? Do these layers exist on gut, for example?

BILLINGTON: The Alcian blue technique has been used on a variety of tissues and they are all positive. Where they differ is in the question of sulphation, in that this dye is extinguished with very low levels of magnesium chloride on all cell types apart from trophoblast. Trophoblast requires very much higher levels of magnesium chloride to achieve the same result, and this indicates a very high degree of sulphation.

ALLISON: Analogies arise with the renal glomerulus, where the epithelial and endothelial cells are in close apposition, and with the lung. In both of these situations a basal lamella with a distinct chemical composition and with immunological properties is present. For example, in the disease called Goodpasture's Syndrome, an autoantibody is formed which reacts specifically with these two types of materials. Has this been tried in the case of the placenta?

BILLINGTON: No, not to my knowledge.

AMOROSO: When Edwards spoke of early stages in the human, showing no evidence of a continuous layer in this so-called fibrinoid in the early human placenta, what was the stage of growth of the embryo: was it before the time when HCG production begins?

EDWARDS: The placental tissues were obtained by menstrual aspiration. This would be approximately 15 days post-implantation. To the best of my knowledge HCG is produced at least a week before the expected time of menstruation.

BAGSHAWE: How sensitive is the method described by Billington for detecting sialic acid? Sialic acid was not detected on trophoblast, yet at least one of the trophoblast's secretory products is a highly sialated molecule, and I am surprised some sialic acid was not detected from that alone.

BILLINGTON: The levels were not elevated compared with other tissues.

BAGSHAWE: You detected some release: what is a significant amount in molecular terms?

BILLINGTON: I would not place much emphasis on our biochemical assays. We were unable to influence any of the histochemical staining by pre-incubation with neuraminidase, which is rather more convincing. However, some of these histochemical techniques are tricky because of interference due to other components of the cell surface.

GARDNER: Some clarification is needed about neuraminidase. In experiments on ectoplacental cone, there are two types of cell which may be quite different, i.e. the diploid stem cell population and the giant cells. In the in-vitro system, the giant cells are probably more persistent, whereas *in vivo* in ectopic transfers the smaller cells are probably more important. Neuraminidase could be more effective on the small cells in ectopic sites which transform into giant cells, whereas *in vitro* you are looking mainly at the surface of giant cells (see also paper by Johnson).

BILLINGTON: Our microcytotoxicity tests are carried out after the ectoplacental cone has been in culture for two or three days. We have looked with the electron microscope at the different types of cells that grow out after that time, and can see two morphologically distinct types (R. F. Searle, unpublished). It seems reasonable to assume that these are the diploid and giant cells, yet there is no effect on any cell in that outgrowth.

GARDNER: If the trophoblast is effectively isolated from the inner cell mass cells, there may not in fact be any significant number of diploid cells present after that period in culture. You could be dealing with giant cells only.

EDWARDS: The association between decidual cells and trophoblast cells is very close (Wynn, 1967*b*; 1969; Martinek, 1970, 1971). There is no evidence of any barrier, or any repulsion, between these types of cell especially in early pregnancy. Such a closely-knit structure does not indicate that the tissues are held apart by differences in charge or similar effects.

BILLINGTON: This may be true of some of the early trophoblast which is in intimate association with the decidual cells. However, cell-surface materials must be coating these cells, even with such close apposition, at least in the haemochorial placenta.

MCLAREN: On this question of intimacy of association between trophoblast and maternal tissues, is it still accepted that in the rabbit, for instance, one gets actual cell fusion between trophoblast and maternal cells?

BILLINGTON: Yes. The evidence for that came from Larsen (1961). There is also some supporting evidence from Glenister (1970).

EDWARDS: Enders & Schlafke (1971) have clear evidence of the fusion of the trophoblastic knobs of the rabbit blastocyst and the uterine epithelium.

AMOROSO: Larsen did not really show cell fusion of the chorio-allantoic placenta, but of the early stages where the abembryonic trophoblast was in contact with the maternal tissues. There was a general breakdown at the site of production of the chorio-vitelline placenta, before the germ layers became completely inverted.

BILLINGTON: Nevertheless, the immunological implications of that are of some interest, although we do not know what happens to the cells afterwards. They probably die.

BODMER: What is the possible significance of the layer of decidual cells in forming a barrier to access by lymphocytes and other components of the maternal immune system to the trophoblastic cell?

BILLINGTON: Decidual tissue may have quite a thick cell-surface coating, and be capable of producing some of this 'fibrinoid' material (Wynn, 1969). Its role in limiting the passage of lymphocytes is not known, although there is evidently a good lymphatic drainage at the placental site. Lymphocytes appear in the space below the uterine epithelium before the decidua is fully established, and there are few in the later stages (see also Discussion after Howe's paper, p. 149).

McLaren: First, one problem about the decidua as a barrier is that, in women, the decidua proper does not develop strongly until a later stage. Secondly, would Bagshawe comment on the attractive idea (Currie & Bagshawe, 1967) of a high negative charge on trophoblast repelling negatively charged lymphocytes? Has anything more been done about that?

Bagshawe: The idea really arose from Kirby, Billington, Bradbury & Goldstein (1964) working on placental fibrinoid, and their observation that it was a highly sulphated material. Observations at that time on various types of tumour cells indicated they had a high net negative charge, and we offered this hypothesis after preliminary testing with various types of tumour cells. But there are some types of tumour where a high net negative charge is not found on the cell surface. The hypothesis could not be tested readily with trophoblast. Neuraminidase may have several actions beside simply clearing sialic acid groups and reducing cell charge. There has evidently been no subsequent work on the surface charge on trophoblast cells.

Curzen: The decidua becomes completely covered by a layer of cytotrophoblast – the cytotrophoblastic shell – which has grown out through the tips of the terminal villi of the villus tree. The cytotrophoblast of foetal origin is in intimate apposition with decidua and this particular area is where the trophoblast becomes invested by what appears to be fibrinoid material, in the sense that it behaves like fibrinoid histochemically. Otherwise the decidua does not appear to constitute any sort of immunological barrier in the human placenta.

Amoroso: Many people have assumed that in the human placenta the important relationship is between foetal trophoblast and circulating maternal blood, where maternal lymphocytes could come in direct contact with foetal tissues. But there is also the cytotrophoblastic shell, which is in direct contact with maternal decidua. If the surface of the microvilli is poor in sialic acid, we must be careful in interpreting the physiological events taking place at the surface of that microvillus. The syncytial trophoblast produces chorionic gonadotrophin, a continuous process, and the large quantity of chorionic gonadotrophin produced at any one moment may be underestimated. Chorionic gonadotrophin may form a membrane on the surface of the microvilli which has to be constantly changed to exert a continuing local effect; when it enters the general circulation it cannot be recycled. More human chorionic gonadotrophin must be produced to keep up the concentration on the microvilli, and this would explain the puzzling phenomenon of the large quantities of HCG secreted during human pregnancy. The work of Bagshawe and his group has shown that HCG may be implicated in immunological phenomena, so I would suggest that the sialic acid in the HCG is the important ingredient at the surface of the placental membrane. It will always be there in small quantities, though its production is high enough to ensure a sufficient concentration on the surface of the microvillus. (See also paper by Borland and Discussion after it p. 165.)

JOHNSON: The question of the yolk sac endoderm in later pregnancy has always worried me. The explanation may be on the lines of Munro's suggestion earlier (p. 9); when these cells are placed in culture all sides of the cells are exposed, whereas only one layer which is devoid of antigens may be exposed *in vivo*. Can you think of a way in which you might be able to test this?

BILLINGTON: Initially, we took the whole membrane by dissection, minced and trypsinized it, and seeded it into culture. The immune cells were then added, and an absolute destruction of the culture was seen. In the next experiment the whole of the intact yolk sac was dipped into trypsin to digest off the outside cells, which are endodermal because of embryonic germ-layer inversion. These were then seeded out. We do not know what happens to the orientation of those cells in culture; they may settle down at random, with different parts of their original surface uppermost.

JOHNSON: I believe Avery & Hunt (1967) grafted whole yolk sac upside down or the right way up and found that the cellular infiltration was slower if the endoderm surface was outwards. Can a short-term organ culture of a dissected membrane, with lymphocytes in a sort of pocket of yolk sac be tested to find out if they enter from the maternal side or the foetal side? In such a system, the junctional complexes would be retained between the endodermal cells.

BILLINGTON: We are approaching this question at the moment. If the uterine wall is split very carefully in the later stages of pregnancy in mice, the yolk sac bulges out. By using either immunized animals or injecting immune lymphocytes, we will assess whether the exposed wall of the yolk sac can be attacked. This operation will retain the normal relationships between the cells and yet put them in a position where lymphocytes could get to them.

BODMER: Could you comment on the specificity control of your cytotoxic assay. You say it is a cell mediated cytotoxic assay. Does it involve K cells or T cells?

BILLINGTON: It involves H-2 and non-H-2 combinations of immune spleen cells in culture with the target monolayers. The effector cells are from mice immunized by skin grafts and intraperitoneal injections of spleen cells.

BODMER: Does the absence of any killing of trophoblast guarantee that the antigen isn't there?

BILLINGTON: The controls used were fibroblasts, and also embryonic tissues other than trophoblast from the same conceptus.

BODMER: I am not familiar with that sort of assay, but difficulties arise with conventional assays of cytotoxic antibodies for the detection of antigens on cells. Many cell types, for instance granulocytes, simply do not react – at least in humans – to anti-HL-A antibodies in a cytotoxic test. Hence unless the antigens are isolated, or inhibition studies are done, the absence of antigens is not convincing (e.g. Miggiano, Nabholz & Bodmer, 1970).

BILLINGTON: I agree. However, we have a conceptus which is composed of two different types of tissue, one is trophoblastic and the other is embryonic. In this test system, the embryonic tissues are destroyed by the immune cells whereas the trophoblast is not. We think that this has some biological significance.

BODMER: But the same sort of differences might arise with respect to other sorts of tissues within the embryo, for example, between phagocytic cells, and non-phagocytic cells. So the significance with respect to the function of the trophoblast itself may not be all that clear?

BILLINGTON: To some degree; except that the trophoblast is the only foetal tissue involved in the placental barrier.

REFERENCES

AVERY, G. B. & HUNT, C. V. (1967). The foetal membranes as a barrier to transplantation immunity. *Transplantation*, **5**, 444.

CURRIE, G. A. & BAGSHAWE, K. D. (1967). The masking of antigens on trophoblast and cancer cells. *Lancet*, **i**, 708.

ENDERS, A. C. & SCHLAFKE, S. (1971). Penetration of the uterine epithelium during implantation in the rabbit. *Am. J. Anat.*, **132**, 219.

GLENISTER, T. W. (1970). Ovo-implantation *in vitro* and its relation to normal implantation. In *Ovo-implantation, Human Gonadotrophins and Prolactin* (eds. P. O. Hubinont, F. Leroy, C. Robyn & P. Leleux); p. 73. S. Karger, Basel, 14.

KIRBY, D. R. S., BILLINGTON, W. D., BRADBURY, S. & GOLDSTEIN, D. J. (1964). Antigen barrier of the mouse placenta. *Nature, Lond.*, **204**, 548.

LARSEN, J. F. (1961). Electron microscopy of the implantation site in the rabbit. *Am. J. Anat.*, **109**, 319.

MARTINEK, J. J. (1970). Fibrinoid and the fetal-maternal interface of the rat placenta. *Anat. Rec.*, **166**, 587.

MARTINEK, J. J. (1971). Ultrastructure of the deciduotrophoblastic interface of the mouse placenta. *Am. J. Obstet. Gynec.*, **109**, 424.

MIGGIANO, V., NABHOLZ, M. & BODMER, W. F. (1970). Detection of HLA and other antigens on fibroblast monolayers using a fluorochromatin cytotoxic assay. In *Histocompatibility Testing 1970* (ed. P. I. Terasaki), p. 623. Munksgaard: Copenhagen.

TEKELIOGLU-UYSAL, M., EDWARDS, R. G. & KISNISCI, H. A. (1975). Ultrastructural relationships between decidua, trophoblast and lymphocytes at the beginning of human pregnancy. *J. Reprod. Fert.* (in press).

WYNN, R. M. (1967*a*). Comparative electron microscopy of the placental junctional zone. *Obstet. Gynec.*, **29**, 644.
WYNN, R. M. (1967*b*). Foetomaternal cellular relations in the human basal plate. An ultrastructural study of the placenta. *Am. J. Obstet. Gynec.*, **97**, 832.
WYNN, R. M. (1969). Non-cellular components of the placenta. *Am. J. Obstet. Gynec.*, **103**, 723.

ANTIGENS OF THE PERI-IMPLANTATION TROPHOBLAST

BY MARTIN JOHNSON

For obvious reasons, discussion on the antigens of peri-implantation trophoblast must be restricted to non-human material and will in fact almost exclusively deal with the mouse. Data on this whole area are thin and in this review I want to establish what few unequivocal observations exist and to attempt a more rational approach to the question of trophoblast antigenicity based on our clearer understanding of the relationship of subclasses of trophoblast cells. Trophoblast differentiation in the mouse over the period of interest is summarised schematically in Fig. 1. Four classes of trophoblast cell are involved, the primary trophectoderm, the primary trophoblastic giant cells, the core proliferative cells of the ectoplacental cone and the secondary giant cells. Whatever the shared properties of cells from any of these four classes may be, the differences in their stage of development and in their developmental histories might be reflected in different antigenic properties.

The general course of the discussion will be first to establish whether or not antigens integral to trophoblast cells can in fact be detected, secondly to attempt the specific definition of these antigens and finally to

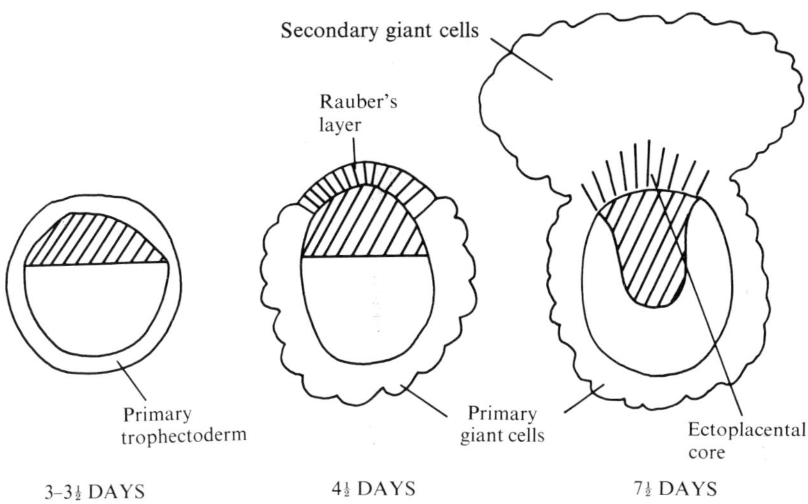

Fig. 1. Schematic representation of mouse trophoblast in the peri-implantation period of pregnancy.

assess briefly their significance in relation to embryo development and survival.

Are antigens integral to the trophoblast detectable?

Simmons & Russell demonstrated in 1967 that an ectopic transplant of rat ectoplacenta to a mouse pre-immunised to rat-species antigens was rapidly and totally rejected. A similar experiment with the blastocyst gave identical results unless the recipient was immuno-suppressed (Dehem, Subis & Vandeputte, 1971). Results of serological tests lead to the expectation of this result.

Thus, xenoantisera raised in either rabbit or rat react with mouse $3\frac{1}{2}$- and $4\frac{1}{2}$-day blastocysts and with the ectoplacental cones derived from $7\frac{1}{2}$-day embryos as revealed both in whole mounts and sectioned material (Gardner & Johnson, 1973; unpublished data with Muggleton-Harris). Cytotoxic tests are clear and unequivocal as are immunofluorescence tests. That the surface xenoantigens visualised by immune fluorescence are integral to the membrane rather than part of an adsorbed or secreted coating layer is rendered likely by the graft rejection and cytotoxic results. Further evidence to this effect is provided by the patching of surface antigens induced by anti-species antibodies on $4\frac{1}{2}$-day blastocysts (Fig. 2). The

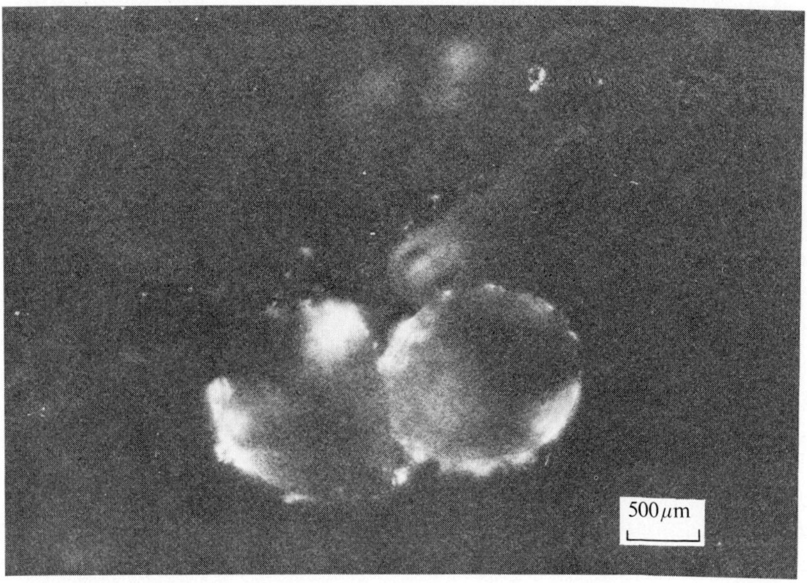

Fig. 2. Two mouse $4\frac{1}{2}$-day blastocysts treated in a two-step immune fluorescence protocol. First layer rabbit anti-mouse species antiserum. Second layer fluorescent antiserum to rabbit IgG. Note patches of stain on surface of trophoblast cells, especially evident on top-right membrane.

patching occurs only slowly at 4 °C when a clear continuous rim of fluorescence is observed. On warming the 4½-day embryo, patching develops on the trophoblast cells within 20 minutes. The patching does not occur, or only occurs slowly, at high and low antiserum dilutions, and is not prevented by 3×10^{-3} M azide. These observations and the clear distinction between phagocytosed material within the cell and the surface fluorescence on the cell make micropinocytosis an unlikely explanation of the results, but we cannot yet formally exclude this possibility (unpublished data with Muggleton-Harris).

The inner cell mass dissected from the 4½-day embryo does not appear to patch with antispecies antisera, neither do unfertilised eggs, blastomeres of the morula or the trophectodermal cells of early cavitating 3½-day blastocysts. Late 3½-day blastocysts do show trophoblastic patching. This difference of ease of patching between cells may represent changing profiles of macromolecular antigens of varying mobility within the membrane, for the xenoantiserum is undoubtedly directed against multiple antigenic specifities. A second explanation of these observations is that during the transformation of trophectoderm to primary giant cells, a change occurs in the membrane properties of the trophoblast which facilitates mobility of determinants within the membrane. Another indication of membrane changes at this time has been described by Yanagimachi & Nicholson who showed a fall in the binding of certain plant lectins immediately prior to implantation (personal communication). Richard Gardner (see p. 51) has also noted changes in the ability of trophoblast to bind and phagocytose melanosomes at the same period. The potential significance of these observations in connection with the vulnerability of the early embryo to immune attack will be discussed later. The important conclusion at this stage, is that integral antigens seem to be susceptible to detection.

Allo-antigenic status of the blastocyst

Conventional embryo-grafting techniques to systemically immunised mice have detected alloantigens on the mouse embryo. Grafts either of 2–8-cell eggs or of 3½-day blastocysts are both totally suppressed without histological trace within one week of transfer. In general, a hyperimmunisation schedule has proved essential, which could indicate a requirement either for a high titre response to strong antigens or for induction of significant immunity to weak antigens. Coisogenic animals were not used, and thus it was not possible to determine which alloantigenic specificities were involved (Kirby, Billington & James, 1966; Simmons & Russell, 1965, 1967). Subsequently 3½-day embryos have been grafted between strains coisogenic for H-2 and between strains differing only at weak, non-H-2 loci. (Searle *et al.*, 1974). Male recipients were immunized by means of two full-thickness male skin-grafts and between two and ten booster i.p. injections of male lymphocytes. Seven days after the final boost, embryo grafts were placed beneath the kidney capsule and were examined one to seven days later. Controls consisted of embryo grafts of the immunised strain to the contralateral side, or of immunising strain to non-immune or 'syngeneically immunised' recipients. The results from these studies are summarised in Tables 1, 2 and 3.

It will be seen first that a difference at H-2 is not required for embryo rejection within seven days. Examination of the grafts at earlier times revealed that the embryo continued to grow ectopically with developing egg cylinder and trophoblast transformation. However, within 48 hours a lymphocytic and polymorph infiltration was evident and this was massive by 72 hours. If the zona pellucida was removed prior to transfer, the lymphocytic invasion was evident within 24 hours. It thus seems highly probable that the antigens were present on the transferred embryo and were not developing subsequently. In contrast to these results when strains differed at H-2 only, no significant rejection or lymphocytic infiltration was observed. Removal of the zona pellucida did not influence this pattern of results. It seems reasonable to assume that any exposed H-2 antigens on blastocyst cells would have been detected in these experiments which yielded rapid and dramatic rejection via non-H-2 antigens. However, until it is shown that grafting between strains coisogenic for a single non-H-2 locus leads to graft rejection, a dosage effect rather than an absolute deficiency of H-2 antigens cannot be formally excluded.

The nature and distribution of the non-H-2 antigens detected was not unequivocally revealed by these experiments. As Richard Gardner has

TABLE 1 *Incompatibility at combined H-2 and non-H-2 loci*

Recipient	Immunised to	Blastocyst donor	Success rate: No. surviving/ No. transferred	Success rate: % age
A (*H-2*^{a})	C57BL (*H-2*^{b})	C57BL (*H-2*^{b})	0/32	0
A (*H-2*^{a})	—	C57BL (*H-2*^{b})	2/3	66.6
C57BL (*H-2*^{b})	A (*H-2*^{a})	A (*H-2*^{a})	0/21	0
C57BL (*H-2*^{b})	—	A (*H-2*^{a})	5/6	83.6
C57BL10 (*H-2*^{b})	C3H (*H-2*^{k})	C3H (*H-2*^{k})	1/14	7.1
C57BL10 (*H-2*^{b})	C3H (*H-2*^{k})	C57BL10 (*H-2*^{b})	7/7	100.0

TABLE 2 *Incompatibility at solely non-H-2 loci (all animals H-2^k)*

Recipient	Immunised to	Blastocyst donor	Success rate: No. surviving/ No. transferred	Success rate: % age
C3H	CBA	CBA	0/10	0
C3H	—	CBA	2/2	100
C3H	C57Br	C57Br	0/19	0
C3H	C57Br	C3H	7/9	77.7

described earlier, the trophoblast is dependent on the presence of an inductive inner cell mass for its proliferation and growth. A primary immune destruction of the inner cell mass might therefore lead secondarily to a failure of trophoblast proliferation and to the erroneous conclusion that the trophoblast possessed target antigens (Gardner, Johnson & Edwards, 1973). In the grafting experiments described above, it was not possible to distinguish a differential rejection of component embryonic tissues. The use of dissected inner cell masses and trophoblastic vesicles in grafting studies and serological analysis is being undertaken currently to examine this question.

Serological tests on whole blastocysts have in fact yielded positive results, but again these have only been unequivocal when differences at weak non-H-2 loci were involved. The mixed anti-globulin and cytotoxic tests were both positive for embryos when antisera directed against combined strong and weak antigens were used (Olds, 1968; Heyner, Brinster & Palm, 1969; James, 1969). When solely H-2 differences were involved, the mixed anti-globulin test, the cytotoxic test and immune fluorescence have all failed to give a clear positive result, despite positive results using

TABLE 3 *Incompatibility at H-2 loci only*

Recipient	Immunised to	Blastocyst donor	Success rate: No. surviving/ No. transferred	Success rate: % age
C57BL10 (*H-2^b*)	C3H (*H-2^k*)	C57BL10 (*H-2^b*)	7/7	100.0
C57BL10 (*H-2^b*)	C3H (*H-2^k*)	B10.Br (*H-2^k*)	10/10	100.0
C57BL10 (*H-2^b*)	B10.Br (*H-2^k*)	C3H + zona (*H-2^k*)	18/21	85.7
C57BL10 (*H-2^b*)	B10.Br (*H-2^k*)	C3H − zona (*H-2^k*)	17/21	80.5

other antisera in the same experimental system (Heyner, Brinster & Palm, 1969; Palm, Heyner & Brinster, 1971; Gardner *et al.*, 1973; Heyner, 1973). In the immunofluorescence tests conducted by Joy Palm and her colleagues an occasional spot positive for H-2 was detected on the surface of one trophoblast cell of the embryo. The origin and significance of this spot is totally obscure. Michael Edidin (1972) has suggested that it might represent residual spermatozoal membrane but the difficulty in successfully staining membranes on spermatozoa themselves for H-2 makes this explanation unlikely (Johnson & Edidin, 1972). Immune fluorescence did, however, give unequivocally positive results on cleaving eggs with known non-H-2 specificities, viz H-3 and H-6 (Palm *et al.*, 1971). Blastocysts were not examined for non-H-2 specificities in this study.

In conclusion, then, whilst non-H-2 antigens are detectable on the preimplantation egg, H-2 antigens have not been detected by any of the many tests used and must therefore be considered quantitatively negligible at this stage.

Before leaving consideration of the blastocyst, three further points are worth making. First, with F_1 embryos, graft rejection of cleaving eggs or of blastocysts occurs even if the only foreign alloantigens are a paternal contribution (Simmons & Russell, 1967; Searle *et al.*, 1974). This cannot be taken as unequivocal evidence of male genomic expression at these early stages, as six to seven days elapsed before analysis. Serological studies along similar lines are needed. In preliminary studies we have found by a cytotoxic test that $3\frac{1}{2}$-day F_1 blastocysts between C3H and C57BL10/ScSn strains express alloantigens of paternal origin. The presence of paternal alloantigens in early F_1 embryos might of course represent a residual paternal membrane component which subsequently dilutes out with further cell division. Quantitative studies or tests which localise antigen will be needed to resolve this question (unpublished data with Muggleton-Harris).

Secondly, the zona pellucida clearly confers some immunological protection. Thus, our own results reported earlier show that lymphocytic infiltration of embryo grafts is delayed in the presence of the zona pellucida. Cytotoxic tests *in vitro* by James (1969), Heyner *et al.*, (1969) and ourselves have all shown that the zona pellucida hinders lysis.

Thirdly, it is our distinct impression that when blastocysts are collapsed prior to incubation with xeno-antisera and complement in cytotoxic tests, they are much more sensitive to lysis. A speculative but intriguing explanation for this observation would be that a polarity of antigen distribution existed, reflecting the polarity of function of the trophoblast cell. Should this prove to be the case, the implications for interpreting data obtained from in-vitro cultures of dissociated trophoblast cells or of any studies involving disruption of trophoblast organisation would be considerable.

Alloantigens of the ectoplacental cone

Three general approaches have been used to investigate the ectoplacental cone, viz grafting to preimmune recipients, induction of transplantation immunity with cone tissue, and absorption of antisera with cone tissue.

In contrast to the blastocyst, the ectoplacental cone appears to be relatively resistant to graft rejection. Simmons and Kirby and their associates failed to suppress the formation of a haemorrhagic vesicle on transplanting ectoplacental cone to recipients pre-immunised across allogeneic differences (Simmons & Russell, 1967; Kirby, *et al.*, 1966). As cross-species grafts resulted in rejection, the failure to suppress ectoplacental growth has been taken as evidence that this tissue, unlike the blastocyst, does not express exposed alloantigens. However, this conclusion may be erroneous, because whereas the survival of blastocystic trophoblast in immune recipients is dependent on the successful proliferation of a very limited area of trophoblast overlying the inner cell mass, the survival of the ectoplacenta requires little or no proliferation after transfer. Indeed, we suspect that little if any proliferation occurs following ectopic transfer of ectoplacental cone even to non-immune recipients, an observation that would not be surprising in view of the removal of the embryonic inductive stimulus to division. Thus, for example, compare the values for the volumes of haemorrhagic nodules formed by ectoplacenta and blastocyst nine days after transfer to normal recipients (Table 4). In those larger grafts derived from blastocysts the embryonic component is quantitatively negligible compared with the considerable numbers of giant cells and the overtly active diploid core of the ectoplacenta. The nodules derived from ectoplacental cone, in contrast, are not only smaller, but possess no core of proliferating cells. Any growth that has occurred appears to be merely the result of giant-cell invasion. It remains therefore quite possible that the dividing cells at the core of the ectoplacenta, like those of the 3½-day blastocyst, express alloantigens. Grafting tests of the type undertaken so far would not detect such a difference. However, the secondary giant cells at least would seem to be resistant to rejection.

The second way in which the question of ectoplacental cone antigenicity has been examined, has involved inducing transplantation immunity

TABLE 4 *Size of trophoblastic nodule 9 days after transfer of ectoplacenta or blastocyst to non-immune syngeneic recipients*

Tissue transplanted	Size-two maximum diameters multiplied (mm^2)
Blastocyst (3½-day)	21.95
Ectoplacental cone (7½-days)	7.47

to skin grafts by sensitising with ectoplacental tissue. Such a procedure, especially after pre-incubation of tissue with neuraminidase, has been claimed to provoke a state of specific immunity to alloantigens as judged by a reduced survival time of skin grafts (Currie, Van Doorninck & Bagshawe, 1968; Taylor & Hancock, 1974, personal communication). However, Simmons (1973) was unable to repeat these results using dissected ectoplacental tissue with or without neuraminidase treatment. Billington (1973) likewise failed to induce transplantation immunity after serial transfer of ectoplacental cone tissue to ectopic sites in allogeneic recipients. In the studies by Currie *et al.*, (1968) the neuraminidase may have been acting as an adjuvant to reveal contaminant maternal cells unwittingly carried over with the cone tissues. But this cannot be the explanation for the results of Taylor & Hancock who elegantly circumvented this problem by using F_1 ectoplacental cone to immunise recipients of maternal strain and induced isoimmunity to paternal alloantigens whether or not neuraminidase pretreatment was performed. The possible contamination of foetal non-trophoblastic cells cannot of course be excluded as a source of immunogen, neither does the experiment distinguish whether or not the cells of the proliferative core of the ectoplacenta or the giant cells were responsible for the induced immunity.

This study by Taylor & Hancock is clearly very important, and it is therefore unfortunate that they chose a strain combination in which the median first-set rejection time was only $8\frac{1}{2}$ days, because this reduces the sensitivity of their test, especially as no control for a non-specific influence of neuraminidase-treated trophoblast injection was included. The alloantigens involved were also not defined. A second-set reaction to multiple non-H-2 alloantigens especially coupled with a first-set reaction against the H-2 antigens on the test graft would be capable of yielding the time-course of graft rejection observed. Perhaps an approach likely to resolve the question of the presence, distribution and type of alloantigens on the ectoplacenta, will be to use the Taylor & Hancock protocol but for strains differing at non-H-2 loci only, when the sensitivity of the test might be improved and incidentally the nature of the alloantigens involved revealed. Should immunity be induced with this experimental approach, it might then prove possible to determine whether or not it is the giant cells or the ectoplacental proliferative core cells that contribute the immunogeneic component by doing a serial passage experiment. Thus by using as immunogen the growth taken from the maternal strain kidney seven days after the transfer there of an F_1 ectoplacental cone, only transformed secondary giant cells will remain as potentially antigenic cells.

A modification of the serial passage approach was used by Schlesinger (1964), in an attempt to detect H-2 on trophoblast by absorption inhibition. Ectopic growths derived seven days after ectopic transfers of blastocysts were used to absorb known anti-H-2 antisera. The system was capable of detecting levels of H-2 up to 1/200 that of adult spleen cells. Results were entirely negative. It is not clear whether just giant cells were present in the surviving ectopic grafts or whether ectoplacental proliferative core cells were also present. However, it seems clear that little if any H-2 antigen is present on at least giant cells.

In summary, the conclusions drawn from studies of transplantation of $7\frac{1}{2}$-day ectoplacental cones must needs be cautious. Serological studies have been limited and the results from grafting and immunisation studies are ambiguous. There is some slender evidence to favour the presence of alloantigens of some sort although probably not H-2 in the region of the ectoplacenta. More important, it is essential to determine what features if any of the dividing ectoplacental cone cells differentiate them antigenically from giant cells, a distinction which has not hitherto been established. Clearly there are two, non-exclusive, possibilities. The cells could show genuine differences in antigenic profile. Alternatively, the cells could be antigenically similar, but have membrane properties rendering them differentially sensitive to immune cells or to antibody. We saw earlier that in respect of ease of patching, phagocytosis and lectin-binding the giant cell differed from the early trophectoderm. A greater ease of patching in giant cells might be construed as a means by which antibody could sweep antigen from the membrane rendering it less vulnerable to the cytotoxic effects of lymphocytes. It has been shown that a dividing cell such as the ectoplacental core cell is more vulnerable to cytotoxic agents than an interphase cell like the giant cell (Graham, Sumner, Curtis & Pasternak, 1973). Possibly, masking layers of sialomucin are also involved. But this explanation seems unlikely, (1) because pre-treatment of ectoplacental cone grafts with vibrio cholera (VC) neuraminidase prior to transfer to immune recipients did not affect graft survival (Simmons, 1973), (2) because the sialomucin layer appears to be incomplete and (3) because its presence does not prevent rejection of, for example, cross-species grafts of ectoplacental cone. What is clear, however, from a number of studies, is that different cells vary remarkably in their sensitivity to immune agents in ways that are not adequately explained in terms of antigenic density alone.

A final consideration that makes us suspect that the proliferative core cells may well differ antigenically from giant cells are experimental results suggesting that core cells may give rise to tissues other than secondary giant cells such as chorionic ectoderm. First, Gardner, Papaioannou & Barton (1974) have shown that the trophoblast layer evidently makes a contribution to some component of the extra-embryo as judged by analysis of glucose phosphate isomerase patterns in embryos derived from reconstituted blastocysts. Secondly, in studying rat:mouse chimaeras, two cases of a rat inner cell mass independently stimulating an overlying ectoplacenta of mouse tissue have been reported. The murine ectoplacental cells, then, seem to contribute to the extra-embryonic ectoderm (Gardner & Johnson, unpublished data). Third, embryonic ectoderm seems to be distinguished from extra-embryonic ectoderm by a surface embryonic phase-specific antigen which may reflect a different origin for the two types of ectoderm (unpublished data).

Carcino-embryonic antigens of the peri-implantation embryo

Kirby (1968) was the first to provide experimental evidence suggestive of an embryo-specific transplantation antigen which he ascribed to the trophoblast. Thus, repeated transfer of allogeneic or syngeneic blastocysts ectopically conferred on the recipient an ability to reject later blastocyst grafts. Subsequent to this analysis both Billington (1973) and we ourselves have provided data which cannot be explained simply in terms of a trophoblast antigen. Thus, repeated immunisation with syngeneic ectoplacental cone did *not* lead to rejection of later grafts of either ectoplacenta or blastocysts. Secondly, it was found that repeated immunisation with syngeneic blastocysts did give some suppression of subsequent blastocyst grafts but *not* of ectoplacental cone (Table 5) (unpublished data). These results suggest either that the tissue antigen was restricted to the diploid cells at the core of the ectoplacenta or that an inner cell mass antigen rather than a trophoblast antigen might be involved. In line with this suggestion, the smaller blastocyst growth obtained after prior sensitisation with blastocyst showed little or no inner cell mass derivatives or diploid core cells. Presumably after giving a transient inductive stimulus to division, the inner cell mass had been destroyed. Thus, extrapolating from Richard Gardner's results, primary immune destruction of the inner cell mass (ICM) via an ICM-specific antigen would lead to failure of blastocyst grafts, without affecting growth of ectoplacental cone.

In line with this suggestion Michael Edidin and his colleagues (Edidin, Pathey, McGuire & Sheffield, 1971) have described an isoantigen present on the mouse ICM and its derivatives of up to $7\frac{1}{2}$ days. The antigen cross-reacts with an undifferentiated testicular teratoma and elicits cellular and

TABLE 5 *Effect on pre-immunisation on survival and size of blastocysts and ectoplacental cone transplanted to ectopic sites*

Immunised with*	Test graft	Success rate	Size of surviving grafts (mm^2)
Ectoplacental cone	Blastocyst	17/20	18.39
Ectoplacental cone	Ectoplacental cone	18/20	5.72
Blastocyst	Blastocyst	11/20	5.70
Blastocyst	Ectoplacental cone	17/20	6.36

*Four immunising transfers of 4–6 embryos or cones.

humoral isoimmunity either after immunization or as a result of multiparity (Edidin, Gooding & Johnson, 1975). The antigen is absent from, or only weakly present on, the trophoblast cells even at $3\frac{1}{2}$-days and is not detected at any time later. Edidin *et al.* (1971) have suggested that the teratoma antigen represents a precursor or embryo-equivalent of the H-2 antigen, and in cultures of various cell lines finds a reciprocal relationship between H-2 and teratoma antigen levels. For purposes of this discussion, it may be significant that trophoblast cells which evidently lack H-2 determinants, also fail to show detectable levels of teratoma antigen.

DISCUSSION AND ASSESSMENT OF SIGNIFICANCE

From the little experimental data available, we can conclude that the cleaving embryo does express both non-H-2 alloantigens and at least one teratoma isoantigen. H-2 antigens cannot be detected. The trophoblast of the $3\frac{1}{2}$-day blastocyst possesses low levels of the teratoma-antigen and again no H-2 antigens have been detected. The ectoplacental cone of $7\frac{1}{2}$ days shows somewhat equivocal evidence of alloantigens of undefined nature and strength, and of unestablished distribution between the subpopulations of diploid core cells and giant cells. The $7\frac{1}{2}$-day ectoplacental tissue lacks the teratoma antigen and evidence as to whether it possesses an organ-specific isoantigen is questionable. All types of trophoblast cell express species antigens seemingly as an integral part of their membranes.

If one observation emerges clearly from the more critical examinations of trophoblast antigenicity, it is the absence or extreme deficiency of H-2 antigens. Trophoblast cells are probably not unique in this respect. In a number of adult tissues, viz brain, human erythrocytes, and muscle, it has proved difficult to detect on individual cells H-2 antigens or their equivalent with any confidence. An absence of H-2 antigens is therefore not incompatible with cellular survival in the adult. In the embryo, there is no evidence for the presence of H-2 antigens on the embryo proper up to about 7 days. Thereafter, whilst H-2 antigens are detectable on whole embryo preparations, the successful demonstration of H-2 specificities on identifiable cell types other than those derived from the haematopoietic system has not been described until late in pregnancy. This evidence, plus the obvious success of intra-or inter-species chimaeras either *in vivo* or *in vitro* in hanging drop cultures, argues against H-2 determinants having an important role to play in developmental processes. Other cell surface moieties are presumably involved in the cell recognition and organization processes. Thus, these lines of reasoning demonstrate that there is no inherent conflict between the absence of detectable H-2 antigens on trophoblast and its success as a tissue.

In contrast, the only known specific and direct involvement of H-2 antigens has been as target antigens for cell-mediated lysis following lymphocyte activation by differences at the mixed lymphocyte response (MLR) gene, or as targets of alloantibodies produced presumably after a similar or related T cell activation of B cells. The first appearance of H-2 antigens at a time when immune responsiveness as judged by graft-versus-host reaction is first demonstrable in the embryo, and the rapid rise in H-2 levels on lymphocytes neonatally at a time when induction of tolerance becomes more difficult and ease of induction of immunity rises, both argue that the role of H-2 determinants is primarily or even solely to be sought in immunological processes. There is thus a most perfect reason for excluding H-2 antigens from trophoblast cells. It would be of interest to determine whether β_2-microglobulin could be detected on the surface of trophoblast cells, as an indication of whether production of H-2 specificities is totally or partially suppressed. A clue as to the nature of H-2 suppression in trophoblast cells might also be sought by examining cells derived from fusion studies between trophoblast and H-2 rich cells.

The role of the non-H-2 transplantation antigens is totally obscure, likewise most other serologically detectable alloantigens have no known function. Possibly like the L-M antigen system in sheep, these alloantigens are involved in basic cellular processes including possibly cell–cell recognition and only fortuitously function as transplantation antigens. They are the antigens which theoretically should put the embryo at greatest risk, particularly between the time of zona-loss and giant-cell transformation, although even in this period it may be only the few cells of Rauber's layer that are at risk. The demonstration by Kirby (1969) that zona-free embryos could survive in the non-decidualised uterus of mice showing delayed implantation, would suggest that even this period is not one of great danger. Indeed, Billingham (1971) has shown that prior intrauterine sensitisation with allogeneic lymphocytes by a protocol known to induce local recall flare on retesting, leads to a state of heightened receptivity to embryos from the allogeneic strain.

In summary, it appears that trophoblast clearly does very well without H-2 alloantigens and possibly does even better with non-H-2 alloantigens at least in peri-implantation stages.

SUMMARY

The evidence concerning the antigenic status of the peri-implantation trophoblast is considered critically in the light of a clearer understanding of the subpopulations of trophoblast cell involved. It is concluded that, on the available evidence, the cleaving embryo does express both non-H-2 alloantigens and at least one teratoma isoantigen. H-2 antigens cannot be detected. The trophoblast of the $3\frac{1}{2}$-day blastocyst possesses only low levels of the teratoma antigen and again no H-2 antigens have been detected. The ectoplacental cone of $7\frac{1}{2}$ days shows somewhat equivocal evidence of alloantigens of undefined nature and strength, and of unestablished distribution between the subpopulations of diploid core cells and giant cells. The $7\frac{1}{2}$-day ectoplacental tissue lacks the teratoma antigen and evidence as to whether it possesses an organ-specific isoantigen is questionable. All types of trophoblast cell express species antigens seemingly as an integral part of their membranes. The possible significance of these conclusions is discussed briefly.

The author wishes to thank Audrey Muggleton-Harris for her permission to quote unpublished joint work and Craig Howe for valuable advice and discussion. The work was supported by an MRC project grant.

REFERENCES

Billingham, R. E. (1971). The transplantation biology of mammalian gestation. *Am. J. Obstet. Gynecol*, **111**, 469.

Billington, W. D. (1973). Does trophoblast express tissue specific antigens? In *Immunology of Reproduction* (ed. K. Bratanov), p. 475, Bulg. Acad. Sci. Press: Bulgaria.

Currie, G. A., Van Doorninck, W. & Bagshawe, K. W. (1968). Effect of neuraminidase on the immunogenicity of early mouse trophoblast. *Nature, Lond.*, **219**, 191.

Dehem, A., Subis, H. & Vandeputte, M. (1971). Effect of immunosuppression on the development of xenogeneic trophoblast under the kidney capsule. *Transplantation*, **11**, 578.

Edidin, M. (1972). Histocompatibility genes, transplantation antigens and pregnancy. In *Transplantation Antigens* (eds B. D. Kahan & R. S. Reisfeld), p. 75. Academic Press: USA.

Edidin, M., Gooding, L. & Johnson, M. H. (1975). Surface antigens of normal early embryos and a tumor model system useful for their further study. Proceedings of the 7th Karolinska Symposium. *Research methods in reproductive physiology. (Immunological approaches to fertility control.)* World Health Organisation, (in press).

Edidin, M., Pathey, H. L., McGuire, G. J. & Sheffield, W. D. (1971). An antiserum to 'Embryoid body' tumour cells that reacts with normal mouse embryos. In *Embryonic and Fetal Antigens in Cancer*, (ed. W. E. Anderson), p. 239. National Tech. Information Service: USA.

Gardner, R. L. & Johnson, M. H. (1973). Investigation of early mammalian development using interspecific chimaeras between rat and mouse. *Nature New Biol.*, **246**, 86.

Gardner, R. L., Johnson, M. H. & Edwards, R. G. (1973). Are H-2 antigens expressed in the preimplantation blastocyst? In *Immunology of Reproduction* (ed. K. Bratanov), p. 480. Bulg. Acad. Sci. Press: Bulgaria.

Gardner, R. L., Papaioannou, V. E. & Barton, S. (1974). Origin of the ectoplacental cone and secondary giant cells in mouse blastocysts reconstituted from isolated trophoblast and inner cell mass. *J. Embryol. exp. Morphol.*, **30**, 561.

Graham, J. M., Sumner, M. C. B., Curtis, D. H. & Pasternak, C. A. (1973). Sequence of events in plasma membrane assembly during the cell cycle. *Nature, Lond.*, **246**, 291.

HEYNER, S. (1973). Detection of H-2 antigens on the cells of the early mouse embryo. *Transplantation*, **16**, 675.

HEYNER, S., BRINSTER, R. L. & PALM, J. (1969). Effect of alloantibody on pre-implantation mouse embryos. *Nature, Lond.*, **222**, 7830.

JAMES, D. A. (1969). Antigenicity of the blastocyst masked by the zona pellucida. *Transplantation*. **8**, 846.

JOHNSON, M. H. & EDIDIN, M. (1972). H-2 antigens on mouse spermatozoa. *Transplantation*, **14**, 781.

KIRBY, D. R. S. (1968). The immunological consequences of extra-uterine development of mouse blastocysts. *Transplantation*, **6**, 1005.

KIRBY, D. R. S. (1969). On the immunologic function of the zona pellucida. *Fertil. Steril.*, **6**, 933.

KIRBY, D. R. S., BILLINGTON, W. D. & JAMES, D. A. (1966). Transplantation of eggs to the kidney and uterus of immunized mice. *Transplantation*, **4**, 713.

OLDS, P. J. (1968). An attempt to detect H-2 antigens on mouse eggs. *Transplantation*, **6**, 478.

PALM, J., HEYNER, S. & BRINSTER, R. L. (1971). Differential immuno-fluorescence of fertilized mouse eggs with H-2 and non-H-2 antibody. *J. exp. Med.*, **133**, 1282.

SCHLESINGER, M. (1964). Serologic studies of embryonic and trophoblastic tissues of the mouse. *J. Immunol.*, **93**, 255.

SEARLE, R. F., JOHNSON, M. H., BILLINGTON, W. D., ELSON, J. & CLUTTERBUCK-JACKSON, S. (1974). Investigation of H-2 and non-H-2 antigens on the mouse blastocyst. *Transplantation* **18**, 136.

SIMMONS, R. L. (1973). Neuraminidase does not unmask histocompatibility antigens on trophoblast. In *Immunology of Reproduction* (ed. K. Bratanov), p. 490. Bulg. Acad. Sci. Press: Bulgaria.

SIMMONS, R. L. & RUSSELL, P. S. (1965). Histocompatibility antigens of fertilized mouse eggs and trophoblast. *Ann. N. Y. Acad. Sci.*, **129**, 35.

SIMMONS, R. L. & RUSSELL, P. S. (1967). Xenogeneic antigens in mouse trophoblast. *Transplantation*, **5**, 85.

DISCUSSION

EDWARDS: We will debate the presence of transplantation and other alloantigens on trophoblast at this point, and reserve discussion on other placental antigens until Faulk's paper.

LOKE: I feel hurt that scepticism persists about transplantation antigens on trophoblast because we have demonstrated not only HL-A antigens but also blood group A antigens on human trophoblast cells. HL-A antigens were demonstrated by using the cytotoxic method on cultured human trophoblast cells taken from hysterectomy cases of 12–14 weeks. Both multi-specific and mono-specific anti-HL-A reagents were used in the presence of complement and the trophoblast cells reacted beautifully with both of them (Loke, Joysey & Borland, 1971). Blood group A could not be demonstrated by the mixed agglutination test, but was shown by the mixed anti-globulin technique and confirmed by using immunoferritin which showed that the antigenic determinants seem to be much further apart on trophoblast cells than on other cells taken from the foetus (Loke & Ballard, 1973). This observation could explain why blood group A antigen was not detected by mixed agglutination, the antigenic determinants being too far apart for sufficient linkage, but could be found by the mixed antiglobulin test. The presence of this antigen at low density confirms or at least supports the theory that immunological enhancement might protect foetal trophoblast against maternal attack. The HL-A antigenic determinants on trophoblast cells are also less dense than on corresponding foetal skin cells, for titration of our anti-HL-A serum against both foetal skin cells and against trophoblast cells showed that the titre against skin cells was much higher – in the region of 1/256 compared to that of the corresponding trophoblast, which is about 1/64 – 1/32. One explanation could be the differential susceptibility of skin cells as opposed to trophoblast cells, but it may equally mean that the antigenic determinants on trophoblast cells are less dense than those on foetal skin (Loke, unpublished data).

The main criticism of most of our results concerns the use of tissue culture and the question of whether we are in fact culturing trophoblast cells. We have confirmatory evidence that the cells are trophoblastic and bear HCG on their surface as shown by the mixed anti-globulin method, for red cells linked with HCG molecules adhered to 85–90 % of the cells in a monolayer culture (Loke, Wilson & Borland, 1972). Trophoblast cells are believed to secrete HCG, hence the monolayer culture consisted predominantly of trophoblast cells. Our results are thus valid and we can conclude that both HL-A and ABO antigens are present but perhaps at a lower density than that of corresponding foetal tissue (see papers of Howe and Borland *et al.* for discussion of whether HCG on surface is indicative of trophoblast, pp. 139 and 157.

FAULK: The argument about antigen density would be useful in the cytotoxic system but we and others have used immunofluorescence, which is a very sensitive technique not necessarily dependent on antigenic density and were unable to find ABO groups on intact trophoblast (McCormick, Faulk, Fox & Fudenberg, 1971).

COOMBS: We also had negative results (unpublished data) with the mixed agglutination test on trophoblast but my worry concerns the mixed anti-globulin test. Could you be sure that there was no other antibody in your anti-A sera?

LOKE: We have tested this anti-A antisera against lymphocytes, and there are no anti-HL-A antibodies in the serum. The serum was absorbed with blood group A substance, and the mixed anti-globulin test with trophoblast was then negative. For those two reasons, we believe that the reaction was due to anti-A reacting with A substance on the trophoblast.

JOHNSON: The density of blood group A substance was low but were you looking at patches rather than a single group A determinant? Without antibody-induced patching, antigenic density might not be low. How many cells in each culture were positive? In your published electron micrographs of trophoblast in culture the cell surface is relatively smooth compared to the convolutions on trophoblast cells fixed *in vivo*. There is also the question of polarity of cells raised by Munro. You could perhaps be exposing determinants which would not otherwise be revealed *in vivo*, and how relevant physiologically is the use of trophoblast in culture?

LOKE: I agree that patching might be an explanation. Trophoblast cells may have a greater ability to form patches compared to the corresponding foetal skin but why that should be I don't know (for relation of patching to antigen susceptibility see Munro's paper).

EDWARDS: Does this evidence of HL-A on late trophoblast not conflict with the evidence presented on the ectoplacental cone stage?

JOHNSON: First these may be different kinds of trophoblast cell, certainly different stages, and anyway some unidentified antigen may be present on the core cells of the ectoplacenta. However, there is no good evidence yet that early giant cells possess H-2 antigens. Three or perhaps four different types of trophoblast are present in later stages, not all at the foeto–maternal interface.

LOKE: In our monolayer cultures there are three types of cells which can be distinguished morphologically. One is a sort of spindle-shaped cell, the second is sort of a small polygonal-type cell, and the third is a multinucleated cell, and we do not know what they represent. If the spindle-shaped cells are fibroblasts from the villus mesenchyme, the small polygonal cells are cytotrophoblast and the multinucleated cells are perhaps the syncytial trophoblast. The polygonal cells and the multinucleated cells secrete HCG and in primary cultures these are the cells that predominate. About 85 to 90% of the cells are of this type, meaning that at that stage the whole culture has not been replaced by spindle-shaped cells.

GARDNER: Is an interval of time in culture essential to see these various antigens expressed, or are they expressed initially? By placing cells in culture you sometimes can get suppression or enhancement of antigen expression. Do your findings really reflect the trophoblast *in vivo*? (see also Discussion after Billington's paper p. 79).

LOKE: I do not know; early tests soon after culture would be advisable, but there are insufficient cells until they have multiplied.

COOMBS: Does the culture medium contain substance A?

LOKE: It contains 20% foetal calf serum.

BODMER: Blood group substance A cannot be present in foetal calf serum, otherwise fibroblast cultures would not be negative. Primary fibroblasts after about the fourth passage are clearly negative for A and they are generally cultured in foetal calf serum (Franks, 1968). Many other kinds of cells are cultured in foetal calf serum and do not express A, so this possibility is unlikely.

JOHNSON: Secretory products from other cells in the cultures such as fibroblasts, or possibly viruses or bacteria could lead to false positives. There is evidence for such cross-reactions with alloantigens (Rapaport, 1972), and if trophoblast cells were picking up such determinants in culture they might appear falsely to be positive. Also, a cross-reactivity between blood group A substance and HCG has been reported (Mätz, Bahl & Mohn, 1973). If confirmed, this observation might cast doubt on the specificity of your identification of A substance.

LOKE: I know of no such examples, but we have not excluded this possibility.

BODMER: Some of our work supports Loke's view. We have been studying the isolation of HL-A antigens mostly from full-term placentae using a simple method of preparing membranes (Snary, Goodfellow, Crumpton & Bodmer, unpublished data). We do not make careful morphological examinations of the tissue but have reason to believe that a large proportion of the material is trophoblastic. The blood is mostly washed away. Membrane preparations are made which are rather free of fibrinoid-type material, the HL-A antigens are solubilised with deoxycholate and the molecule purified to a rather high degree of purity following a previously-published procedure (Snary *et al.*, 1974). We find lots of HL-A antigen there, and can quantitate the amount of HL-A in terms of cell equivalents of antigen. Preliminary data indicate that we obtain from a normal sort of placenta the equivalent of up to ten litres of blood, hence any doubt about the extract coming wholly from contaminating blood is out of the question. The HL-A antigen could be maternal, or foetal.

On the basis of several placentae so far examined, only one of the maternal haplotypes is in effect present and this is, as expected, the same haplotype as expressed by the foetus. Since such substantial amounts come from placentae, the antigen must be present on trophoblast and be foetal in origin, for otherwise both maternal haplotypes would be detected equally. The question of the expression of the paternal antigen is problematical. Some preliminary evidence suggests lower expression of paternal than maternal antigen but more work is needed here. Perhaps the paternal antigens have been covered over, and Loke, using cell cultures, might be better able to detect the relative expression of both maternal and paternal antigens. However, it seems clear that we are finding substantial amounts of HL-A on plasma membrane from placentae.

EDWARDS: You have been able to detect so much HL-A antigen that there should be no difficulty in finding it by other tests. Yet we have heard from others that these antigens are not detectable on trophoblast. Judged from your results, it is very difficult to understand how the idea has grown that histocompatibility antigens are absent from this tissue.

BODMER: I agree. Immunofluorescence for H-2 and HL-A, for example, is not an easy technique, and there are limitations with other techniques. The possibility that HL-A may be present on the placenta is an old idea; Van Rood (1962) made crude placental extracts and certainly found HL-A antigens present. We believe that placenta is a rich and convenient source of material for preparation of HL-A antigens.

CURZEN: How pure are the tissues you use? The preparations of cell membranes from crude placental homogenates, however carefully washed free of maternal and foetal blood cells, have not come exclusively from trophoblast. Probably less than 50 % of the membranes are of trophoblastic origin. The rest would be stromal cells and possibly even contributing membranes from foetal capillaries.

BODMER: I would not make any claims for purity, and our material is certainly very crude. However, we do try to excise the remaining pieces of the decidua. We argue on the basis of the amount of material extracted, and on the question of foetal or maternal origin, and do not know whether the outer or inner trophoblast layers differ in amount of expressed HL-A antigen. Our cell preparations appear to be mostly syncytial trophoblast, and do not seem to contain much material from vessels as all the fibrous tissue is left behind.

FAULK: There was only half of the maternal HL-A haplotype in placenta, and you believe this is quite reasonable. I wonder if, within the framework of the HL-A dogma, there is any room for a foetal type of HL-A which would type out in your sera in an unexplained way.

BODMER: The specificity of the antigens is defined by an inhibition assay using antibodies previously characterised by their activity on lymphocytes. Hence the HL-A antigen is that typed on lymphocytes.

ALLISON: Was there any maternal antibody present against the paternal haplotype in the particular cases where you had apparent suppression?

BODMER: No, I think not in that case. In the four or five placentae so far examined, we have only detected activity in one so far, perhaps because of the level of detection.

FAULK: Among our eluted antibody the occasional cytotoxic reaction occurs in the Terasaki system, but when run through the Eurotransplant HL-A computer system it does not correlate with the expected HL-A activity, and it actually seems to have anti-cell-membrane activity (Faulk *et al.*, 1974).

BODMER: Was the antibody polymorphic and did it associate with an antigen? Family studies would show if it was linked to HL-A. Our antibody has a high titre.

FAULK: Antibody-mediated cytotoxicity would suggest an association with an antigen. Family studies would of course be very useful.

JOHNSON: This question is very important. It has been implied that placental cells might be particularly sensitive for some reasons to suppression of the expression of their paternal haplotype by maternal antibody. But take the alternative case in which the foetus might suppress expression of the alien maternal haplotype on decidual cells. If sufficient decidua was present in the extracted placental material, the resultant haplotype ratios would be as described by Bodmer, provided the trophoblast itself was neutral for alloantigens.

DAVIES: I think it is obvious enough that HL-A (human) or H-2 (mouse) or Ag-B (rat) antigens are present on trophoblast, but I am not convinced that their immunogenicity matters. In Johnson's work with transplantation, the recipients were all pre-immunised not, however, with trophoblast or embryo, but with adult lymphoid cells. The absolute immunogenicity of the embryo for rejection purposes thus seems not to have been established.

The immunogen for graft rejection is not what we thought it was and I would like to describe that kind of experiment that takes two years to do but two minutes to describe! It is an enhancement experiment. Passive enhancement is easier to manipulate than active enhancement; it is difficult to achieve with skin grafts but easy with tumours, and for our purposes best shown with kidney or heart transplants in rats. We actually used auxiliary hearts, which were passively enhanced using an allo-antiserum of the usual kind containing Ag-B antibodies. Ag-B is the major histocompatibility system of the rat. Table 1 shows the two types of rats

TABLE 1 *Enhancement of rat auxiliary heart allografts*

Serum	A/gus anti Wag (or A/gus anti F_1)
Graft	Wag heart → A/gus (or F_1) (Ag-B2 = $H\text{-}1^w$) → (Ag-B1 = $H\text{-}1^l$)

that we used – A/gus is one kind of rat and Wag is the other kind of rat – the transplant is from an F_1 hybrid to A/gus. The hearts are sewn in abdominally, aorta to aorta, and pulmonary artery to inferior vena cava, basically to give a coronary flow. In Fig. 1, group (*a*) are controls treated with normal serum, and they all reject their heart transplants by about ten days. The enhancing serum was used in group (*b*) and you will see that it is very effective; a number of rats grafted at one year of age eventually died of old age without rejecting their transplant, and they were only treated with this serum in the first week after transplantation. The serum had a cytotoxic titre of about 1/3000.

We then absorbed the serum with red cells, (we have also used platelets) as a 'tissue' expressing Ag-B but perhaps not the product of other genes in the same major histocompatibility region, for they do not stimulate lymphocytes to DNA incorporation (MLR). After full absorption we no longer have any cytotoxic titre in the sense that there is no longer a full release of radioactive label, and only a subpopulation of the lymphocyte suspension used as target remains susceptible to this antibody. Yet we can

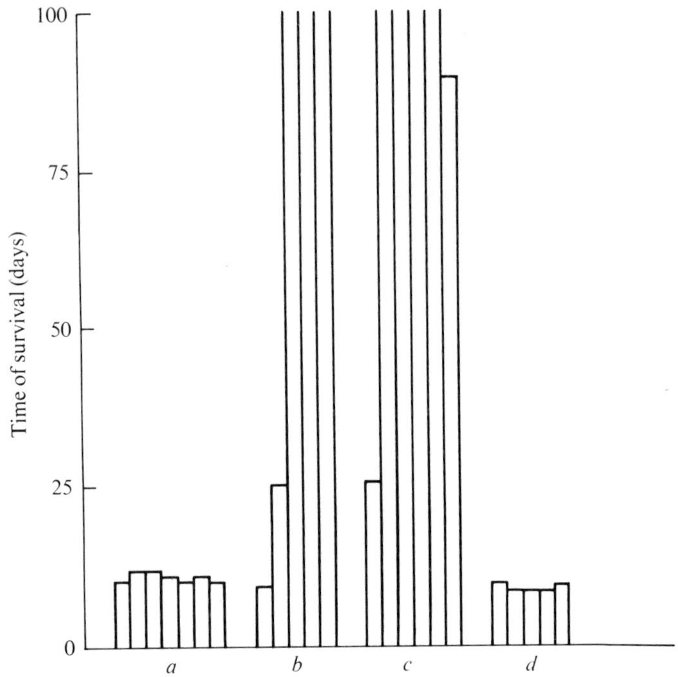

Fig. 1. Survival of rat auxiliary heart transplants from (A/gus × Wag) F_1 to A/gus. Series *a*, controls injected with normal rat serum, 1 ml (intravenous) on day of grafting and days, 1, 3, 5, 7, 9 thereafter. Series *b*, injected (the same dose schedule) with enhancing alloantiserum; series *c*, treated with the same serum but fully absorbed with Wag erythrocytes; series *d*, the same serum fully absorbed with Wag erythrocytes and subsequently absorbed with Wag spleen/thymus cells (Davies & Alkins, 1974).

still enhance fully (Fig. 1, group (*c*)), even after removing these Ag-B antibodies to which everyone has attributed the enhancement phenomenon in the past. In fact, enhancement is perhaps even better than before absorption. If this absorbed serum is now absorbed again with lymphocytes, then the enhancing activity is removed, as shown in Fig. 1 group (*d*).

There are a number of reasons for thinking that this result is not due to formation of immune complexes. To begin with, absorbing with red cells or platelets should produce complexes in just the same way as absorbing with white cells. We know, of course, that lymphocytes shed antigen: red cells evidently do not, and it seems rather unlikely that platelets do. There are a number of other reasons why Ag-B complexes are probably not involved. By simply taking rat IgG labelled with ^{125}I and mixing this, as a marker, with the enhancing alloantiserum, we find no evidence for complexes on gel filtration.

We do not know much yet about the distribution of the new immunogen. The open circles in Fig. 2 show the original serum, and open triangles are that serum absorbed with red cells and tested against spleen cells. The squares are tests against lymph node target cells. There is about 30% killing of lymphocytes, no reaction with bone marrow, no reaction with thymocytes. My colleague Staines has confirmed these findings by similar

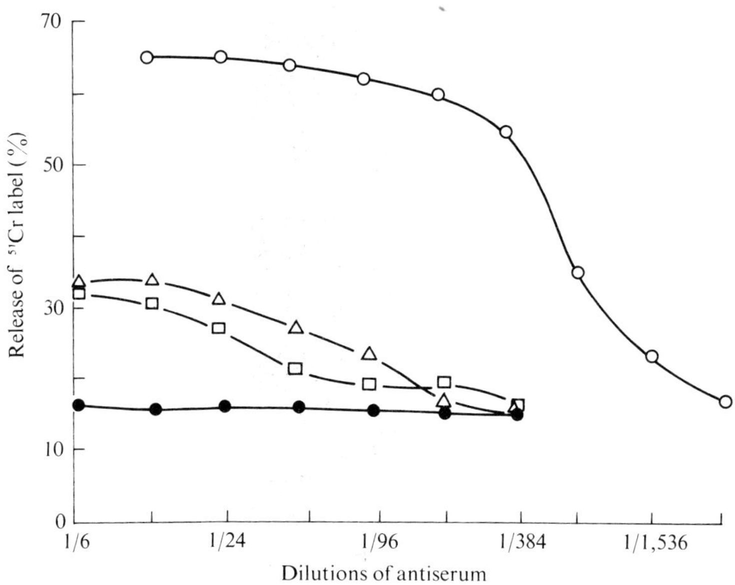

Fig. 2. Complement-mediated cytotoxicity of enhancing alloantiserum. Original titre against Wag lymph node target cells, ○. After maximal absorption with Wag erythrocytes, titre against spleen cells, △, against lymph node cells, □. There was no measurable reaction (●) using as target cell, bone marrow, thymocytes, peripheral leukocytes (unfractionated) or normal peritoneal exudate cells (Davies & Alkins, 1974).

experiments in mice with skin grafts, although these are enhanced for only about a week. Nevertheless, this enhancement of mouse skin grafts is highly significant and using the small residual cytotoxic titre to monitor columns a new peak – which we believe to be the product of Ir (immune responsiveness) genes – is separable from H-2. The Ir gene product sensitises the recipient against a transplant, and a role for Ag-B (H-2 or HL-A) as currently recognised is no longer clear. There may be some involvement in what we like to call the 'effector arc', so in the case of trophoblast or the foetus the question arises as to whether the immunogen leading to rejection is present at all. It is not self-evident by looking for H-2 in the mouse embryo or HL-A in the human trophoblast, or HL-A antibodies in sera of multiparous women (Davies & Alkins, 1974; Staines, Guy & Davies, 1974; Archer, Smith, Davies & Staines, 1974; Davies & Hess, 1974).

ALLISON: Is there any relationship to the MLC system?

DAVIES: We would like to think so, yes, but the relationship of Ir and MLC is not yet clear. One would perhaps hope that MLC inhibition would be a measure of these new antibodies.

ALLISON: The difference between H-2 and H-3 differences described by Johnson could be very interesting because there is now evidence, as far as T cells are concerned, of two separable events. The first is an MLC difference – this also applies in the human situation – which generates a blast, but this by itself is not enough, for a second difference is needed for cytotoxicity. In mice the difference at H-2, and at certain of the HL-A haplotypes in man, brings about lysis. The distinction between having an H-2 or an H-3 antigen on the surface of the trophoblast could thus be important from the functional point of view in terms of immunisation at the effector level.

DAVIES: I should add that in the mouse, we have obtained similar results in congenic situations. So the sensitising antigen is determined by a gene in the major histocompatibility region.

BODMER: Did you absorb with any other tissue, because after immunisation with spleen, you could be detecting these immunogens on other types of tissue as well.

DAVIES: So far we have absorbed with red cells, platelets and liver – it is difficult to absorb out with liver. We have not tried fibroblasts.

ADINOLFI: Coggin & Anderson (1974) used either foetal extracts or a single-cell suspension from foetuses and injected them into adult animals. When these animals were challenged with tumours, their lymphocytes were able to kill the tumour cells *in vitro*. This is evidence that foetal tissues, not necessarily trophoblast, may have immunogenicity.

DAVIES: I was talking about immunogenicity in terms of effecting allo-rejection. I suppose the Coggin & Anderson experiment concerns CEA (carcinoembryonic antigen).

ALLISON: This has to be contrasted with my presentation, showing that multiparous animals frequently do react against tumours. Events during pregnancy lead to immunisation of the mother, therefore the tissue is immunogenic.

DAVIES: Yes, but this presents no danger for the foetus. It does not matter whether HL-A antibodies are made by the mother because they will not effect the foetus, even if they traverse the placenta. The question is what the barrier must do to protect the foetus. The only kind of barrier needed may be to prevent white cells from the foetus entering and immunising the mother, but in fact a barrier is necessary only if the foetus possesses cells expressing Ir gene products. Our impression is that Ir is expressed only on switched-on cells; the foetus may not have switched-on cells. In that case it is not immunogenic in relation to rejection and no barrier is required.

BODMER: Surely, a response leading to rejection can be invoked without using immune cells as part of the stimulus.

DAVIES: That is not self-evident. One would predict from this that skin graft rejection is due to passenger and resident lymphocytes.

BODMER: You certainly can do it with tumours.

DAVIES: Which tumours? Lymphomas may well express Ir, and which tumour cell suspension is free from lymphocytes?

BODMER: Then you are suggesting that lots of non-lymphoid cells express Ir. It is hard to believe that the foetus during all of its time – especially in man, which is a very long time – is not expressing any of the cell recognition antigens which may be involved in rejection.

DAVIES: You mentioned that tumour cells would even produce HCG. Why should they not make Ir products?

JOHNSON: Could I make three points about Dr Davies' remarks. First, whilst I agree that a difference at Ir is needed to stimulate for major transplantation antigens such as AgB, H-2 or HL-A, a difference at AgB, etc. is also needed to kill. In his experiments all blocking with anti-Ir antibody – if that's what it is – would be blocking stimulation. If, however, the animal is pre-stimulated by lymphocytes and skin grafts, as in our experiments, the test grafts of trophoblast should have been rejected were they to express H-2. Therefore we can conclude that the cells lack the major transplantation antigens. Second, we have no data that bears on the expression of Ir by trophoblast cells. No one, to my knowledge, has succeeded in inducing transplantation immunity by sensitising only with trophoblast across strains coisogenic for the H-2 region. Foetal lymphocytes, in contrast, from as early as seven days in the mouse presumably do express Ir, for they are capable of giving a graft-versus-host reaction (Hofman & Globerson, 1973), although in this experiment the authors could not strictly exclude the possibility that a few contaminating maternal lymphocytes were not responsible for their results. This could easily be tested however. Third, could I ask Davies and Allison whether a difference at M locus or some other Ir locus is also necessary in the case of minor loci such as H-3? Or is an H-3 difference alone adequate for both stimulation and killing?

DAVIES: I would guess that *H-3* alone would not be sufficient.

BODMER: Could I confirm that Johnson's data might be consistent with H-2 antigens being absent from the giant cells but present on the dividing cells in the early trophoblast stages.

JOHNSON: Yes, this would be consistent, but I doubt the possibility.

BODMER: Studies on H-2 and HL-A have shown that these molecules are subject to turnover (see e.g. Bernoco *et al.*, 1973). The lack of antigen in giant cells might be a relatively non-specific and uninteresting result because the cell surface is not turning over and so the antigen is absent. It may seem surprising that some cells have no H-2 or HL-A, but this really should not be the case. Many cells have little or no H-2 or HL-A, e.g. tumour cells, and HeLa cells have maybe W-28 but very little else (Miggiano, Nabholz & Bodmer, 1970, Kennett, Bodmer and others, unpublished results). Cells can be induced to shed all their antigen and be quite viable for a time. There is no basis for believing that H-2 or HL-A must be present for the viability of the cell. Cell–cell recognition demands this sort of function.

JOHNSON: The heavy phagocytic and secretory activity of trophoblast might indicate a lot of instability in the cell surface, in contrast to your assumption. A belief exists that H-2 does occupy a very fundamental role in cell to cell relationships. My point is that I find little or no evidence to suggest H-2 is important in anything except immunological processes.

BODMER: You may be right about it being involved in mainly immunological processes.

GARDNER: Could you say more about the fact that H-2 or HL-A might be important in certain cell–cell interactions? Most cells in the body have these surface antigens, so how could they be important in cell–cell interactions unless there is a different density of these antigens?

BODMER: It is not the HL-A or H-2 serological determinants, that are present on just about all cells, which would be involved but products of other genes present in that region. There may be hundreds or thousands of such genes including loci such as Ir, MLC and for graft rejection. So far, the defined products controlled by genes in the region which have tissue specificity are connected with the immune system. However tissue-specific antigens which are not concerned with immunological processes may be carried in that chromosomal region (Bodmer, 1972).

DAVIES: With respect to a tumour cell being apparently devoid of H-2. If a cell is not an adequate target cell for H-2 or HL-A antibody, it does not mean that the antigens are not there. If such cells are extracted, the antigen is there. A good example is shown by immunoselection where an

F_1 tumour grown in the parental strains may lose its antigen as judged by surface expression but it is really only covered up in some way, e.g. by additional neuraminic acid residues on the surface glycoproteins.

BODMER: A negative variant of HL-A2 has been produced by the immunoselection *in vitro* of a lymphoid cell line (Pious, Hawley & Forrest, 1973). When extracted these cells clearly do not have any antigen (Pious, Bodmer & Bodmer, 1974). This is not really a contradiction: there just may be different ways in which you get immunoresistance.

ALLISON: I would also like to mention Pious's work, for these cells are resistant to immune cytolysis. This could be relevant to the trophoblastic cells. If the antigens are not expressed on the surface of the cells, they could not be recognised by antibody or by lymphocytes.

BODMER: There may be different ways to obtain immune resistence. The TA-3 Haushka is an immune-resistant cell and quite clearly the antigen is present. In the case of Pious's lines, the antigen is not present, so there has been a switching-off mechanism at the genetic level, and not some sort of epigenetic masking. There may be different ways in which the overt expression of an antigen can be eliminated with respect to serological antigens or cell-mediated antigen.

BAGSHAWE: In view of the analogy with the TA-3 cells, it is interesting that Sanford showed the immunogenicity of the TA-3 cell to be greatly increased by neuraminidase (Sanford, 1967).

BODMER: Yes, some of these suppression systems could be explained simply by increased activity of an enzyme placing sialic acid on various molecules. HeLa cell derivatives cannot be typed in the normal way for HL-A; after absorption, W-28 and virtually nothing else can be found. Hybrids between HeLa and peripheral blood lymphocytes express the HL-A antigens of the peripheral blood lymphocytes, but nothing new from HeLa, and nothing is exposed by neuraminidase. This illustrates another type of suppression which could be genetic, although we do not consider it is a deletion in the simple sense. The chromosome is still there and the linked markers are there (Kennett *et al*, unpublished data).

REFERENCES

ARCHER, J. R., SMITH, D. A., DAVIES, D. A. L. & STAINES, N. A. (1974). A skin-graft enhancing antiserum which recognizes two new B-cell alloantigens determined by the major histocompatibility locus of the mouse. *Immunology* (in press).

BERNOCO, D., CULLEN, S., SCUDELLER, G., TRINCHIERI, G. & CEPPELLINI, R. (1973). HL-A molecules at the cell surface. In *Histocompatibility Testing 1972* (eds. J. Dausset & J. Colombani) p. 327. Munksgaard: Copenhagen.

BODMER, W. F. (1972). Evolutionary significance of the HL-A system. *Nature Lond.*, **237**, 139.

COGGIN, J. H., Jr & ANDERSON, N. G. (1974). Cancer, differentiation and embryonic antigens: Some central problems. *Adv. Cancer Res.*, **19**, 105.

DAVIES, D. A. L. & ALKINS, B. J. (1974). What abrogates heart transplant rejection in immunological enhancement? *Nature, Lond.*, **247**, 294.

DAVIES, D. A. L. & HESS, M. (1974). Isolation and partial characterization of a new alloantigen genetically linked to the major histocompatibility locus of the mouse. *Nature, Lond.*, (in press).

FAULK, W. PAGE, JEANNET, M., CREIGHTON, W. D., CARBONARA, A. & HAY, F. (1974). Immunological studies of the human placenta: characterization of immunoglobulins on the trophoblastic basement membrane. Proceedings of the 2nd Internat. Planned Parenthood Federation Cambridge Workshop *J. Reprod. Fert.*, (in press).

FRANKS, D. (1968). Antigens as markers on cultured mammalian cells. *Biol. Rev.*, **43**, 17.

HOFMAN, F. & GLOBERSON, A. (1973). Graft-versus-host response induced *in vitro* by mouse yolk sac cells. *Eur. J. Immunol.*, **3**, 179.

LOKE, Y. W. & BALLARD, A-C. (1973). Blood group A antigens on human trophoblast cells, *Nature, Lond.*, **245**, 329.

LOKE, Y. W., JOYSEY, V. C. & BORLAND, R. (1971). HL-A antigens on human trophoblast cells, *Nature, Lond.*, **232**, 403.

LOKE, Y. W., WILSON, D. V. & BORLAND, R. (1972). Localisation of human chorionic gonadotrophin in monolayer cultures of trophoblast cells by mixed agglutination, *Am. J. Obstet. Gynec.*, **113**, 875.

MÄTZ, L., BAHL, O. P. & MOHN, J. F. (1973). Blood-group activity of human chorionic gonadotrophin. *Biochem. biophys. Res. Comm.*, **55**, 717.

MCCORMICK, J. N., FAULK, W. PAGE, FOX, H. & FUDENBERG, H. H. (1971). Immunohistological and elution studies of human placentae. *J. exp. Med.*, **133**, 1.

MIGGIANO, V., NABHOLZ, M. & BODMER, W. F. (1970). Detection of HL-A and other antigens on fibroblast monolayers using a fluorochromatin cytotoxicity assay. In *Histocompatibility Testing 1970.* (ed. P. I. Terasaki) p. 623. Munksgaard: Copenhagen.

PIOUS, D., BODMER, J. G. & BODMER, W. F. (1974). Antigenic expression and cross-reactions in HL-A variants of lymphoid cell lines. *Tissue Antigens* (in press).

PIOUS, D., HAWLEY, P. & FORREST, G. (1973). Isolation and characterization of HL-A variants in cultured human lymphoid cells. *Proc. natn. Acad. Sci. USA*, **70**, 1397.

RAPAPORT, F. T. (1972). The biological significance of cross-reactions between histocompatibility antigens and antigens of bacterial and/or heterologous mammalian origin. In *Transplantation Antigens* (eds. B. D. Kahan & R. A. Reisfeld), p. 182. Academic Press: New York.

SANFORD, B. H. (1967). An alteration in tumour histocompatibility induced by neuraminidase. *Transplantation*, **5**, 1273.

SNARY, P., GOODFELLOW, P., HAYMAN, M. J., BODMER, W. F. & CRUMPTON, M. J. (1974). Subcellular separation and molecular nature of human histocompatibility antigens (HL-A). *Nature, Lond.*, **247**, 457.

STAINES, N. A., GUY, K. & DAVIES, D. A. L. (1974). Passive enhancement of mouse skin allografts: Specificity of the antiserum for major histocompatibility complex antigens. *Transplantation*, (in press).

VAN ROOD, J. J. (1962). Leukocyte grouping: a method and its application. Ph.D. Dissertation, University of Leiden.

ANTIGENS ON POST-IMPLANTATION PLACENTAE

BY PAGE FAULK, PRESLOV TRENCHEV, JACK DORLING AND JOHN HOLBOROW

The full-term human placenta may be considered as a type of homograft that has remained healthy and intact for approximately nine months. Other types of untreated homografts do not usually enjoy this degree of success, and it has been suggested that a great deal could be learned about transplantation if the mechanisms responsible for the successful placental graft were understood (Billingham, 1964). Medawar (1953) has suggested that these mechanisms are immunological. Several experiments have supported this concept, and Beer & Billingham (1971) have reviewed much of this experimental work.

Youtananukorn & Matangkasambut (1972) have reported cell-mediated immunity (CMI) to placental antigens by maternal lymphocytes, but the placental antigens were not identified. Direct immunohistological evidence that the placenta may be a site of immunological activity in humans was presented by McCormick *et al.* (1971). They demonstrated immunoglobulins (Ig), complement, and fibrinogen on the trophoblastic basement membranes (TBM), and were able to elute IgG from TBM. Such placental eluates have subsequently been shown to inhibit several blastogenic reactions by human lymphocytes in culture (Faulk, 1972; Bonneau *et al.*, 1973; Faulk *et al.*, 1974). However, the antigens towards which these reactions are directed have also not been identified.

It is obviously important to identify the trophoblast antigens that are responsible for the immunological reactions described above. This report describes the results of our efforts to identify several placental antigens. We have included immunoproteins, fibrillar proteins, and contractile proteins. Results of this investigation suggest that actin-like proteins are found on trophoblasts, and that these antigens are also found on lymphocytes. Antibodies to such antigens could block and protect trophoblasts from cell-mediated immune damage, and might be very important in the immunobiology of pregnancy.

MATERIALS AND METHODS

Tissues

Normal full-term placentae were studied. These were obtained upon delivery and were usually perfused with phosphate buffered saline (PBS). Blocks of tissue were cut from the mid-portion of the central cotyledon and frozen in liquid nitrogen, and 2 μm sections were prepared in a cryostat for immunofluorescence studies.

Antigen, antisera and conjugates for immunofluorescence

Antisera to immunoproteins. Globulin fractions of rabbit antisera monospecific for IgG, IgA, IgM, C3 and C4 were conjugated with fluorescein isothiocyanate (FITC) by the dialysis technique (Clark & Shepard, 1963) which produces uniform labelling of protein and less non-specific reactions. Unreacted FITC was removed by Sephadex G25 gel filtration. Fluoresceinated IgG fractions with the optimal fluorescein: protein ratios were prepared from the crude globulin conjugates by elution from DEAE-cellulose columns with 0.08 M phosphate buffer, pH 6.3 according to the procedure of McDevitt *et al* (1963). The pH of each labelled IgG fraction was adjusted to pH 7.2 and heat-inactivated normal rabbit serum 20% v/v) was added to stabilize the dilute protein solution and to inhibit any subsequent nonspecific binding of the labelled material in sections of placenta. These conjugates were subjected to staining reactions and specificity checks as described by Faulk & Hijmans (1972). Sections were examined using transmitted light from a halogen lamp and FITC-interference filters in a Reichert microscope.

Antisera to fibrillar proteins. Collagen was extracted from guinea pig skin at acid pH according to the method in Hawley & Faulk (1970). Rabbits received three subcutaneous injections of native acid-soluble collagen each month, and a sample was obtained a fortnight after the last injection. This antiserum had a passive haemagglutination titre of 1:1000 using tanned sheep erythrocytes coated with native guinea-pig collagen. The anti-fibrinogen/fibrin serum was purchased as a FITC-conjugate from Hyland Laboratories (California, USA).

Antisera to contractile proteins. Pregnant and non-pregnant human uteruses were obtained freshly from the operating theatre and kept frozen at $-70\,^{\circ}$C. They were thawed immediately before use, cut into pieces and homogenized.

Extraction procedures

Actin was extracted in de-ionized distilled water at 4 °C from acetone dried tissue powder prepared from the homogenate (Iyengar & Weber, 1963) and purified by passage through a column of Sephadex G200 (Adelstein, Godfrey & Kielley, 1963). The second peak contained the actin, which aggregated readily in 0.1 M KCl and bound added myosin specifically.

Tropomyosin was extracted with 1.0 M KCl according to the procedure of Bailey (1948) but with two modifications: firstly, isoelectric point precipitation was performed at pH 4.1; and secondly ammonium sulphate precipitation was repeated three times, the fraction precipitating between 53 and 50% saturation being collected (Carsten, 1968; Carsten, 1971).

Myosin was extracted with KCl at low ionic strength (10.075) (Huriaux, Hamoir & Oppenheimer, 1967). Crystallized trypsin (Seravac Laboratories) was added in a weight ratio 1:200. The digestion period was prolonged to 40 minutes because of the greater resistance of smooth muscle myosin to trypsin digestion (Hamoir, 1973). Heavy meromyosin was isolated from the digest by ammonium sulphate precipitation and recovered between 45 and 50%.

The various extracted protein solutions were run electrophoretically in 6–8% sodium dodecyl sulphate polyacrilamide gel (Weber & Osborne, 1969) and the separated protein bands stained with Coomassie Blue (25%). Only one band was observed with heavy meromyosin, tropomyosin and actin; the other preparations gave two or more bands.

Immunization and specificity of antisera

Immune sera against the extracted proteins were raised in rabbits by means of a single subcutaneous injection of 2 mg protein per rabbit emulsified in incomplete Freund's adjuvant. Animals were bled 21 days later.

The sera were tested by double diffusion in agar containing 0.6 M KCl, and by indirect immunofluorescence as previously described (Farrow *et al.*, 1971). For the latter, 6 μm cryostat sections of unfixed snap-frozen rat organs were used, and the sera were screened for antibodies at a dilution of 1/30. The sera were also tested on monolayer tissue culture preparations of chick embryo fibroblasts and of HeLa cells taken after 24 hours growth. A goat anti-rabbit Ig FITC conjugated antiserum (Nordic) of tested specificity was used at a dilution of 1:30, which gave no background staining of substrates when normal rabbit serum was used as the middle layer.

The specificity of antisera was evaluated by double diffusion in agar. The anti-tropomyosin sera gave a single line of precipitation against tropomyosin, and no precipitate against the other proteins. Anti-myosin gave a single line with both actomyosin and myosin. Anti-heavy meromyosin gave a single line with heavy meromyosin only. The precipitation lines took 8–10 days to develop at 4 °C, and several refillings of the antiserum wells were necessary. With the actin-anti-actin system no precipitates were obtained. However, the anti-actin serum stained smooth muscle, and it stained epithelial cells in renal glomeruli (this is the subject of another report).* The anti-actin sera also stained chick embryo fibroblasts, mouse L cells, and HeLa cell monolayers. Specific fluorescence was seen as a fine filamentous network in the cytoplasm. All of the actin staining was removed by absorption of the anti-actin serum with actin. The ability of actin to absorb specific reactivity was abolished if the actin solution was pre-treated with myosin, and the resulting insoluble complex removed by centrifugation. Pre-treatment of actin with tropomyosin did not have this effect. Absorption with human serum or with saline extracts of human uterus produced no effect. Absorptions were done by adding one drop of protein solution containing 0.125 mg to one drop of antiserum and eight drops of PBS.

Immune electron microscopy

Placentae were fixed for 30 minutes at 4 °C in 1% paraformaldehyde in 0.1 M PO_4 pH 7.4 containing 7.5% sucrose. Uptake of antibodies was studied at the ultrastructural level by the use of peroxidase labelled antiglobulin (Avrameas, 1969) in an indirect technique as previously described (Webb & Dorling, 1972).

RESULTS

Immunofluorescence

Immunoproteins. Direct immunofluorescence staining was done with FITC-labelled monospecific antisera to human IgG, IgA, IgM, C3 and C4. As shown previously by McCormick *et al.* (1971), we found most of the staining of placental villi to be limited to the TBM and the basal portions of trophoblasts. These results are listed in Table 1. Areas of fibrinoid necrosis also stained heavily for IgG and complement. Portions

* Localization of smooth muscle contractile proteins in kidney by immune-electron microscopy, by Janet Webb, J. Dorling, P. Trenchev and E. J. Holborow, to be published.

TABLE 1 *Immunoproteins in human normal term placentae identified by immunofluorescence*

Immunoprotein antigens	Location of antigens		
	Trophoblasts	TBM	Stroma
IgG	±	+	−
IgA	−	−	−
IgM	−	−	−
C3	−	+	−
C4	−	+	−

of TBM in some villi were not stained, and other areas were densely stained. The staining was both smooth and granular. The elution studies of Faulk *et al.* (1974) showed that placental IgG was not limited to any one subclass, and immunogenetic typing showed the IgG to be of maternal origin. Placental eluate-IgG has also been shown to inhibit blastogenic reactions of human lymphocytes that have been stimulated by old tuberculin or phytohaemagglutinin (PHA), as well as by the mixed-lymphocyte (MLC) reaction (Faulk, 1972; Faulk, Carbonara & Jeannet, 1973; Bonneau *et al.*, 1973; Faulk *et al.*, 1974).

Fibrillar proteins. Direct immunofluorescence staining was done with a FITC-labelled antiserum to fibrinogen, and indirect staining was done with rabbit anti-guinea pig collagen sandwiched with a FITC-labelled sheep anti-rabbit Ig serum. Fibrinogen staining was identified on TBM, and intense staining was observed in areas of fibrinoid necrosis. Collagen staining was diffusely and randomly scattered about the stroma of villi. It was also noted on TBM and faint staining was identified in some trophoblasts. These results are listed in Table 2.

TABLE 2 *Fibrillar proteins in human normal term placentae identified by immunofluorescence*

Fibrillar protein antigens	Location of antigens		
	Trophoblasts	TBM	Stroma
Collagen	±	+	+
Fibrinogen	−	+	+*

* Positive staining in areas of fibrinoid necrosis.

Contractile proteins. All staining for contractile proteins was done by indirect immunofluorescence. Intense staining of villi stroma was routinely and reproducibly seen with the anti-myosin serum (Fig. 1). Staining was in ordered whorls about foetal stem vessels and throughout the stroma, but TBM and trophoblasts were always negative. The

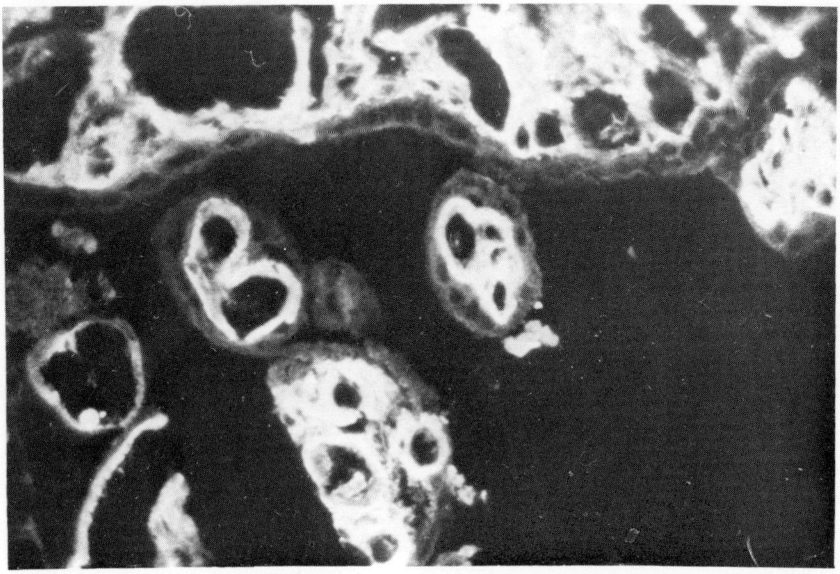

Fig. 1. Pattern of anti-myosin staining of human normal, term placenta.
Note staining of the villus stroma with sparing of trophoblasts. The staining patterns for troponin and tropomyosin are similar to myosin but anti-myosin staining is more intense.

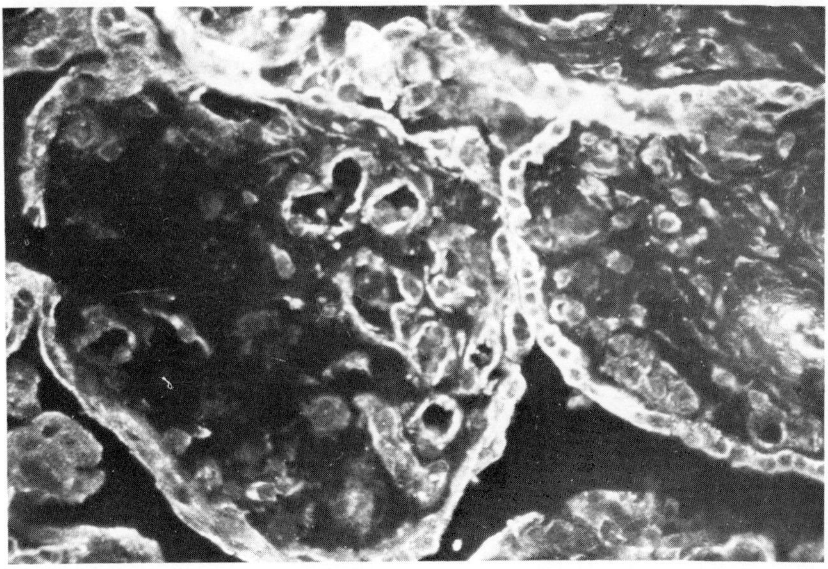

Fig. 2. Pattern of anti-actin staining of human normal, term placenta.
Note staining of trophoblasts with sparing of villus stroma. Staining within stroma is due to fibroblasts, macrophages, and foetal stem vessel endothelium. This pattern is similar to that seen for actinin and heavy meromyosin. In our experience, staining of vascular endothelium is not seen in non-inflammed adult vessels, but is unique to foetal tissue, placenta, and inflammed adult tissue such as rheumatoid synovium.

anti-actin serum routinely and reproducibly stained trophoblasts (Fig. 2), being most intense at the apical and basal aspects of these cells. Staining within the stroma was limited to the cytoplasm of cells, thought to be fibroblasts, and to the cytoplasm of pericytes and endothelial cells of foetal stem vessels. Areas of staining within TBM were noted, but see paragraph below on immunoelectron microscopy. Results of staining reactions with antisera to contractile proteins are listed in Table 3. Adult endothelium did not stain with anti-actin.

TABLE 3 *Contractile proteins in human normal term placentae identified by immunofluorescence*

Contractile protein antigen	Location of antigens			
	Trophoblasts	TBM	Stroma	Fibroblasts
Actin	+	–*	–	+
Actinin	+	–	–	+
Heavy meromyosin	+	–	–	+
Myosin	–	–	+	–
Troponin	–	–	±	–
Tropomyosin	–	–	±	–

* see Figs. 4 and 5.

Immuno-electron microscopy

Positive antibody-peroxidase staining for anti-IgG generally confirmed the results obtained by immunofluorescence (Table 4) with the additional observation that nonperfused tissues tended to stain endothelial cells of foetal stem vessels. The anti-collagen serum stained positively along fibres that were ultrastructurally identified as collagen (Fig. 3*b*), and staining was observed in the adventitia of foetal stem vessels.

Positive antibody-peroxidase staining for myosin generally confirmed the results obtained by immunofluorescence. The overall pattern of anti-myosin staining was different from the pattern of anti-collagen staining,

TABLE 4 *Protein antigens in human normal term placentae identified by immuno-electron microscopy*

Antigens identified	Trophoblasts	TBM	Stroma
IgG	–	+	–*
Collagen	±	±	+
Myosin	–	–	+
Actin	+	–	–†

* Foetal stem vessels stain in non-perfused placentae, and areas of fibrinoid necrosis stain in perfused specimen.

† Foetal stem vessel endothelial cells stain as do macrophages and fibroblasts.

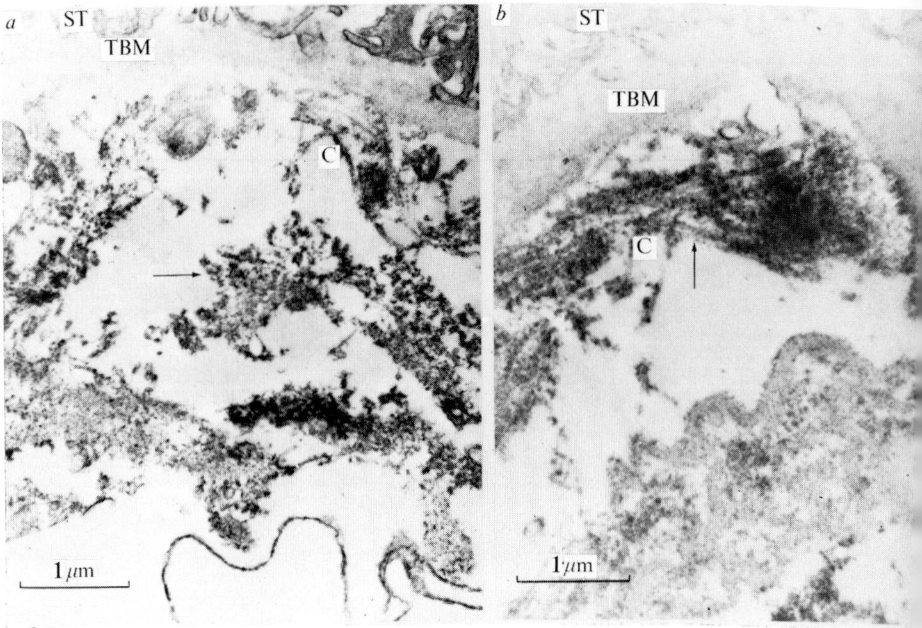

Fig. 3. Immuno-electron microscopy of human normal, term placenta: Antibody-enzyme staining for myosin and collagen.
(a) Anti-myosin pattern below TBM. A granular and stippled distribution (arrow) independent of collagen.
(b) Anti-collagen pattern below TBM. Distribution is along collagen bundles (arrow). ST is syncytial trophoblast and C is collagen.

even though the two were often seen in close proximity (Fig. 3*a*). Anti-actin staining also confirmed the results of immunofluorescence staining. Actin staining seemed to be more prominent in the basal areas of syncytial trophoblasts, and this appeared to be fibrillar with areas of positive staining seeming to continue into the TBM (Fig. 4). The basal aspects of syncytial trophoblasts were folded and appear to be inserted into TBM (Ashley, 1965), but this was not true for cytotrophoblasts (Fig. 5).

DISCUSSION

The results of this study indicate that at least three different groups of antigens are present in mature placentae: immunoproteins, fibrillar proteins, and contractile proteins. The immunoproteins seem to be specifically deposited within placentae, and some evidence has accrued that (*a*) they are of maternal origin (Faulk *et al*., 1974), (*b*) they have specificity for sites either on or in TBM (McCormick *et al*., 1971), and (*c*) they inhibit certain blastogenic reactions of human lymphocytes *in*

the transport of maternal IgG across the 'placental barrier'. This would provide a reasonable basis for the transport of antibody across cells proposed by Brambell (1966).

Placental eluates are known to inhibit MLC reactions and it has been proposed that this represents maternal blocking antibodies that protect trophoblasts from maternal CMI (Faulk, 1972; Faulk *et al.*, 1973; Faulk *et al.*, 1974). It is thought that the specificity for eluate-IgG is a 'trophoblast antigen' that is shared by lymphocytes, and evidence for cross-reacting antigens between trophoblasts and lymphocytes has been reported (Beer, Billingham & Yang, 1972). The nature of this antigen is not known. We have been impressed with the similarity of staining patterns for IgG and actin on trophoblasts, and we have recently observed positive staining of lymphocytes with anti-actin sera consistent with the earlier report of Fagraeus *et al.* (1973) for cortisone-sensitive lymphocytes. The similarity of staining patterns for actin and IgG on trophoblasts, the capacity of placental-IgG to inhibit MLC, and the presence of actin-like antigens on lymphocytes suggests that actin or an actin-like protein may be one of the trophoblast antigens responsible for common antigencity between lymphocytes and trophoblasts. This possibility is presently being studied. If it proves to be true it will put forward a useful approach to the study of the placenta as a homograft.

SUMMARY

This is a study of antigens in the post-implantation human normal term placenta. Antigens were identified by immunofluorescence and immunoelectron microscopy. The protein antigens were arbitarily grouped as immunoproteins, fibrillar proteins, and contractile proteins. IgG, C3 and C4 were identified on the basal aspects of trophoblasts and on the trophoblastic basement membrane (TBM). The fibrillar proteins collagen and fibrinogen were identified in villus stroma, and on TBM as well as areas of fibrinoid necrosis, respectively. Six contractile proteins, actin, actinin, heavy meromyosin, myosin, troponin and tropomyosin were identified. Actin, actomyosin and heavy meromyosin were noted in trophoblasts, and myosin, troponin and tropomyosin were noted in the villus stroma. Basal folds were identified only on syncytial trophoblasts, and immunoelectron microscopical studies identified actin-positive filaments extending into the basal folds and through TBM. Actin-like structures were identified both on trophoblasts and lymphocytes and it is suggested that maternal immunity to these structures could generate blocking antibodies to protect trophoblasts from maternal cell-mediated immunity. Such specificity would offer a basis for previously observed cross-reactions between trophoblasts and lymphocytes.

We thank Mrs Janet Webb for assistance with the antibody enzymes studies, Dr L. Conochie for characterization of the anti-collagen serum, and Dr G. Offer and Mr R. Starr of the MRC Muscle Biophysics Unit for myosin.

REFERENCES

ADELSTEIN, R. S., GODFREY, J. E. & KIELLEY, W. N. (1963). G-actin: preparation by gel filtration and evidence for a double-stranded structure. *Biochem. biophys. res. Commun.*, **12**, 34.

ALLISON, A. C. (1973). The role of microfilaments and microtubules in cell movement endocytosis and exocytosis. In *Locomotion of Tissue Cells*, (eds. Ruth Porter and D. W. Fitzsimons), Ciba Foundation Symposium p. 109. Elsevier: Amsterdam.

ASHLEY, A. (1965). Study of the human placenta with the electron microscope. *Arch. Path.*, **80**, 377.

AVRAMEAS, S. (1969). Coupling of enzymes to proteins with glutaraldehyde. Use of the conjugates for the detection of antigens and antibodies. *Immunochemistry*, **6**, 43.

BAILEY, K. (1948). Tropomyosin: a new asymmetric protein component of the muscle fibril. *Biochem. J.*, **43**, 271.

BEER, A. E. & BILLINGHAM, R. E. (1971). Immunobiology of mammalian reproduction. *Adv. Immunol.*, **14**, 1.

BEER, A. E., BILLINGHAM, R. E. & YANG, S. L. (1972). Further evidence concerning the autoantigenic status of the trophoblast. *J. exp. Med.*, **135**, 1177.

BILLINGHAM, R. E. (1964). Transplantation immunity and the maternal–fetal relation. *New Eng. J. Med.*, **270**, 667.

BONNEAU, M., LATOUR, M., REVILLARD, J. P., ROBERT, M. & TRAEGER, J. (1973). Blocking antibodies eluted from human placenta. *Transplant. Proc.*, **5**, 589.

BRAMBELL, F. W. R. (1966). The transmission of immunity from mother to young and the catabolism of immunoglobulins. *Lancet*, **ii**, 1087.

CARSTEN, M. E. (1968). Tropomyosin from smooth muscle of the uterus. *Biochemistry*, **7**, 960.

CARSTEN, M. E. (1971). Uterine smooth muscle: troponin. *Arch. Biochem. Biophys.*, **146**, 597 and **147**, 353.

CLARK, H. F. & SHEPARD, C. C. (1963). A dialysis technique for preparing fluorescent antibody. *Virology*, **20**, 642.

FAGRAEUS, A., THE, H. & BIBERFELD, G. (1973). Reaction of human smooth muscle antibody with thymus medullary cells. *Nature New Biol.*, **246**, 113.

FAULK, W. PAGE (1972). See Discussion after chapter by Beer, A. E. & Billingham, R. E. In *Ontogeny of Acquired Immunity* (eds, R. Porter and Julie Knight), Ciba Foundation Symposium, p. 172. Elsevier: Amsterdam.

FAULK, W. PAGE, CARBONARA, A. & JEANNET, M. (1973). Immunobiological studies of human placentae. In *Immunology of Reproduction* (eds K. Bratanov, R. G. Edwards, V. H. Vulchanov, V. Dikov and B. Somlev) p. 405. Bulgarian Academy of Sciences: Sofia.

FAULK, W. PAGE & HIJMANS, W. (1972). Recent developments in immunofluorescence. *Progr. Allergy*, **16**, 9.

FAULK, W. PAGE., JEANNET, M., CREIGHTON, W. D., CARBONARA, A. & HAY, F. (1974). Immunological studies of the human placenta: Characterization of immunoglobulins on the trophoblastic basement membrane. *J. Reprod. Fert. Suppl.*, (in press).

FOX, H. (1968). Fibrinoid necrosis of placental villi. *J. Obstet. Gynaecol. Brit. Commonw.*, **75**, 448.

HAMOIR, G. (1973). Extractibility and properties of the contractile proteins of vertebrate smooth muscle. *Phil. Trans. R. Soc. London B*, **265**, 169.

HAWLEY, P. R. & FAULK, W. PAGE (1970). A circulatory collagenase inhibitor. *Brit. J. Surg.*, **57**, 900.

HURIAUX, F., HAMOIR, G. & OPPENHEIMER, H. (1967). Low molecular weight fragment from succinylated myosin of bovine carotids. *Arch. Biochem. Biophys.*, **120**, 274.

IYENGAR, M. R. & WEBER, H. H. (1964). The relative affinities of nucleotides to G-actin and their effects. *Biochim. biophys. Acta*, **86**, 543.

LUSTIG, L., COSTANTINI, H. & MANCINI, R. E. (1969). Serological and immunofluorescent study of soluble collagen in tissue culture of fibroblasts. *Proc. Soc. exp. Biol. Med.*, **130**, 283.

MCCORMICK, J. N., FAULK, W. PAGE., FOX, H. & FUDENBERG, H. H. (1971). Immunohistological and elution studies of the human placenta. *J. exp. Med.*, **133**, 1.

MCDEVITT, H. O., PETERS, J. H., POLLARD, L. W., HARTER, J. G. & COONS, A. H. (1963). Purification and analysis of fluorescein-labelled antisera by column chromatography. *J. Immunol.*, **90**, 634.

MEDAWAR, P. S. (1953). Some immunological and endocrinological problems raised by evolution of viviparity in vertebrates. *Symp. Soc. exp. Biol.*, **11**, 320.

TRENCHEV, P., SNEYD, P. & HOLBOROW, E. J. (1974). Immunofluorescent tracing of smooth muscle contractile protein antigens in tissues other than smooth muscle. *Clin. exp. Immunol.*, (in press).

WEBB, J. A. & DORLING, J. (1972). The use of peroxidase-labelled antiglobulin for ultrastructural localization of tissue antigens reacting with serum antibodies. *J. Immunol. Methods*, **2**, 145.

WEBER, K. & OSBORN, M. (1969). The reliability of molecular weight determinations by dodecyl sulfate polyacrilamide gel electrophoresis. *J. biol. Chem.*, **244**, 4406.

YOUTANANUKORN, V. & MATANGKASAMBUT, P. (1972). Human maternal cell mediated immune reaction to placental antigens. *Clin. exp. Immunol.*, **11**, 549.

DISCUSSION

BODMER: We have also worked with antibody or immunoglobulin eluted by low pH from placenta, and have some evidence for polymorphic activity (Goodfellow & Bodmer, unpublished data). Would you accept that when a pool of material is eluted from several placentae, a heterogeneous mixture of polymorphic antibodies may result, so accounting for the relatively non-specific inhibition of MLC (mixed lymphocyte culture)? Material eluted from a single placenta, careful matching of the HL-A types of the test cells, and particular attention to the maternal and paternal antigens would give a clearer result.

FAULK: Regarding your first point, Van Rood in Leiden has studied 25 sets of matched mother-baby-eluate samples for anti-HL-A, and found no anti-HL-A activity by either agglutinating or cytotoxic assays. Identical results were also obtained for the same samples by Jeannet in Geneva. Occasional reactions were observed, but showed no relation to HL-A. This may be analogous to your observations of polymorphic activity. Regarding your second point, eluates prepared from pooled placentae would probably yield a heterogeneous mixture of polymorphic antibodies, as you suggest. However, most of our data were obtained from single placenta, thus disallowing the possibility for such a mixture (Faulk *et al.*, 1974).

JOHNSON: Do you know the subclass of IgG involved in the placental eluate?

FAULK: In the placental eluates, measurements by Frank Hay have shown all four subclasses of IgG to be represented, the proportion of each being roughly similar to that amount in the maternal serum.

JENKINS: Even spontaneous activity seems to be inhibited as judged by the data you have presented, for levels are less than unity. Do you know why the immunoglobulins in your placental sections are confined to isolated villi?

FAULK: Spontaneous activity is inhibited, as you say. I have no explanation for the location of immunoglobulins.

BODMER: If the control IgG is taken from females it could possess maternal HL-A antibodies. Another possibility is that inhibiting antibody could be present in levels too low for detection by cytotoxic tests.

FAULK: These are certainly possibilities. Our control sera were not always from females, and they were usually typed and found to contain no anti-HL-A activity. Inhibiting antibody may have been there in low titre.

COOMBS: Faulk and Bodmer obtain a material in placental extracts which interferes with the stimulation of lymphocytes by lectins. Could this material also be interfering with antibodies reactive with lymphocytes, and so undermine the basis of your testing of the HL-A extracts we were discussing earlier? (see Discussion after Johnson's paper, p. 103).

BODMER: If you mean the purified antigen obtained from placenta, this has been characterized chemically and purified in the same way as that obtained from cell lines (Snary *et al.*, 1974). It is free from immunoglobulin as far as we can tell. The antibody we eluted may inhibit an MLC reaction but so far we have only tested it in serological tests.

FAULK: We find only trivial interactions of eluate with labelled PHA (phytohaemagglutinin). In any case, this would not account for inhibition of MLC with eluate.

CURZEN: The distribution of the immunoproteins, complement and IgG is interesting and were described as being related to the basement membrane. But I would be inclined to call them subtrophoblastic because they were evidently not confined to the basement membrane. The photomicrographs of these deposits seem to be of a sufficiently high magnification to permit deciding on the distribution of the material in villi of different sizes. The deposits appeared to be present in main stem villi or possibly first-branch villi, rather than in terminal villi.

FAULK: I agree. You perhaps noticed in some of my photographs that a few villi were unstained. A definite decision whether staining is limited to main-stem villi cannot be made but, judging from their size, I doubt that this is the case.

CURZEN: The pictures are in many ways similar to ours, using a far cruder antiserum – just an anti-human globulin with FITC label. Our material was unperfused tissue and we used snap-frozen cryostat sections of full-term placentae (Curzen, 1970).

FAULK: Many difficulties arise with unperfused tissue. The trophoblast is heavily engaged in the transport of maternal IgG, so it would not be surprising to find staining.

EDWARDS: Have you found a similar situation in the placenta of other species?

FAULK: No, but Voisin (1974) has found almost exactly the same thing in the mouse. He also found inhibition of MLC with the mouse placental eluates.

COOMBS: The phenomenon you describe could be a part of the normal transport of immunoglobulin in other species. There should be no great surprise at finding immunoglobulin, complement, and other proteins also there.

FAULK: That is true. However, in the preparation of these eluates, the tissues are perfused and homogenized and then extensively washed in physiological saline until all soluble and/or unbound materials are washed free; then the pH of the washing fluid is dropped to a value of 2.7 and it is only then that one observes an outpouring of IgG. This would seem to indicate a specific binding.

BODMER: Our elution step utilizes purified membranes and a low pH, which indicates a fairly specific binding rather than a non-specific association with immunoglobulins.

COOMBS: But would the low pH lyse the cells so releasing the immunoglobulin?

ALLISON: Part of Brambell's hypothesis implies that immunoglobulin being transported is actually associated with membranes, and you may be merely eluting that material. (See Discussion after Adinolfi's paper, p..211.)

BODMER: The placenta may be just a large soak for immunoglobulin from the mother, and just stopping them from getting through to the foetus.

FAULK: I cannot agree that the placenta is a big soak, for the transport of maternal IgG across the human placenta is rather specific. According to the Brambell hypothesis (1966), IgG is intracytoplasmically transmitted in phagolysosomes. If the transporting cells are disrupted one might expect that the transport vesicles would be similarly destroyed.

ALLISON: If IgG is bound to membrane, the loss of vesicles would not matter and the binding might be reversed at low pH.

FAULK: Yes, that is a valid criticism. However, IgG prepared from maternal sera only weakly inhibits the MLC reaction, if at all. Placental IgG eluate inhibits quite well.

ALLISON: Your suggestion that the antibody is anti-actin is a little unlikely because as far as we know this material is not present on the cell surface, but is found inside the plasma membrane. I prefer the notion that the antibody is directed against some other polymorphic antigen – either the MLC locus itself, or something like it, and the fact that specific reactions occur should be followed up.

FAULK: I did not mean to imply that anti-actin is responsible for the inhibition of MLC. On the contrary, it was simply mentioned as a possibility that merits consideration! I am very sympathetic with the concept that our eluates may have specificity for something like the MLC locus, but there is presently no data to support this.

MUNRO: The eluted immunoglobulin was fluorescently labelled: does it stain tissues other than placenta? Lymphocytes would be particularly interesting.

FAULK: This eluate does not stain thyroid, kidney, gut, lung, or basement membranes, nor does it stain lymphocytes. Some preliminary work indicates that it stains fibroblasts.

MUNRO: It is absorbed by lymphocytes, and the fact that this material will react against lymphocytes in MLC reaction is very important. Does absorption with lymphocytes stop the staining of placenta?

FAULK: This has not been done. We have shown that lymphocytes treated with eluate and then washed extensively were still inhibited in the MLC reaction, but in this case we were absorbing on to the responder (Faulk *et al.*, 1974).

ALLISON: Some lymphocytes carry surface receptors for immunoglobulin, and their structure modified by acid elution, leading to a non-specific uptake of immunoglobulin by lymphocytes, would inhibit MLC.

FAULK: This is a possibility, but the maternal serum IgG is treated in exactly the same way, and it inhibits poorly.

MUNRO: You used an Fab_2 fraction, and also blocked the MLC reaction, hence I doubt that the last point is relevant.

LOKE: In our work using sections of young unperfused trophoblast, the IgG appeared as little globules of the cytoplasm in the syncytial trophoblast and not in linear distribution that you found (Loke, unpublished). Could there be a difference between young and full-term trophoblast?

FAULK: Perfused tissues should be used wherever possible. We have looked at trophoblast as early as 12 weeks gestation and these are nicely positive. However, perfusion is very difficult at 12 weeks, and we have doubts about the results: it may be simply a maternal IgG in the progress of being transported across the placenta.

ALLISON: The comment on the presence of actinomyosin in the endothelium is interesting, because it also occurs in regenerating endothelium, or in granulation tissue. Actinomyosin may be a feature of this particular kind of endothelium, and would indicate that the endothelial cells can move, although this property may be irrelevant to any barrier effect. I am concerned about the difference in specificity between the myosin serum and the heavy meromyosin serum. One is meant to be a fragment of the other.

BODMER: In relation to the specificity of the anti-collagen antibodies, is there any possibility that some of this specificity could be directed to attached sugar residues? One way to rule out this possibility might be to try and inhibit the reaction using specific sugars.

FAULK: In granulating tissue and indeed in foetal tissue, an immature or loosely cross-linked form of collagen is present (Robins, Shimokomaki & Bailey, 1973). I imagine that the heterophilic antibody is recognizing either immature collagen, or a foetal type of procollagen (Von der Mark, Coick & Bornstein, 1973). However, there may be some carbohydrate as you suggest. In any case, homologous antisera (i.e. rabbit anti-human collagen) do not recognize trophoblast or TBM, but intensely stain connective-tissue stroma of placental villi (Faulk *et al.*, 1974).

REFERENCES

BRAMBELL, F. W. R. (1966). The transmission of immunity from mother to young and the catabolism of immunoglobulins. *Lancet*, **ii**, 1087.

CURZEN, P. (1970). The antigenicity of the human placenta. *Proc. R. Soc. Med.*, **63**, 65.

FAULK, W. PAGE, JEANNET, M., CREIGHTON, W. D., CARBONARA, A. & HAY, F. (1974). Immunological studies of the human placenta: characterization of immunoglobulins on the trophoblastic basement membrane. *J. Reprod. Fert.*, (in press).

ROBINS, S. P., SHIMOKOMAKI, M. & BAILEY, A. J. (1973). The chemistry of collagen cross-links. *Biochem. J.*, **131**, 771.

SNARY, P., GOODFELLOW, P., HAYMAN, M. J., BODMER, W. F. & CRUMPTON, M. J. (1974). Subcellular separation and molecular nature of human histocompatibility antigens (HL-A). *Nature, Lond.*, **247**, 457.

VOISIN, G. (1974). Demonstration, nature and properties of material antibodies of term placenta and directed against paternal antigens. *J. Reprod. Fert.* (in the press).

VON DER MARK, K., COICK, E. M. & BORNSTEIN, P. (1973). The immunology of procollagen. *Arch. biochem. Biophys.*, **156**, 356.

LYMPHOCYTE PHYSIOLOGY DURING PREGNANCY: IN-VIVO AND IN-VITRO STUDIES

BY CRAIG HOWE

The lymphocyte is capable of both recognizing and destroying cells bearing non-self antigenic determinants. This fact, one of the tenets of the theory of immunological surveillance (Burnet, 1971), should be relevant to the maternal–foetal immunological relationship. That foetal-specific (Volkova & Maysky, 1969; Allison, p. 19), sex-linked (Polackova, 1970) and paternally-derived alloantigens (Searle *et al.*, 1974) arise during the development of the mammalian embryo is no longer a matter for conjecture. Less certain however, are the factors preventing successful immune recognition of, and/or tissue damage to, components of the conceptus expressing these foreign determinants.

Other contributions to this symposium have reviewed the evidence for and against the expression of several classes of antigens on the trophoblast. As a counterpart to these antigenic analyses, this review will be an attempt to define immunological factors with the potential for disrupting the maternal–foetal unit.

Although the success of the foetus as an allograft cannot be solely explained on the basis of blunted maternal immune competence, pregnancy-induced changes in lymphocyte distribution and traffic do occur. The role of the trophoblast in these phenomena is ill-defined as is the nature of the putative aggressor cell. The experiments summarized here include: (1) Studies of lymphocyte populations in the pregnant animal; (2) the use of in-vitro assays to elucidate the nature of the immune response to antigens of the trophoblast; (3) the ability of lymphoid cells sensitized to trophoblast hormones to interrupt pregnancy; (4) new directions for further research.

Studies of lymphocyte populations in the pregnant animal

Early studies of the immunological components of maternal–foetal relations concentrated on the functional criteria of sensitization or tolerance (Billingham, Palm & Silvers, 1965; Breyere & Barrett, 1960). Thus it was found that pregnancy led either to sensitization or tolerance depending on how immune competence was assessed. Transplantation of tumours to mice previously pregnant by the tumour-donating strain suggested an enhanced tumour growth (Currie, 1969). This apparent depression of cell-mediated immunity was subsequently found to be due not to any modification of the effector cell *per se*, but rather to the production of a humoral factor(s) capable of interfering with lymphocyte recognition and/or destruction of the foreign cells (Hellström, Hellström & Brawn, 1969). Although the above experiments left many questions unanswered, it was not until the allograft reaction was more fully understood and some of the subtleties of the tumour–host relationship were appreciated that the immunological factors operating during pregnancy were re-investigated. Rather than concentrating on the humoral aspects of maternal sensitization as did the earlier investigations, these later studies emphasized the changes in the central and peripheral lymphoid organs accompanying pregnancy. Miller, Mills & Baines (1973), Beer & Billingham (1971) and Maroni & DeSousa, (1973) found marked changes in lymphoid tissue histology and in the incorporation of DNA precursors into these same organs. Some of these alterations were common to all murine pregnancies while others were unique to animals carrying allogeneic foetuses, implying that stimulation by paternally derived alloantigens accounted for the differences. Whether it is the trophoblast which serves as the antigenic source causing the lymphoid changes accompanying pregnancy is not yet known, but study of the uterine transfer of pure trophoblastic vesicles (Gardner, 1972) might answer this question.

The recognition of two major classes of lymphocytes on the basis of their differing origins and cell-surface characteristics has led to a re-evaluation of a variety of immunological phenomena (Raff, 1973). Murine thymus-derived (dependent) lymphocytes (T cells) express an alloantigen on their surfaces designated θ (or Thy 1). Antisera to θ in the presence of heterologous complement are cytotoxic for this class of lymphocyte and such in-vitro treatment is singularly useful for depleting a heterogeneous lymphoid population of its T cell component. The other major class of lymphocyte is derived from the bone marrow (B cell) and in the case of the mouse can be recognized either by a xenoantigen designated 'mouse bone-marrow lymphocyte antigen' (MBLA) or by its high-density display of surface-incorporated immunoglobulin. Antisera to immunoglobulin light-chain determinants (K chains) in the presence of complement are cytotoxic to this population (Stobo, Rosenthal & Paul, 1973). Solid supports to which anti-immunoglobulins have been coupled have also been used to remove B cells from mixed lymphoid populations by retaining the immunoglobulin bearing cells. Conversely, the cells not retained by these supports are highly enriched in T lymphocytes. Although the techniques used to distinguish human T and B cells are somewhat different from those used for the mouse (Bentwich & Kunkel, 1973) there is no doubt that the concept of two lymphocyte populations is a general one.

Utilizing antisera to the θ antigen and a heterologous antiserum (rabbit)

TABLE 1 *Nature of lymphocyte redistribution engendered by pregnancy*

Lymphocyte population	Nature of change	Relative T Cell contribution*	Relative B Cell contribution†	Unique to allogeneic pregnancy
Thymus	Involution	N.C.‡	—	No
Mesenteric lymph node	Transient increase in weight	↑	↓	No
Spleen	Transient increase in weight	↑	↑	No
Peyers patches	N.C.	N.C.	N.C.	No
Bone marrow	N.C.	—	N.C.	No
Lymph nodes draining the uterus	Increase in weight	↑	N.C.	Yes
Peripheral blood	N.C.	↓	↑	No
Peritoneal exudate	Increase in cell number	↑	N.C.	Yes

* Determined by sensitivity to anti-θ serum.

† Determined by sensitivity to anti-MBLA or anti-mouse Ig.

‡ N.C. No change.

↑ Rise in relative contribution.

↓ Fall in relative contribution.

Lymphoid tissues were evaluated throughout allogeneic or syngeneic pregnancies. In most of the tissues studied a population of lymphoid cells insensitive to both of the aforementioned antisera was observed.

to MBLA, it has been possible to follow the changes in T and B lymphocytes during pregnancy. Table 1 is a compilation of the results of a study comparing syngeneic and allogeneic matings. Several findings emerge from such a study. Both syngeneic and allogeneic pregnancies led to changes in the composition of lymphoid tissues and in the case of the thymus and peripheral nodes these changes were qualitatively similar. It is likely that such redistribution may be explained by the changing hormonal milieu engendered by pregnancy. The situation is however not without complications, in that pseudopregnancy has only a very minor effect on murine lymphoid populations (Howe, unpublished). The peritoneal exudate and the lymph nodes draining the uterus (i.e. the lumbar and renal nodes) showed a marked increase in the number of T cells present during the later half of a first allogeneic pregnancy. These T cells may represent cells activated by paternal alloantigens or alternatively an influx of non-sensitized lymphocytes attracted to the draining node by chemotactic factors. The peripheral blood lymphocytes are also of interest in terms of the declining proportion of cells sensitive to anti-θ treatment. Studies of lymphoid cell traffic during pregnancy have indicated that the nodes draining the uterus are capable of trapping lymphocytes (Howe, unpublished) by an as yet undefined mechanism and it may be this preferential sequestration of peripheral T cells that accounts for the redistribution characteristic of allogeneic pregnancies. Not until the functional correlations of these lymphoid structural changes have been defined, and an examination of lymphocyte traffic to and through the uterus and associated foetal membranes carried out, will the significance of these various observations be clear.

Although sometimes observed after a single pregnancy, maternal sensitization in both the human and murine cases is more frequently detected in multiparous individuals. This may imply that repeated doses of antigen are necessary or that the commonly used assays of haemagglutination (Goodlin & Herzenberg, 1964), leucoagglutination or antibody-mediated cytotoxicity (Kaliss, 1973) are not sufficiently sensitive to detect the weak immune responses induced by a single pregnancy. In an effort to explain some of the findings of lymphocyte redistribution and to detect specific changes in the sera of primiparous mice, the assay outlined in Fig. 1 was adopted. This test is based on the ability of certain humoral factors to interfere with the homing of lymphocytes to the various lymphoid tissues (Ford, 1966). Paternal and, as a control, maternal lymphocytes were radiolabelled, incubated in the test serum or serum fraction obtained from mice carrying syngeneic or allogeneic foetuses. Following incubation, the cells were administered intravenously to syngeneic recipients and the fate of the cells determined by counting the tissues in a gamma spectrometer. The data in Table 2 represent a summary of experiments in which it was found that sera from females bearing allogeneic, but not

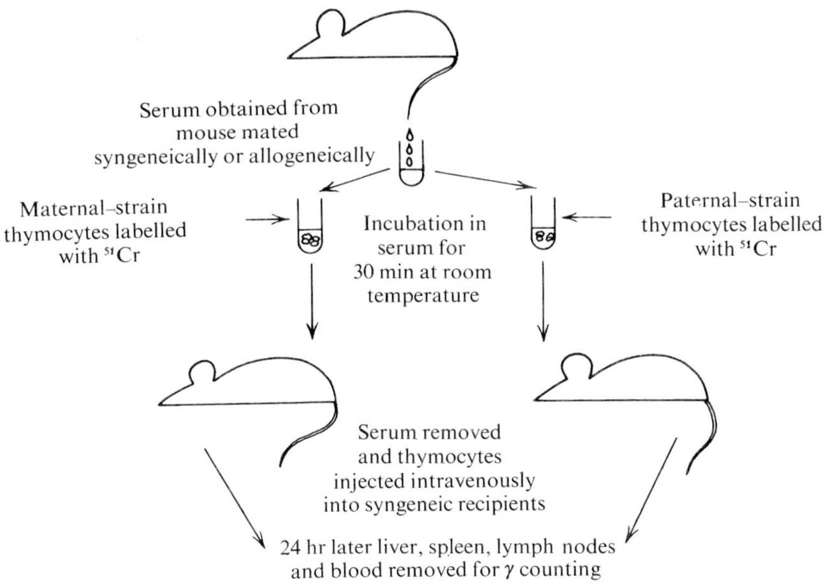

Fig. 1. Details of an assay to detect pregnancy-engendered serum factors capable of modifying lymphocyte homing.

syngeneic, foetuses inhibited lymphocyte homing. A detailed investigation of this phenomenon revealed that the serum factor was specific; it occurred in 85% of primiparous CBA mice pregnant by C57B1/10ScSn males and was present in significant amounts for only five days postpartum. Characterization of the factor has not been attempted, but preliminary experiments indicated that while none of the positive sera contained cytotoxic antibody (lymphocyte viability > 95% at time of injection), the activity responsible for lymphocyte redistribution was

TABLE 2 *Effects of various sera on thymocyte distribution*

	Distribution of radiolabelled cells (% of injected radioactivity)			
Source of serum	Blood	Peripheral nodes†	Spleen	Liver
Multiparous – allogeneic	0.34*	2.28	17.32	15.88
Uniparous – allogeneic	0.27	2.89	14.18	15.42
17th day of pregnancy – allogeneic	0.08	4.26	9.19	10.55
Virgin female	0.09	4.05	10.11	11.43
Multiparous – syngeneic	0.11	3.94	8.64	13.79
Uniparous – syngeneic	0.07	3.88	10.36	12.71

*Mean ± (SD) of four animals.
†Inguinal, brachial and axillary nodes.

removed by the passage of serum over an anti-mouse IgG column. It would be of interest to determine if the redistribution factor is similar to known antibodies (Schlesinger, Shlomui-Korzash & Israel, 1973) and if the lymphocyte surface components being masked or redistributed are shared with the trophoblast. Testing of the material eluted from the placenta (see paper by W. Page Faulk *et al.*) should be carried out.

Several problems concerning lymphocyte physiology in the pregnant animal remain. First and foremost is the identification of antigens leading to maternal sensitization. The importance of various antigens in triggering lymphocytes to specific response pathways is not yet clear. While it is commonly held that lymphocyte–allograft interaction leads to activation of T cells, the recent awareness of non-T cell cytotoxic mechanisms (Perlmann, O'Toole & Unsgaard, 1972), the identification of suppressor lymphocytes (Katz & Benacerraf, 1973) and the recognition of blocking and unblocking antibodies (Sjøgren & Bansal, 1972) are all indicative of the complexity of the cell mediated immune response, especially when considered in the context of the maternal–foetal unit. Ideally studies of cell–cell interaction should be carried out under in-vitro conditions but, as the next section will reveal, even these supposedly defined systems are not without complications.

In-vitro techniques

Numerous techniques have evolved for the measurement of lymphocyte responses *in vitro*. Whether all of these assays have in-vivo correlates or whether in-vivo events can be accurately reproduced *in vitro* are questions still in need of answers. The success of the mixed lymphocyte reaction (MLR) as a predictor of donor–recipient histocompatibility is an excellent testimonial to the utility of in-vitro techniques, but even more importantly, observations of this test have served as a stimulus for a re-examination of the fine structure of the major histocompatibility locus (Dupont *et al.*, 1974).

Several of these in-vitro assays have been applied to the study of maternal–foetal relations in an attempt to answer two questions: (1) What is the overall level of maternal lymphocyte reactivity during pregnancy? (2) Do maternal lymphocytes show specifically heightened or depressed response to foetal antigens?

The first question has been commonly assessed by quantitating lymphocyte responses to non-specific stimuli (Janossy, Greaves, Doenhoff & Šnajdr, 1973). Lectins and bacterial products are most often used and are especially valuable because they allow separate evaluation of T and B cell reactivity. Unfortunately, studies of non-specific lymphocyte stimulation yield conflicting results. While some investigators have claimed that lymphocytes obtained from pregnant women are less reactive than the

TABLE 3 *In-vitro assays in which serum from pregnant or post-partum animals inhibits lymphocyte reactivity*

Assay	Nature of inhibition	Reference
Mitogen induced lymphocyte transformation	Non-specific	St Hill, Finn & Denye (1973)
Mixed lymphocyte reaction	(1) Non-specific (2) Specific and non-specific components	Kasakura (1973) Howe (1974, unpublished)
Production of migration inhibition factor	Specific	Youtananukorn & Matangkasambut (1973)
Colony inhibition	Specific	Edidin, Gooding & Johnson (1974)
Allorosette formation	Non-specific	Howe (1974, unpublished)
Cell-mediated microcyto-toxicity	Specific	Hellström, Hellström & Brawn (1969)

cells of non-pregnant individuals (Purtilo, Hallgren & Yunis, 1972), other reports have found no difference between the groups (Harrison, 1972), and in at least one case a heightened reactivity to low lectin concentrations was noted (Carr & Stites, 1972). The part that serum factors play in this assay and in in-vitro assays of all kinds further complicates the interpretation of results. Serum taken from pregnant animals interferes with the assays of cell-mediated immunity listed in Table 3. Again, while these observations are of interest, their significance awaits characterization of the serum factors.

The mixed lymphocyte reaction has been employed by several investigators to study specific maternal reactivity toward foetal or paternal antigens (Ceppellini, 1971). This reaction, known to be the in-vitro correlate of the recognition phase of the allograft reaction, is a measure of T cell reactivity toward products of the major histocompatibility complex (MHC). Although the nature of the reaction precludes measurement of pre-existing sensitization or accurate determination of lymphocyte reactions toward antigens specified by genes outside the MHC, the assay is a valuable method for generating specific cytotoxic T cells *in vitro* and is a sensitive system for screening humoral factors capable of abrogating immune recognition (Gordon, Stinson, Souther & Oppenheim, 1971).

In one report, serum factors arising from maternal sensitization and capable of inhibiting the MLR were detected in 16 of 45 sera from multipara (Robert, Betuel & Revillard, 1973). This same survey demonstrated that blocking activity was probably due to IgG alloantibody, but was dissociated from the presence of cytotoxins. As in an earlier report (Ceppellini *et al.*, 1971) activity of these sera appeared, at least in some cases, to be directed toward responder as well as stimulator cells. The

MLR should be the single most powerful in-vitro technique for exploring the immunological relationship between mother and foetus. Before this potential can be realized however, a precise knowledge of the cell-surface recognition units must be sought. When these components are ultimately defined, determination of their presence or absence on the trophoblast may explain some of the ways in which mammalian pregnancy appears to contravene the basic tenets of transplantation biology.

The application of in-vitro techniques to the study of lymphoid cell-trophoblast interactions has received little attention. The difficulties in maintaining trophoblast *in vitro* and the problem of unambiguous identification of trophoblast as opposed to non-trophoblast cells are probable explanations for the infrequent use of this type of system (see Discussion after Johnson's paper, p. 101). Normal mouse and normal human trophoblast can be cultured, but typical trophoblast cells are difficult to identify after 7–12 days *in vitro* and stable cell lines have only been derived from human choriocarcinomas (Kohler & Bridson, 1971). Newer techniques, such as the perfusable capillary system (Knazek, Kohler & Gullino, 1974), may prove advantageous by providing suitable growth conditions.

Studies by Currie (1968) and by Douthwaite & Urbach (1971) have examined the effects of allogeneic and syngeneic lymphoid cells on cultures of human trophoblast. Both reports revealed that trophoblast *in vitro* was destroyed by syngeneic maternal cells as well as by allogeneic lymphoid populations. In the later study, lymphocytes purified by the removal of adherent cells failed to lyse trophoblast cells, encouraging the authors to speculate that cells other than small lymphocytes were required for cytotoxicity. Unfortunately, these early studies gave no indication of the trophoblast antigens involved nor were they helpful in identifying the effector cells. Several conclusions based on the two investigations are incompatible with a classic T-cell-mediated cytotoxicity. (1) The kinetics of the reaction (in both studies) are difficult to reconcile with the known mechanisms of effector cell–target cell interaction. (2) The ratio of lymphoid cells to trophoblast cells in the case of the Douthwaite study, although not explicitly stated, was implied to be remarkably low. (3) Purified lymphocytes failed to manifest cytotoxicity. The current recognition of several means of cell-mediated cytotoxicity (McLennan, 1973) suggest that non-T cell mechanisms may be operating in the aforementioned experiments. Further studies should identify the aggressor cell by selective removal of T or B lymphocytes according to the methods alluded to in the first section of this review. Careful removal of phagocyte cells and the use of anti-immunoglobulin reagents as potential inhibitors of cytotoxicity may lead to a definition of the effector cell. Trophoblast cultures might also be employed as cellular immunoadsorbents, the resulting non-adherent cells being tested for their ability to kill paternal lymphoblasts or fibroblasts. Conversely, lymphocytes activated in maternal–paternal MLRs might be evaluated for their ability to bring about trophoblast lysis. By the very nature of these reactions, basic information about trophoblast cell surface antigens will be gleaned. For example, if antibody-dependent monocyte killing was operative, it would seem likely that the antigens being recognized were not products of the MLR. Although beyond the scope of this review, one cannot fail to notice the manipulative advantages of utilizing choriocarcinoma cell lines, especially those which differentiate in response to progesterone precursors (Pattillo *et al.*, 1972), in studying the antigen dynamics accompanying the cytotrophoblast–syncytiotrophoblast transition. A final caveat regarding the dangers of generalizing from the results obtained in any single in-vitro assay is necessary. A recent analysis, comparing two assays previously thought to measure a single parameter, revealed that each technique was in fact a measure of separate populations active in the cell-mediated immune response (Scolland & Fitch, 1974). Therefore comparisons between assays are at best hazardous.

Immune response to trophoblast hormones

The difficulties inherent in identifying trophoblast antigens should be apparent from the studies summarized in this volume. The trophoblast is an endocrine tissue as well as a protective filter and it is the former capacity which may provide an immunological marker. Several reports have, by immunofluorescent or immunoperoxidase techniques, identified cells of the human placenta involved in the synthesis and secretion of human chorionic somatomammotrophin (HCS) (also known as human placental lactogen) and human chorionic gonadotrophin (HCG) (de Ikynakoff & Cedard, 1973; Midgely & Pierce, 1962). Heteroantibodies to HCG were cytotoxic to syncytial trophoblast cells cultured in the peritoneal cavity of rabbits (Morisada, Yamaguchi & Iizuka, 1972), but in another study employing rabbit anti-HCG in the absence of complement, the in-vitro growth of a choriocarcinoma cell line was not altered (Nisula & Kohler, 1974). The later report did, however, show that these same antibodies were effective in reducing the size and gonadotrophin production of the tumour adapted for growth in the hamster cheek pouch. These results taken together with earlier findings of enhanced growth of human choriocarcinoma in animals pre-treated with anti-lymphocyte serum (ALS) (Lewis, Davis & Ross, 1969) suggest that in the xenogeneic system, at least, antibodies to HCG may require an ALS-sensitive cell in order for tumour destruction to occur. At the same time it should be noted that mechanisms operative in these xenograft studies may bear little or no resemblance to the syngeneic model. The preceding results might be taken as evidence for considering HCG to be a structural component of the trophoblast cell surface, a suggestion also supported by data from serological techniques, including mixed haemadsorption (see paper by R. Borland).

A recently developed assay tests the ability of spleen cells, taken from an animal immunized with a soluble protein antigen, to induce cytolysis of ^{51}Cr-labelled chicken red blood cells coated with the sensitizing antigen (Golstein, Schirrmacher, Rubin & Wigzell, 1973). Mice immunized with small doses of either HCG or HCS in adjuvant developed spleen cells specifically cytotoxic for the correspondingly coated chicken red blood cells. The results of such an assay are given in Table 4. Although the levels of red cell lysis appear to be low, they were not unlike previous studies using other antigens. Thus both of the hormones were immunogenic in this heterologous system. Because of the known capacity of passively transferred antibodies to HCS to interrupt pregnancy in the rat (Gusdon, 1972), the transfer of sensitized spleen cells was evaluated for its effect on murine pregnancy. At various times during the second half of gestation 4×10^7 syngeneic nucleated spleen cells, obtained from animals previously immunized with HCG (a crude commercial preparation, 3550 i.u./mg), HCS (a purified commercial preparation) or an irrelevant antigen, keyhole limpet haemocyanin (KLH), were administered (intraperitoneally) to C3H mice. Several animals were killed at varying intervals after receiving spleen cells in order that a gross assessment of maternal and embryo pathology could be made. Most of the mice in each experimental group were allowed to carry on through the time of projected parturition. The results of this preliminary investigation are summarized in Table 5.

Two points emerge from an examination of the data. First, the transfer of spleen cells from animals immunized with HCS, but not HCG or KLH, was able to interrupt pregnancy. Second, these effects were to a degree dependent on the stage of pregnancy. The five animals receiving HCS sensitized cells killed prior to term revealed embyros which were dead and partially resorbed. Apart from the obvious histopathological study

TABLE 4 *Cytotoxicity of immune spleen cells towards chicken red blood cells coated with HCG or HCS*

Immunogen	Target cell	Inhibitor	Cytotoxicity index*
HCG	CRBC-HCG	—	37
HCG	CRBC-HCG	HCG	–4
HCG	CRBC-HCS	—	8
HCS	CRBC-HCS	—	44
HCS	CRBC-HCS	HCS	–9
HCS	CRBC-HCG	—	6

The assay was carried out over an 18-hour period at a ratio of 100 immune cells for every one target cell. Spontaneous release varied between 4 and 10%.
*Index of specific release. Mean of 3 determinations.
CRBC: chicken red blood cells.

TABLE 5 *The effect of transferring cells sensitized to human placental hormones on the outcome of pregnancy in the mouse*

Series and sensitizing antigens	Day of gestation when cells were transferred	Number of animals receiving cells*	Number of animals producing offspring	Number of live offspring produced
I. HCG	8	8	8	49
	11	6	6	40
	14	8	8	39
	16	8	7	43
II. HCS	8	6	6	42
	11	7	5	35
	14	7	2	11
	16	8	1	4

* Matings were C3H × C3H and syngeneic cells were transferred.

required to examine the nature and quantity of lymphoid infiltration, several other questions are raised by these results. Is there a structural component of the murine placenta (trophoblast) which cross-reacts with HCS? And is the experiment essentially different from the one previously reported (Gusdon, 1972)? By injecting spleen cells, enough primed B lymphocytes and plasma cells may have been transferred to have accounted for the findings on a purely humoral basis. Alternatively, the antibody-dependent monocyte (K cell) killing mechanism (Allison, 1973) observed to be active in the in-vitro killing of chicken red blood cells may have been operative *in vivo*. Finally, autonomous T-cell-mediated tissue destruction cannot be ruled out. The judicious use of in-vitro cell separation techniques before transfer should be informative. The possibility that endocrine products of the trophoblast, both proteins and steroids (Smith & Brush, 1973), may be structural antigens as well as secretory products requires further examination.

New directions

Several recent theories contend that cell mediated immunity constitutes a destructive threat to the maternal–foetal unit. Countering this threat are the placental filter, the low-density antigenic display on tissues forming the maternal–foetal interface and humoral blocking factors, all acting in concert to maintain homeostasis. While these concepts are consistent with observations of certain allograft and tumour–host models, such theories should not be adopted without a thorough consideration of views which might at present appear to be less conventional. For example, the immunostimulation theory of Prehn and colleagues has been proven to operate with both a cellular (Prehn & Lappé, 1971) and humoral contribution (Shearer, Philpott & Parker, 1974) and, in at least one intriguing experiment (Lappé & Schalk, 1971), has been applied to the problem of immune recognition of pregnancy. Clearly there is a need for further studies which take as their starting point the possible beneficial role that lymphocyte recognition of foetal antigens might play in development.

A second promising system is that of teratocarcinoma, a neoplasm arising spontaneously in the ovary or testis of humans and susceptible strains of mice (Stevens, 1967). The tumour, consisting of heterogeneous foetal elements, has already received attention as a model for differentiative processes. Inchoate studies of teratoma cell surfaces have yielded interesting results (Artzt *et al.*, 1973). Especially fascinating is a recent report of teratocarcinoma-like outgrowths arising from ectopically transplanted mouse egg cylinders. Host factors, presumably the cellular and humoral immune systems, were shown to be of primary importance in determining the success or failure of these grafts (Damjanov & Solter, 1974). The potential for separating and isolating various stages of foetal development, either by in-vivo or in-vitro manipulations of these unique tumours, should provide an excellent adjunct to more conventional studies.

Having considered the lymphocyte throughout this review, it is only fair to suggest that, as in certain tumour systems (Alexander, Evans & Grant, 1972), cells other than lymphocytes might be the determining factors in maintaining the balance characteristic of maternal–foetal relations. Studies to date on the immunological interruption of pregnancy (Beer & Billingham, 1971) have not positively identified the effector cells. An experiment carried out several years ago (Currie, 1969), for which there was no adequate explanation at the time, considered in the light of recent evidence implied that macrophages, probably with lymphocyte co-operation, could specifically influence the outcome of an allogeneic murine pregnancy. The discovery of macrophage-activating agents and the delineation of steps leading to specific macrophage arming suggest that a study of this cell's physiology and distribution during gestation would be highly rewarding.

Current knowledge of gestational physiology, the complexity of cell surfaces comprising the foetal–maternal interface and the multiple effector pathways of immune surveillance caution against any single explanation for the success of mammalian pregnancies in a potentially hostile environment. Perhaps the concept of the 'foetus as an allograft' is too restrictive to be of further use in planning experiments to test the various hypotheses.

SUMMARY

The lymphocyte as effector cell *in vivo* and *in vitro* is considered in terms of the maternal–foetal unit. The organ distribution of murine T and B cells during pregnancy is presented and an assay for a humoral factor

MARONI, E. S. & DESOUSA, M. A. B. (1973). The lymphoid organs during pregnancy in the mouse. A comparison between a syngeneic and an allogeneic mating. *Clin. exptl. Immunol.*, **13**, 107.

MIDGELY, A. R. Jr, & PIERCE, G. P. (1962). Immuno-histochemical localization of human chorionic gonadotrophin. *J. exp. Med.*, **115**, 289.

MILLER, K. G., MILLS, P. & BAINES, M. G. (1973). A study of the influence of pregnancy on the thymus gland of the mouse. *Am. J. Obstet. Gynec.*, **117**, 913.

MORISADA, M., YAMAGUCHI, H. & IIZUKA, R. (1972). Toxic action of anti-HCG antibody to human trophoblast. *Int. J. Fertll.*, **17**, 65.

NISULA, B. C. & KOHLER, P. O. (1974). Effect of antiserum to HCG on growth and function of choriocarcinoma *in vivo* and *in vitro*. *Cancer Res.*, **34**, 512.

PATTILLO, R. A., STORY, M. T., HERSHMAN, J. M., DELFS, E. & MATTINGLY, R. F. (1972). Hormone control of differentiation and embryonic antigens in human placental tumor cells *in vitro*. In *Embryonic and Foetal Antigens in Cancer*, vol. **2**, (eds N. G. Anderson, J. H. Coggin, E. Cole and J. W. Hollerman), p. 45.

PERLMANN, P., O'TOOLE, C. & UNSGAARD, B. (1972). Lymphocyte-mediated tumor cells destruction *in vitro*. Some general principles and reactivities in human urinary bladder carcinoma. In Conference on Immunology of Carcinogenesis, *National Cancer Institute Monograph*, **35**, 223. Dept. of Health Education and Welfare: Washington D.C.

POLACKOVA, M. (1970). Ontogenetic development of the male specific antigen in mice. *Folia Biologica*, **16**, 12.

PREHN, R. T. & LAPPÉ, M. A. (1971). An immuno-stimulation theory of tumor development. *Transplant. Rev.*, **7**, 26.

PURTILO, D. T., HALLGREN, H. M. & YUNIS, E. J. (1972). Depressed maternal lymphocyte response to phytohaemagglutinin in human pregnancy. *Lancet*, **i**, 769.

RAFF, M. C. (1973). T and B lymphocytes and immune responses. *Nature, Lond.*, **24**, 19.

ROBERT, M., BETUEL, H. & REVILLARD, J. P. (1973). Inhibition of the mixed lymphocyte reaction by sera from multipara. *Tissue Antigens*, **3**, 39.

SCHLESINGER, M., SHLOMUI-KORZASH, Z. & ISRAEL, E. (1973). Antigenic differences between spleen-seeking and lymph node-seeking thymus cells. *Eur. J. Immunol.* **3**, 335.

SCOLLARD, D. & FITCH, F. (1974). T cell participation in cell mediated cytotoxicity *in vitro* (Abstract). *Fedn Proc.*, **33**, 779.

SEARLE, R. F., JOHNSON, M. H., BILLINGTON, W. D., ELSON, J. & CLUTTERBUCK-JACKSON, S. (1974). Investigation of H-2 and non H-2 antigens on the mouse blastocyst. *Transplantation*, **18**, 136.

SHEARER, W. T., PHILPOTT, G. W. & PARKER, C. W. (1974). Humoral immunostimulation I. Increased uptake of [^{125}I]iododeoxy-uridine and [^{3}H]thymidine into TNP cells treated with anti-TNP antibody. *J. exp. Med.*, **139**, 367.

SJØGREN, H. O. & BANSAL, S. C. (1972). Relationship between blocking and unblocking tumor growth *in vivo*. In Conference on Immunology of Carcinogenesis, *National Cancer Institute Monograph*, **35**, 231. Dept. of Health Education and Welfare: Washington, D.C.

SMITH, N. C. & BRUSH, M. G. (1973). Progesterone-macromolecular interactions. In Preparations of human placental villous surface membrane (Abstract). *J. Endocr.*, **59**, xvii.

STEVENS, L. C. (1967). The biology of teratomas. *Adv, Morphol.*, **6**, 1.

ST HILL, C. A., FINN, R. & DENYE, V. (1973). Depression of cellular immunity in pregnancy due to a serum factor. *Br. med. J.*, **3**, 513.

STOBO, J. D., ROSENTHAL, A. S. & PAUL, W. E. (1973). Functional heterogeneity of murine lymphoid cells v. Lymphocytes lacking detectable surface– or immunoglobulin determinants. *J. exp. Med.*, **183**, 71.
VOLKOVA, L. S. & MAYSKY, I. N. (1969). Immunological interaction between mother and embryo. In *Immunology and Reproduction* (ed. R. G. Edwards) p. 211. International Planned Parenthood: London.
YOUTANANUKORN, V. & MATANGKASOMBUT, P. (1973). Specific plasma factors blocking human maternal cell-mediated immune reaction to placental antigens. *Nature New Biol.*, **242**, 110.

DISCUSSION

EDWARDS: You have alluded to the distribution of HPL (human placental lactogen) and HCG using the peroxidase technique; is HCG produced by the syncytial trophoblast or the cytotrophoblast, a point which would be rather important in view of your last results?

HOWE: The studies of de Ikonicoff & Cedard (1973) implicated the syncytiotrophoblast as the chief source of HCG. The Langhans cells were negative as determined by their failure to stain with the peroxidase conjugate.

CURZEN: In the human, the evidence is equivocal. Fluorescent tracing studies have shown HCG to be largely localized in the cytotrophoblast in the hands of some workers (Fox & Kharkongor, 1970). In agreement with this observation, HCG concentration falls in the peripheral blood after about 14 weeks, and this fall could be related to the declining number of cytotrophoblast cells in the human placenta from that time onwards (Mishell, Wide & Gemzell, 1963).

MUNRO: I would like to raise a general point about the use of HCG to identify trophoblast cells; how readily will HCG bind to the surface of trophoblast cells or, for that matter, other types of cells? Does the appearance of HCG on the cell surface give evidence of the synthesis of HCG by that cell, or could it simply indicate that the cells had taken up the hormone from another source? This point could be relevant to the detection of the types of cells synthesizing HCG in the placenta.

BAGSHAWE: The binding of HCG to cells has evidently not been studied formally. The hormone does not bind readily to untreated red cells (Bagshawe, unpublished observations). (see Discussion after Borland's paper, page 167.)

CURZEN: The transfer of sensitized spleen cells is a fascinating experiment but I have difficulty in interpreting its possible physiological significance, because HPL was used to sensitize mouse lymphocytes or spleen cells. These sensitized cells then interrupt mouse pregnancy. Some evidence by Contractor & Davies (1973) showed that HPL dramatically inhibits the human mixed lymphocyte reaction between unrelated adult lymphocytes; hence it might have been a non-specific inhibiting effect. I also query the significance of this study when a human hormone is being used across a species barrier.

HOWE: HCS is known to inhibit MLR, but the finding may not be relevant to my experiments in which HCS is an antigen. HCS could have a variety of effects on the lympho-reticular system; Arezzini, De Gore, Tarli & Neri (1972) found an increase in weight of lymphatic tissue when it was given to Dwarf mice, and Arrenbrecht & Sorkin (1973) suggested that human growth hormone, which bears structural and functional resemblance to HCS, induced T cell differentiation in mice. We should be careful about interpreting the effects of HCS on lymphoid populations.

There are difficulties in using antibodies against a human hormone to influence pregnancy in the mouse. The basis for attempting such experiments was the observation of Gusdon (1972), for the histology presented in his study was compatible with a type of cell mediated immune response, and led me to carry out the cell transfers.

EDWARDS: An important point for your interpretation is to know that cells capable of producing HPL or an HPL-like substance exist in the mouse placenta. The mouse needs ovarian hormones for the maintenance of pregnancy, and any placental hormones are presumably luteotrophic. The role of prolactin in mice as a hormone stimulating the formation of the corpus luteum has been well studied in contrast to the role of HCG as a possible luteotrophin in man. Such species differences might indicate that protein hormones are not produced by the mouse placenta. Is there any evidence to show that in fact the mouse placenta does produce either HCG or HPL-like molecules?

AMOROSO: Yes, I think there is a good deal of evidence, especially the work of Lyons and his group from California. I think it is well documented that the mouse does produce placental prolactin comparable to the one described by Astwood & Greep nearly forty years ago.

BILLINGTON: More recent experiments by Zeilmaker (1968), where the effect of ectopic transfers of mouse blastocysts on the endocrine state of the recipient was assessed, indicated that certain effects could have been mediated directly by the developing trophoblast. They gave strong evidence of luteotrophic hormone (LTH) production by the trophoblast.

EDWARDS: I do not wish to go deeply into endocrinology, but caution is necessary in interpreting such results. If trophoblast is transplanted, the recipient must be hypophysectomized, or other methods must be used to ensure that any effect of the pituitary is quite separate from the effects of the transplanted trophoblast.

HOWE: Recent studies of Kasakura (1971, 1973) on the blocking of the mixed leucocyte reaction by serum from pregnant women indicates that the active factor is non-specific. It does not seem to be cortisol, because there was no correlation between serum cortisol levels and ability of sera to block the MLC reaction.

JENKINS: The transfers of HCG-sensitized cells were carried out rather late in pregnancy. Would you expect a different effect if the transfers were carried out earlier?

HOWE: This could well be.

BAGSHAWE: Can animals that have been actively sensitized to either HCG or HPL become pregnant?

HOWE: I have not examined many of these animals. Sensitization to

HCG does not suppress pregnancy, and insufficient data exist on sensitization to HPL.

COOMBS: Differences in hormone output between syngeneic and allogeneic matings could affect your results.

HEAP: What was the appearance of the placenta in those animals treated with anti-HPL cells where foetuses were being resorbed?

HOWE: Gross examination revealed no great difference from animals receiving normal lymphoid cells. If gross structural changes occur they are not easily seen.

MUNRO: You are using the killing of hormone-coated red cells as an indication of the cell-mediated part in the immune response. The killing of chicken cells is mediated by antibody and the K cell system (Greenberg, Hudson, Shen & Roitt, 1973). This possibility should be tested before claiming that your experiments are really any different from simply passively transferring antibody to HPL.

HOWE: In many ways these may be identical phenomena. From what we know about the autonomy of T cell killing, I doubt that we are working with a T cell phenomenon.

JOHNSON: I agree. If mice are immunized with HPL, then presumably T cells recognize only those HPL determinants absent from mouse placental lactogen and will not, by themselves, kill cells with MPL on their surface. They will, however, stimulate B cells which will react with both MPL and HPL. A K cell mediated killing is therefore more likely. This is in line with the idea of tolerance that was discussed earlier.

ALLISON: The exact analogy is in thyroiditis. In that case, there is no T cell sensitization against the autoantigen but K cell damage occurs. An easy method of deciding is to mix aggregated gammaglobulin with the effector cells before they are added to the target cells, to find out if killing is inhibited.

BAGSHAWE: Do the cells that you inject home on to the conceptus?

HOWE: Perhaps not, but this is being investigated in detail at the present time.

EDWARDS: Many workers have shown an astonishing migration of lymphocytes into the uterine stroma early in pregnancy, but others have expressed doubt and debate about the meaning of this phenomenon. Several reports have thus appeared on the movement of lymphocytes, monocytes and granulocytes towards the deciduotrophoblastic zone during the early stages of human and animal placentation especially (e.g. Hamilton & Boyd, 1960; Kirby, 1970; Dallenbach-Hellweg & Nette, 1964). Would this possibly be one explanation of the loss of lymphocytes during pregnancy or is this idea complete nonsense?

HOWE: The idea is not complete nonsense at all. Tracing of lymphoid cells to the uterus and, at later stages of pregnancy, to the mature placenta certainly needs to be done before any comprehensive theory of lymphocyte traffic during pregnancy can be discussed. The lymph nodes draining the uterus and the peritoneal exudate are the only sites where an increase in θ-positive lymphocytes can be attributed to the presence of allogeneic embryos. A loss of θ-bearing cells occurs from the peripheral blood but we do not know if the uterus is acting as a sink for these θ-bearing lymphocytes.

CURZEN: One of the difficulties in the human situation is obtaining very early implantation stages. My impression from the few available is that the large numbers of lymphocytes near the blastocysts are a very transient phenomenon, and few lymphocytes are found after the first trimester of pregnancy (Hertig & Rock, 1961). Hardly any are found in relation to the placenta and its basal plate.

AMOROSO: Many people have studied implantation but very few have looked for lymphocyte infiltration in the human. The stages between seven days and 14 days are the time when infiltration would be expected. There is an abundance of lymphocytes in the basal plate of this early implantation. There is a traffic of lymphocytes in the blood sinuses, so these cells obviously enter the uterus through the blood pathway. These will get to the surface of the cell and attack the basal plate.

BAGSHAWE: According to Park (1971) lymphocytes are found throughout pregnancy in close association with the trophoblast. He attributes contrary statements on this point to Hertig's studies which were apparently restricted to embryos in the first two weeks after nidation, so resulting in the view that lymphocytes were not seen.

EDWARDS: The studies quoted above showed in general that large numbers of lymphocytes colonized these regions in early pregnancy. Later on, the number of lymphocytes declined. Information from Howe's work could illustrate the stage of pregnancy when lymphocyte traffic occurs.

HOWE: Changes take place throughout pregnancy but most of those described in my paper occurred late in mouse pregnancy, i.e. about fifteen or sixteen days. In the case of the spleen, the most dramatic changes take place in the spleen at ten days. Preferential changes were not observed during early pregnancy in the mouse.

BILLINGTON: Lymphocyte traffic in rodents does occur in the early stages of pregnancy, since significant numbers accumulate in the subepithelial space at the site of blastocyst attachment just prior to implantation. Once implantation has occurred the lymphocytes move away (Potts, 1965).

JOHNSON: The data shown in Fig. 1 represent incidental observations which may or may not have anything to do with lymphocytes. Mouse implantation sites of six to nine days were stained with anti-immunoglobulin antisera and at the edge of the decidua the cytoplasm of some cells stained very heavily. Staining was absent with anti-albumin, very

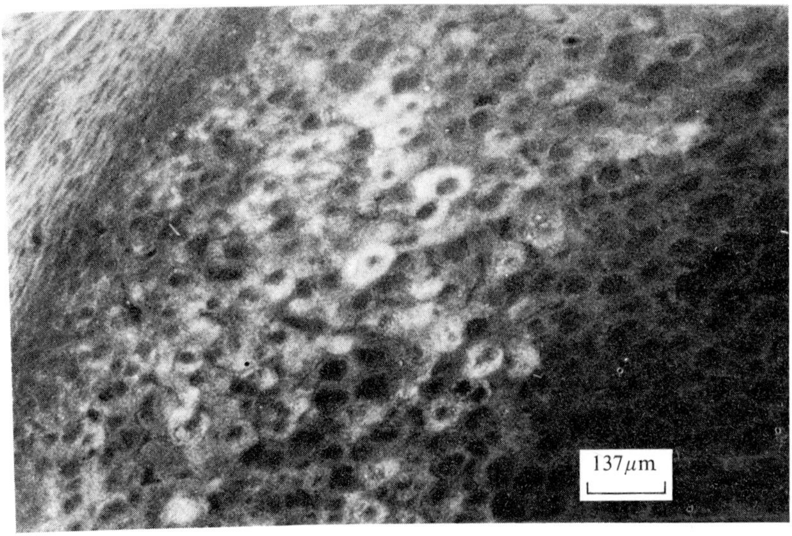

Fig. 1. Section through the edge of the decidua from a $7\frac{1}{2}$-day, pregnant mouse uterus stained directly with fluorescein-conjugated antiserum to mouse IgG. $\times$ 110. (unpublished data).

weak with IgM but strong with anti-γ-chain. Presumably these are B-cell derivatives.

COOMBS: Are viruses likely to be located in these placental sites?

ALLISON: This is certainly possible, for such sites are likely places for certain types of infections: brucellosis and malaria, for examples.

GARDNER: Concerning the cause of the accummulation of lymphocytes about the implantation site in early stages of pregnancy, I suspect that the effect is non-specific. They may accumulate because of the vascular changes occurring as a part of decidualization, and perhaps become trapped there through stromal odema and changes in capillary permeability. Might not this lead to a non-specific accumulation in these sites?

EDWARDS: In that case, various blood components and even erythrocytes should accumulate there as well. If the accumulation was simply due to changes in capillary permeability, then the oedematous picture would be complex.

McLAREN: Red cells are not seen at the site of implantation, as far as I know. There is extensive oedema, with attendant capillary permeability changes and many white cells.

ALLISON: This sounds exactly like inflammatory sites. An accumulation of lymphocytes occurs there and specifically of young T lymphocytes.

McLAREN: Histamine is also involved in the development of the decidual reaction.

GARDNER: This question could be answered if lymphocyte accumulation was measured when means other than pregnancy had been used to induce decidualization, such as intraluminal oil or traumatic damage (Finn, 1965). The implications would be the appearance of a similar pattern regardless of the type of decidualization stimulus. In my experience oil-induced deciduomata produce a rather similar accumulation of lymphocytes.

EDWARDS: Would anyone care to comment about the Kurloff cell mentioned by Howe during his presentation?

COOMBS: The Kurloff cell looks very much like a lymphocyte – in fact it is thought to be a lymphocyte – and is only found in the guinea-pig. This cell migrates to the uterus in pregnancy and is responsive to oestrogen. The production of these cells is stimulated, presumably from the spleen, as a result of the injection of oestrogen (Marshall, Swettenham, Vernon-Roberts & Revell, 1971). It is a highly sulphated glycoprotein which has been characterized by Dean & Muir (1970), and is very toxic to macrophages.

EDWARDS: In Faulk's data the amount of lymphocyte suppression with placental extracts was so considerable that the lymphocytes in pregnancy would be almost virtually non-reactive. Do you find a similar degree of generalized lymphocyte depression?

HOWE: Faulk is using a very special source of immunoglobulin. The reason why we do not find such a drastic suppression of lymphocytes may be because most of the serum immunoglobulin reactive in lymphocyte suppression is associated with the placenta rather than free in the serum. I do not find the same type, or degree of suppression in mixed leucocyte cultures as he obtained. Faulk was using only 250 μg to achieve this very marked suppression.

EDWARDS: The traffic of lymphocytes through the placenta must be quite astonishing, and during a brief period of time, a large proportion of the lymphocytes might be exposed to this placental immunoglobulin. After these lymphocytes leave the placenta, do they then become fully reactive in terms of the MLC reaction? Once these lymphocytes are blocked, is it possible that the serum factors are removed, and the lymphocytes are again reactive?

HOWE: Faulk implies that the phenomenon was a fairly irreversible process. In an evaluation of the mixed leucocyte culture by assaying specific protein synthesis, I find that certain sera from pregnant animals will inhibit protein synthesis for 12 hours, but if the sera are then removed and the lymphocytes cultured for a further period in a neutral serum, they then respond quite normally. At least in this in-vitro system, therefore, the effect seems to be reversible.

BODMER: Surely, if an antibody binds to a lymphocyte in an antigen–antibody type of reaction, a procedure like elution would be needed to remove it. The process of antibody binding to lymphocytes is probably irreversible.

ALLISON: But there is a turnover of membrane constituents.

BODMER: Turnover occurs only over a relatively long period of time (e.g. Bernoco *et al.*, 1973). However, I do not see how a lymphocyte passing through the placenta would pick up immunoglobulin.

EDWARDS: In the conditions described by Faulk, lymphocytes might well pick up immunoglobulin. The situation is very different to that in serum, for there is a mass of surrounding placenta.

BODMER: From our data, it seems clear that any antibody extracted is attached to membrane, because the membranes are purified (see earlier discussion).

EDWARDS: The origin and destination of this immunoglobulin of Faulk's and how it concentrates in the plasma, are important questions. It is certainly maternal, from what he said. We are sorry he has had to leave, although some of these points were answered in an earlier meeting (Faulk *et al.*, 1974).

Could you tell us more about the MLC tests in pregnancy? What is your opinion of the various responses reported by different workers: some showing heightened responses, some depressed, and still others unchanged. It is all very confusing.

HOWE: There is less confusion about the MLC data than there is about the data regarding stimulation of lymphocytes by mitogens. Hopefully, some of the confusion about the mitogen data can be explained away by changes in the various lymphoid cell populations. In MLC, a non-specific component may be involved but its nature is still unclear.

EDWARDS: Several authors have studied changes in the size of lymph nodes during pregnancy. Hetherington's data appear to be in opposition to those reported by others.

HOWE: This difference may be explained in terms of methodology, for he was examining uniparous animals whereas Maroni & de Sousa (1973) used multiparous females. Too much importance should not be placed on lymph node size; studies with cell populations and the functional capabilities of lymphocytes are the most important aspects.

BILLINGTON: Is there any good evidence on the use of the rosette-forming assay to detect sensitization to paternal strain antigens in the mouse?

MCLAREN: We have found some evidence for an increase in rosette-forming cells in multiparous mice after allogenic pregnancies (Baines, Ansell, McLaren & Micklem, unpublished data), but not during the course of a first allogeneic pregnancy.

COOMBS: There are at least four or five different types of rosetting reactions. Which one do you mean?

MCLAREN: I mean lymphocytes forming allo-clusters with paternal-strain erythrocytes using the technique of Micklem, Asfi, Staines & Anderson (1970).

BILLINGTON: This evidence fits with data showing the detection of sensitization to be difficult during the first pregnancy, but to be more easily found after three or four pregnancies by the MIF test (Eldred, unpublished results) or by GVH reactivity (Maroni & Parrott, 1973).

HOWE: In some studies, Currie was able to interrupt pregnancy by administration of the adjuvant, *Corynebacterium parvum* (Currie, 1969). What type of cells are activated by this organism?

ALLISON: *Corynebacterium parvum* does not increase T cell activity, and unlike Freund's complete adjuvant, it seems to act chiefly on macrophages and to some extent on increasing antibody production.

JENKINS: Taylor, Hancock & Scott (1973) do not believe the destruction of trophoblast *in vitro* is allogeneic inhibition. There are three reasons for their doubt: destruction does not take place until 72 hours after introduction of lymphocytes into the culture; transformation of the lymphocytes occurs before lysis of the cells; and the reaction depends on the presence of non-lymphocytic leucocytes, all of which stresses the difference between these observations and allogeneic inhibition. Taylor believes her results to be similar to those of Ginsberg, Ax & Berke (1970).

HOWE: Her data on the transformation of lymphocytes are very interesting, but the kinetics of killing she describes are not really compatible with the Ginsberg model.

REFERENCES

AREZZINI, C., DE GORE, V., TARLI, P. & NERI, P. (1972). Weight increase of body and lymphatic tissue in dwarf mice treated with human chorionic somatomammotrophin (HCS). *Proc. Soc. exp. Biol. Med.*, **141**, 98.

ARRENBRECHT, S. & SORKIN, E. (1973). Growth hormone-induced T cell differentiation. *Eur. J. Immunol.*, **3**, 601.

BERNOCO, D., CULLEN, S., SCUDELLER, G., TRINCHIERI, G. & CEPPELLINI, R. (1973). HL-A molecules at the cell surface. In *Histocompatibility Testing 1972*. (eds. J. Dausset & J. Colombani), p. 327. Munksgaard: Copenhagen.

CONTRACTOR, S. F. & DAVIES, H. (1973). Effect of human chorionic somatomammotrophin and HCG on phytohaemagglutinin-induced lymphocyte transformation. *Nature New Biol.*, **243**, 284.

CURRIE, G. A. (1969). The foetus as an allograft: the role of maternal 'unresponsiveness' to paternally derived foetal antigens. *Foetal Autonomy*, (eds G. E. W. Wolstenholme, M. O'Connor & A. Churchill), p. 32, Ciba Foundation Symposium: London.

DALLENBACH-HELLWEG, G. & NETTE, G. (1964). Morphological and histochemical observations on trophoblast and decidua of the basal plate of human placenta at term. *Am. J. Anat.*, **115**, 309.

DEAN, M. F. & MUIR, HELEN (1970). The characterization of a protein-polysaccharide isolated from Kurloff cells of the guinea pig. *Biochem. J.*, **118**, 783.

De Ikonicoff, L. K. & Cedard, L. (1973). Localisation of human chorionic gonadotrophic and somatomammotrophic hormones by the peroxidase immunohistoenzymologic method in villi and amniotic epithelium of human placenta (from six weeks to term). *Am. J. Obstet. Gynec.*, **116**, 1124.

Faulk, W. Page, Jeannet, M., Creighton, W. D. Carbonara, A. & Hay, F. (1974). Immunological studies of the human placenta: characterization of immunoglobulins on the trophoblastic basement membrane. *J. Reprod. Fert.* (in press).

Finn, C. A. (1965). Oestrogen and the decidual cell reactions of implantation in mice. *J. Endocr.*, **32**, 223.

Fox, H. & Kharkongor, F. N. (1970). Immunofluorescent localisation of chorionic gonadotrophin in the placenta and in tissue cultures of human trophoblast. *J. Path.*, **101**, 277.

Ginsberg, H., Ax, W. & Berke, G. (1970). Functional implications of cellular immunity. An analysis of graft reaction in cell cultures. In *Pharmacological treatment in organ and tissue transplantation*, p. 85. Excerpta Medica Foundation: Amsterdam.

Greenberg, A. H., Hudson, L., Shen, L., & Roitt, I. M. (1973). Antibody dependent cytotoxicity due to a 'small' lymphoid cell. *Nature New Biol.*, **242**, 111.

Gusdon, J. P. (1972). An immunological method of pregnancy destruction and contraception. *Am. J. Obstet. Gynec.*, **112**, 472.

Hamilton, W. J. & Boyd, J. D. (1960). Development of the human placenta in the first three months of gestation. *J. Anat.*, **94**, 297.

Heatherington, C. M. (1973). The absence of any effect of maternal/fetal incompatibility at the *H-2* and *H-3* loci on pregnancy in the mouse. *J. Reprod. Fert.*, **33**, 135.

Hertig, A. T. & Rock, J. (1961). Two human ova of the pre-villous stage, having an ovulation age of about eleven and twelve days respectively. *Contrib. Embryol. Carnegie Inst.*, **29**, 129.

Kasakura, S. (1971). A factor in maternal plasma that suppresses the reactivity of mixed leucocyte cultures. *J. Immunol.*, **107**, 1296.

Kasakura, S. (1973). Is cortisol responsible for inhibition of MLC reactions by pregnancy plasma. *Nature, Lond.*, **246**, 496.

Kirby, D. R. S. (1970). Immunological aspects of implantation. In *Ovoimplantation, Human Gonadotrophins and Prolactin* (eds P.O. Hubinont, F. Leroy, C. Robyn & P. S. Leleux). Karger: New York.

Maroni, E. S. & deSousa, M. A. B. (1973). The lymphoid organs during pregnancy in the mouse. A comparison between a syngeneic and an allogeneic mating. *Clin. exp. Immunol.*, **13**, 107.

Maroni, E. S. & Parrott, D. M. V. (1973). Progressive increase in cell-mediated immunity against paternal transplantation antigens in parous mice after multiple pregnancies. *Clin. exp. Immunol.*, **13**, 253.

Marshall, A. H. E., Swettenham, K. V., Vernon-Roberts, B. & Revell, P. A. (1971). Studies on the function of the Kurloff cell. *Int. Arch. Allergy*, **40**, 137.

Micklem, H. S., Asfi, C., Staines, N. A. & Anderson, N. (1970). Quantitative study of cells reacting to skin allografts. *Nature, Lond.*, **227**, 947.

Mishell, D. P. Wide, L. & Gemzell, C. A. (1963). Immunologic determination of human chorionic gonadotrophin in serum. *J. clin. Endocr.*, **23**, 125.

Park, W. W. (1971). *Choriocarcinoma. A study of its pathology.* Heinemann: London.

POTTS, M. (1965). Implantation: an electron microscopic study with special reference to the mouse. Ph. D. Thesis, University of Cambridge.

TAYLOR, P. V., HANCOCK, K. W. & SCOTT, J. S. (1973). The possible role of serum factors in masking the antigenicity of human trophoblast *in vitro.* In *International Congress on Immunology in Obstetrics and Gynaecology*. Excerpta Medica no. **281** (18).

ZEILMAKER, G. H. (1968). Luteotrophic activity of the ectopic mouse trophoblast. *J. Endocr.*, **41**, 455.

IMMUNOLOGICAL PRIVILEGE RESULTING FROM ENDOCRINE ACTIVITY OF TROPHOBLAST *IN VIVO*

BY ROBERT BORLAND, YUNG WAI LOKE AND DONALD WILSON

The foetus, the foetal membranes and the placenta are derived from genetic material of both parents and thus are genetically different from the maternal environment in which they survive and develop for varying periods of time – depending upon the species. It would be logical to assume that the genetically unique conceptus will possess potentially antigenic material not present in the mother and thus should be subject to maternal immunological reaction directed against it. If such a maternal response does take place, it certainly does not appear to affect the foetus adversely.

We are thus faced with the problem of attempting to understand the mechanisms involved which would account for maternal reluctance to reject the 'foreign' foetus, foetal membranes and placenta as she would any other allograft. Numerous hypotheses have been proposed to explain the above phenomenon, many of which fall by the wayside as new work comes to light, but a current and widely held opinion is that there exists some form of 'foetal–maternal barrier' which acts to prevent rejection of the allogeneic foetus from the maternal host during pregnancy. (Billingham, 1964, Currie & Bagshawe, 1967; Bradbury, Billington, Kirby & Williams, 1969 and Loke, Joysey, Valerie & Borland, 1971).

In the placenta, the trophoblast cells provide the unbroken 'barrier' between the maternal blood system and the other foetal tissues, and in mammals with a haemochoric placenta this 'barrier' may be only two to three cells thick. It would seem then that the barrier must be provided by the cells of the trophoblast and/or some non-cellular material closely associated with the trophoblast in this region of the placenta. For the trophoblast cells themselves to constitute such a barrier would necessitate the cells being *non-antigenic* – this seems an unlikely explanation since these cells have been shown to possess HL-A antigenic determinants although they may not necessarily elicit an immune response in the mother. (Loke *et al.*, 1971).

Thus, we must consider the possibility that there is some form of a non-cellular barrier closely associated with the trophoblast layer and that such an immunologically inert barrier would have to function in either of two ways:

(a) By blocking the *afferent arc* of the immunological circuit, i.e. by preventing the trophoblast antigens reaching and stimulating the maternal immunological mechanisms or

(b) By blocking the *efferent arc* of the immunological circuit and preventing a maternal response reaching the trophoblast cells.

There now appears to be considerable evidence to suggest that sensitization of the pregnant female against foetal antigens does take place (Bonneau *et al.*, 1973). Consideration will now be given to the possible nature of a barrier which prevents the maternal immunological attack from reaching the trophoblast and foetal cells.

The first suggestion as to the nature of this barrier came from Kirby and his colleagues (1964) who suggested that such an extracellular barrier did exist and consisted of a periplacental fibrinoid material shown subsequently to be a sialomucoprotein. (Bradbury *et al.*, 1969; Bradbury, Billington & Kirby, 1965). The interesting immunological properties of other mucoprotein-rich sites such as the hamster cheek pouch (Billingham & Silvers, 1962) and the anterior chamber of the rabbit eye (Billingham & Boswell, 1953) add further weight to the suggestion that the sialomucoprotein coat may be responsible, at least in part, for the unusual foetal–maternal immunological relationship since there is an abundance of evidence for prolonged graft survival in the above sites. (See paper by Billington, p. 67.)

So far there has been little discussion on the possible physiological status of this sialomucin layer associated with the trophoblast and our attention to this was focussed by two findings, first that the human chorionic gonadotrophin (HCG) hormone is produced by the trophoblast cells (Loke & Borland, 1970) and second, that HCG has been shown to be a glycoprotein with sialic acid residues linked to it (Bahl, 1969*a*, 1969*b*, and Canfield *et al.*, 1971). In the light of the above findings it did not seem unreasonable to examine the possibility that HCG in addition to its gonadotrophic effects, may act as an immunological barrier substance in man.

Before discussing the more specific aspect of HCG acting as a possible immunological barrier, there has in the past been considerable interest in the general immunological effects of altered hormonal levels during pregnancy. For example Heslop, Krohn & Sparrow (1954) demonstrated the prolonged survival of skin allografts in pregnant rabbits where transplantation was carried out at the peak of enhanced hormone secretion (three weeks gestation). These grafts survived twice as long as those transplanted earlier or later in pregnancy. As against this, however, certain animal species do not show any significant influence of pregnancy on the development of skin allograft sensitivity, e.g. mice (Medawar & Sparrow, 1956) and cattle (Billingham & Lampkin, 1957). The increased survival of allografts in the rabbit was only observed when transplants were carried out at the peak of cortico-steroid production which is a relatively brief period when compared with the length of gestation.

The chorionic gonadotrophins have been described so far as naturally occurring only in primates and equines although HCG does have profound physiological effects when injected into a wide range of experimental animals. Park (1958) demonstrated that in normal mice receiving premedication with HCG, intravenous inoculations of suspensions of placenta, liver and kidney cells had an increased survival time whereas in this context, oestrogens and progesterone were relatively ineffective. Also working with mice Pearse & Kaiman (1967) demonstrated that daily injections of HCG delayed skin allograft rejection by an average of two days. Working with male, female, mature, immature, and castrated or hypophysectomized rats, Jolly (1960) found that allogeneic trophoblast transplants survived only ten days whereas when experiments were carried out using pregnant rats, the trophoblast transplants survived for an average of eighteen days. Other studies in rats have shown that following treatment with exogenous HCG, oestrogens and progesterone, only HCG and oestrogen were able to produce thymic atrophy and HCG alone was capable of reducing spleen weight in treated animals (Nelson *et al.*, 1967).

Studies in humans have shown that trophoblast allografts in non-pregnant individuals survived 30–40 days only, whereas in pregnant recipients, or those with endogenous or exogenous hyperoestrogenism, there was a prolonged survival of the trophoblast with proliferation and HCG production (Lajos *et al.*, 1964). A later study from the same laboratory concluded that both trophoblast and foetal skin allograft survival in such hyperoestrogenic individuals may be mediated by an anterior pituitary substance (Lajos *et al.*, 1967).

The experiments which shall now be described were designed to examine the *local effect of HCG* on the transplantation of trophoblast and other foetal cells and the possible role of HCG in conferring immunological privilege on trophoblast cells *in vivo.*

Cell suspensions were prepared from primary cultures of human trophoblast and foetal skin cells taken from the same conceptus (Loke & Borland, 1970) by treating the monolayer cultures with trypsin (0.1% in calcium and magnesium-free phosphate buffered saline) at room temperature. The cells were washed in warm Hanks balanced salt solution (BSS) and exposed to 1600 I.U. HCG/ml in Hanks BSS for two hours at 37 °C. Controls were exposed to Hanks BSS only. Five intradermal injections of 0.1 ml aliquots (10^6 cells) of each cell suspension were placed into marked sites on the shaved back of each of twelve inbred virgin female guinea-pigs. The transplantation sites were removed surgically under general anaesthesia, serial histological sections were made and

TABLE 1 *Survival of human foetal cells following transplantation into guinea-pig skin*

Cells and treatments	Surviving human foetal cells at various times following transplantation				
	5 Days	10 Days	15 Days	20 Days	25 Days
Trophoblast	++++	+++++	++++	+++	+++
Trophoblast + HCG	++++	++++++	++++++	++++	++++
Skin	++	++	+	0	0
Skin + HCG	++	++++++	+++++	++++	+++

stained with haematoxylin and eosin. Sections were examined and photographed on a Zeiss Ultraphot microscope and comparative estimates made of the surviving transplanted cells at various times up to 25 days following transplantation (Table 1). From Table 1 it can be seen that the untreated foetal skin cells were rejected more rapidly than the corresponding trophoblast cells which is in agreement with earlier work done with guinea-pig trophoblast (Borland, Loke & Oldershaw, 1970). HCG-treated skin cells survived longer and gave a transplant survival pattern similar to that of the trophoblast cells. Significant numbers of healthy foetal skin cells were still present in the guinea-pig subcutis 25 days after transplantation.

Addition of HCG to trophoblast cells made little difference to the observed survival rate of the cells which is not unexpected if one accepts that HCG is being produced by these cells. Injections of equivalent volumes of Hanks BSS and HCG produced no visible pathology in the guinea-pig subcutis even at five days. A similar series of experiments comparing HCG-treated, washed and unwashed cells revealed no significant differences in their behaviour on transplantation. A tenfold increase in the concentration of HCG made no difference to the survival pattern of treated trophoblast or skin cells as shown in Table 1. Thus, although the precise mechanism is not yet clear, it seems that HCG can act locally and interfere with the normal transplantation rejection process.

It was then deemed necessary to attempt an evaluation of the role of HCG in human trophoblast survival by enzymically altering the sialoglycoprotein using neuraminidase (VCN) replacing the HCG *in vitro* and comparing the *in-vivo* survival times of these cells with normal untreated trophoblast cells. Monolayer primary cultures of human trophoblast from three isolates were prepared as described previously and treated as follows:

Four cultures from each isolate were exposed to 50 units of VCN in 1 ml Hanks BSS for 30 minutes at 37 °C and washed in five changes of warm Hanks BSS. A further four cultures from each isolate were treated as above, but after washing with Hanks BSS each monolayer was exposed to 1600 units of HCG in Hanks BSS for 30 minutes at 37 °C and washed as

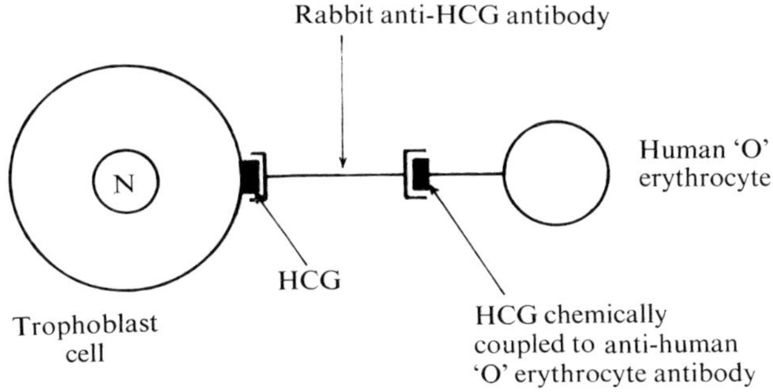

Fig. 1. A schematic representation of the mixed agglutination test. The figure shows a human 'O' erythrocyte with its attached HCG molecule, linked by means of an anti-HCG antibody molecule, to an HCG molecule on the surface of a trophoblast cell.

before. A further four dishes from each trophoblast isolate were retained as controls and treated with Hanks BSS only. Following washing, the cells were removed from the dishes, washed twice in warm Hanks BSS and finally resuspended to give a concentration of 10^6 cells/ml.

Six inbred virgin female guinea-pigs each received 15×0.1 ml (10^5 cells) intradermal injections on the dorsum in three rows of five each, comprising five from untreated trophoblast cultures, five from VCN-treated cultures and five from VCN- and HCG-treated cultures. Each inoculation site was clearly marked and one site was removed surgically from each of the three groups on each animal every five days as described previously.

Prior to setting up the above tests, primary trophoblast cultures before and after treatment with VCN and/or HCG, were tested for the presence of HCG on the cells using HCG-linked human O cells as indicators. A schematic representation of the mixed agglutination test is shown in Fig. 1.

When the primary monolayer cultures were tested, most of the untreated trophoblast cells were producing HCG *in vitro* as shown by the rosettes of HCG-linked red cells around them. Following prior treatment

TABLE 2 *The presence of HCG on human trophoblast cells in primary monolayer culture as determined by the mixed agglutination test*

Cultures	Treatment	% Cells showing +ve mixed agglutination for presence of HCG	
Primary monolayer cultures of human trophoblast cells	Controls	60–70	65
	Neuraminidase	0	0
	HCG	40–50	45
	Neuraminidase + HCG	40–50	45

TABLE 3 *Survival of human trophoblast cells following transplantation into guinea-pig skin*

Days post-transplant	Mean no. of surviving human trophoblast cells following transplantation		
	Neuraminidase-treated	Controls-untreated	Neuraminidase + HCG-treatment
5 days	+	+++	++
10 days	+	++++	+++++
15 days	0	++	+++
20 days	0	+	+++
25 days	0	+	+++

of trophoblast cells with VCN no rosette formation was seen, but subsequent exposure of the treated cells to HCG prior to carrying out the mixed agglutination test resulted in the reappearance of rosette formation in the cultures (Table 2). These findings show that HCG present on untreated trophoblast cells can be removed or enzymatically altered and replaced again by treatment of the cells with commercial HCG.

The results of transplanting the VCN and HCG-treated human trophoblast cells are shown in Table 3. The untreated trophoblast cells were present for at least 25 days after transplantation whereas the VCN-treated cells were not found beyond ten days. However, when the VCN-treated cells were exposed to HCG and thoroughly washed prior to transplantation, the cells were found to survive for at least 25 days.

From these findings it would appear that VCN-treatment of human trophoblast cells prior to transplantation results in the rapid rejection of these cells by the host as described previously for murine trophoblast by Currie, van Doornick & Bagshawe (1968). However, this ability of trophoblast cells to survive longer than other cells in a foreign environment, can be restored by treating the cells with HCG. It is concluded that HCG, in addition to its widely recognized gonadotrophic role, may have an important part to play in protecting the human trophoblast cells from attack in a foreign environment and as such may constitute an important part of the foetal–maternal immunological barrier (see also paper by Allen, p. 217).

ABSTRACT

It has been a widely held concept that the trophoblast pericellular mucoprotein is in some way responsible for protecting the foetus from immunological attack. In this study it has been shown that the treatment of trophoblast cells with neuraminidase results in the loss of their immunological immunity, but that subsequent treatment of the cells with human chorionic gonadotrophin (HCG) restores protection to the trophoblast cells.

The authors wish to acknowledge the excellent technical assistance of Mrs Anne Ballard and Miss Gillian Thomas. The study was supported by a grant from the Cancer Research Campaign.

REFERENCES

BAHL, O. P. (1969*a*). Human chorionic gonadotrophin. I. Purification and physicochemical properties. *J. biol. Chem.*, **224**, 567.

BAHL, O. P. (1969*b*). Human chorionic gonadotrophin. II. Nature of the carbohydrate units. *J. biol. Chem.*, **244**, 575.

BILLINGHAM, R. E. (1964). Transplantation immunity and the foetal–maternal relation. *New Eng. J. Med.*, **270**, 667.

BILLINGHAM, R. E. & BOSWELL, T. (1953). Studies on the problem of corneal homografts. *Proc. R. Soc. B.*, **141**, 392.

BILLINGHAM, R. E. & LAMPKIN, G. H. (1957). Further studies in tissue homo transplantation in cattle. *J. Embryol. exp. Morphol.*, **5**, 551.

BILLINGHAM, R. E. & SILVERS, W. K. (1962). Studies on cheek pouch skin homografts in the syrian hamster. *C.I.B.A. Foundation Symposium* on *Transplantation*, p. 90.

BONNEAU, M., LATOUR, M., REVILLARD, J. P., ROBERT, M. & TRAEGER, J. (1973). Blocking antibodies eluted from human placenta. *Transpl. Proc.*, **5**, 589.

BORLAND, R., LOKE, Y. W. & OLDERSHAW, P. J. (1970). Sex difference in trophoblast behaviour on transplantation. *Nature, Lond.*, **228**, 572.

BRADBURY, S., BILLINGTON, W. D. & KIRBY, D. R. S. (1965). A histochemical and electron microscopical study of the fibrinoid of the mouse placenta. *J. R. Micros. Soc.*, **84**, 199.

BRADBURY, S., BILLINGTON, W. D., KIRBY, D. R. S. & WILLIAMS, F. A. (1969). Surface mucin of human trophoblast. *Am. J. Obstet. Gynec.*, **104**, 416.

CANFIELD, R. E., MORGAN, F. J., KAMMERMAN, SANDRA, BELL, JENNIFER, J., & AGOSTO, GLADYS, M. (1971). Studies of human chorionic gonadotrophin. *Recent Prog. Horm. Res.*, **27**, 121.

CURRIE, G. A. & BAGSHAWE, K. D. (1967). The antigenicity of normal and malignant trophoblast: some implications. *Advances in Transplantation*, **34**, 523.

CURRIE, G. A., VAN DOORNINCK, W. & BAGSHAWE, K. D. (1968). Effect of neuraminidase on the immunogenicity of early mouse trophoblast. *Nature, Lond.*, **219**, 191.

HESLOP, R. W., KROHN, P. L. & SPARROW, F. M. (1954). The effect of pregnancy on the survival of skin homografts in rabbits. *J. Endocr.*, **10**, 325.

JOLLY, W. P. (1960). The persistence of trophoblast on extra-uterine tissues in the rat. *Am. J. Anat.*, **106**, 109.

KIRBY, D. R. S., BILLINGTON, W. D., BRADBURY, S. & GOLDSTEIN, D. J. (1964). Antigen barrier of the mouse placenta. *Nature, Lond.*, **204**, 548.

LAJOS, L., CSABA, I., DOMANY, S., SZEKELY, J. & SAARY, Z. (1967). Effect of placentotropin on trophoblast homografts. *Acta Morph. Acad. Sci. Hung.*, **15**, 185.

LAJOS, L., GORES, J., SZEKELY, J., CSABA, I. & DOMANY, S. (1964). The immunologic and endocrinologic basis of successful transplantation of human trophoblast. *Am. J. Obstet. Gynec.*, **89**, 595.

LOKE, Y. W. & BORLAND, R. (1970). Immunofluorescent localization of chorionic gonadotrophin in monolayer cultures of human trophoblast cells. *Nature, Lond.*, **228**, 561.

LOKE, Y. W., JOYSEY, VALERIE, C. & BORLAND, R. (1971). HL-A antigens on human trophoblast cells. *Nature, Lond.*, **232**, 403.

MEDAWAR, P. B. & SPARROW, E. M. (1956). Effects of adrenocortical hormones, ACTH and pregnancy on skin transplantation immunity in mice. *J. Endoc.*, **14**, 240.

NELSON, J. H., HALL, J. E., LIMSON, G. M., FREIBERG, H. & O'BRIEN, F. J. (1967). Effect of pregnancy on the thymolymphatic system. *Am. J. Obstet. Gynec.*, **98**, 895.

PARK, W. W. (1958). Experimental trophoblastic embolism of the lungs. *J. Path. Bact.*, **75**, 257.

PEARSE, W. H. & KAIMAN, H. (1967). Human chorionic gonadotrophin and skin homograft survival. *Am. J. Obstet. Gynec.*, **98**, 572.

DISCUSSION

EDWARDS: May we start by considering the purity and potency of the HCG. Was it a highly purified placental preparation?

BORLAND: In the case of the first experiment, it was a commercial preparation of HCG (5000 i.u./mg) which was fairly crude. In the second series of experiments, however, we did obtain a much more highly purified preparation of HCG. It was tested for contaminant immunoglobulin and found to be negative.

EDWARDS: Immunoglobulin would be only one of the possible contaminants influencing your results, for as Howe mentioned, other placental extracts suppress lymphocyte activity. In none of your experiments is HCG unambiguously identified as the immunosuppressive factor. Attempts to prove that HCG is responsible by neutralizing it with antibody must utilize specific antibody against the hormone. A similar requirement is needed in the agglutination tests.

LOKE: Care must be taken when using commercial HCG, as shown by some of my results. HCG was being used in tests designed to show whether it could inhibit the cytotoxic activity of anti-HL-A antisera. It did, for both trophoblast and lymphocytic target cells. How did it work? It does not stick on to the cell surface and prevent the combination of antibody and antigen, as judged by immunofluorescence. It might have been anti-complementary activity, and Pepys and I started a series of experiments (Loke & Pepys, 1974). Using the standard haemolytic complement assay, we examined two commercial preparations of HCG, each of about 3000 i.u./mg, as well as the International Standard HCG from Mill Hill, for specific activity. All these three preparations inhibited haemolysis in the presence both of human and guinea pig complement. The inhibition was linear, and preferentially affected the initial stages of complement fixation – that is the generation of EAC.142 from guinea-pig R.3, and EAC.43 from human R.3. An effect on the terminal stages was also observed. The preparation also converted C.3. We suspected that some of the observations might be explained by contaminants of these preparations. The HCG preparations were therefore incubated with an anti-HCG antiserum, the precipitate was centrifuged, but the supernatant still inhibited haemolysis. Aggregated Ig could be the contaminating factor, and an antiserum to human Ig light chains revealed that all three preparations contained Ig molecules. When all the immunoglobulin contaminants were absorbed out, the HCG preparations were no longer anti-complementary. This was confirmed by using a highly purified HCG that was obtained from Serono, Rome which also failed to inhibit haemolysis. We suggest, therefore, that previous results should be cautiously interpreted in the light of our findings.

EDWARDS: When gonadotrophins are highly purified, they may be less effective than impure samples in evoking physiological responses. Could you have altered the nature of the molecule by highly purifying the HCG? One way to resolve this question would be to measure its biological activity after purification.

LOKE: We did not assay the endocrine activity of the HCG.

ALLISON: In your experiments, it is unlikely that HCG was involved, but immune complexes or aggregated gammaglobulin might have been. A repeat of your experiments, coating the skin cells with aggregated gammaglobulin, could be revealing.

BILLINGTON: Tumour tissues cultured *in vitro* have been shown to manifest prolonged survival on transfer to allogeneic recipients (Jacobs & Huseby, 1967), and survival of certain other tumours can also be increased by pre-incubation in substances like chondroitin sulphate and a variety of other quite unrelated acid mucopolysaccharides (Lippman, 1968).

BORLAND: Undoubtedly one could test a whole range of glycoproteins, preferably those linked to sialic acid. However, trophoblast is known to produce HCG, and it does not produce chondroitin sulphate.

BODMER: In relation to the point made by Allison, I believe that most gammaglobulin does not contain sialic acid residues, and yet neuraminidase seemed specifically to remove the ability to protect.

ALLISON: The specificity of the enzyme must be tested, because many preparations of neuraminidase also contain protease activity.

BODMER: If the neuraminidase was acting specifically to remove sialic acid, the removal of HCG from the cell by this enzyme indicates either a binding of HCG to cells dependent on sialic acid or the detection by antibody of an antigen of HCG dependent on sialic acid.

BORLAND: I believe that the neuraminidase is in fact removing the HCG *in toto* by cleaving a sialic acid binding group.

BAGSHAWE: There was a theory at one time that the sialic acid component acted as a passport for HCG binding to receptor sites but this is now unacceptable. Desialation of HCG renders it biologically inactive *in vivo* simply because of a greatly shortened half-life but it remains immunologically recognizable by xenogeneic antisera.

HOWE: Bagshawe's comment about desialation and biological activity does not hold true for in-vitro studies. Recent experiments by Channing & Kammerman (1973) using isolated ovarian cells have shown that desialated HCG is equally effective in producing various responses.

BORLAND: This means a loss of activity or just an increased plasma removal of desialated HCG *in vivo*. A recent paper suggested that the half-life of HCG is markedly reduced after desialation (Morell *et al.*, 1971).

ALLISON: It is taken up rapidly into the liver, like many other proteins after removal of their sialic acid groups.

BAGSHAWE: Neuraminidase may be acting as an adjuvant, binding to the surface and modifying the antigenicity of the cells. You may be measuring the addition of a masking antigen rather than the removal of the protective mechanism.

BODMER: Under these circumstances, a complete block to the detection of the molecule on the surface by mixed agglutination would be unlikely, hence this suggestion is doubtful.

COOMBS: Whatever happens, there is a very interesting phenomenon here. Bagshawe showed that neuraminidase did not interfere with the antigenicity of HCG, as judged by reactions with the antibody used in radio-immunoassay. It must presumably be a very specific anti-gonadotrophic antibody. Borland's antibody may not be directed against that same determinant but against one sensitive to removal of sialic acid.

After treatment with neuraminidase and re-addition of HCG, the number of HCG-positive cells did not return fully to the previous level, but did so after treatment with papain. This observation can be explained if neuraminidase removes a sialic acid molecule, and does not affect the rest of the molecule, so that the uptake of further material will be blocked. With papain, the whole molecule could be hydrolysed away, thus making the space available again.

JOHNSON: A positive result for HCG after treating cells with neuraminidase, then re-adding HCG, could imply that the receptor for HCG was still there and capable of binding the hormone, the whole HCG molecule having been cleaved away by the enzyme.

BORLAND: We obtained a positive result on immediately re-adding HCG after neuraminidase, so the receptor must still be there.

GARDNER: The survival of xenogeneic grafts of foetal skin cells by coating them with HCG is fascinating, since, unlike trophoblast cells, they are presumably incapable of producing HCG. How long does HCG remain bound to these cells *in vivo*, and is it lost with turnover of the membrane constituents?

BORLAND: This work has not been done yet. We were concerned about a non-specific local effect of HCG, and the time interval must be extended beyond 25 days if possible. In the mixed agglutination tests, foetal cells treated with HCG appear to react more weakly than trophoblast cells.

JOHNSON: The turnover of HCG could be measured *in vitro* by finding out how long fibroblasts exposed to HCG remain positive in the mixed agglutination tests.

JENKINS: Are you certain that the cells surviving after 20 days are derived from trophoblast? Evidence from Taylor & Hancock (1973) indicates that after nine days the cultures are principally fibroblasts rather than trophoblast or giant cells.

BORLAND: We are convinced ourselves that they are trophoblast. There is no reason to assume that the cell is fibroblastic because it has become bipolar *in vitro*. We observe giant multinuclear cells and also small polygonal cells although there are also some binuclear cells too.

JOHNSON: You are casting doubt on the identification of trophoblast cells and also on Loke's data presented earlier (p. 101) by showing that HCG combines with fibroblasts or skin cells. Your assumption that a cell with HCG on its surface must be trophoblast might be misleading. Two or three trophoblast cells producing HCG in a culture of a hundred fibroblasts might coat all or many of your cells in culture to make them look as though they were in fact HCG-producing trophoblast cells.

BORLAND: What you suggest is possible. This would not of course affect the basic tenet that HCG is the mechanism by which trophoblast cells (and maybe other cells acquiring HCG) have prolonged survival on transplantation.

HOWE: Is HCG in fact an abnormally sticky molecule? Looking at preparations of HCG I never found it to bind to red cells, lymphocytes or various other cells. I again wonder if HCG is being detected.

BAGSHAWE: We have no evidence to suggest HCG is sticky. The idea of HCG being involved in an immunosuppressive role is novel. Although the endocrinological function of HCG is anything but clear, it is certainly a hormone and the idea of a hormone with a primary action on the immunological system is interesting. We have some relevant data on a substantial group of patients with choriocarcinoma (Bagshawe & Golding, 1970). If a skin graft from the patient's husband or from her child is placed on her, it is rejected more slowly than normal, but grafts from unrelated donors tend to be rejected at the normal time. This would seem to be inconsistent with a general immunological suppression. All of these patients had high circulating levels of gonadotrophins.

BORLAND: I would like to emphasize that we were dealing with a local effect of HCG as a possible 'immunoprotective' coat on the trophoblast cell surface, and not with the general parenteral effect of HCG as a possible immunosuppressant which has been alluded to in the past. A general effect of HCG could have a central effect on the lymphoreticular system, whereas in our system the results were a direct comparison between treated and control cells all in the same animal and any differences could not be accounted for by a general parenteral effect of HCG.

CURZEN: You scored the survival time for your grafted cells but did you also score local inflammatory response? This might show an inverse relation to the survival of cells.

BORLAND: We did score the inflammatory response, but it was not marked. The reaction was purely mononuclear: neither polymorphs nor eosinophils were present.

CURZEN: I do not suggest that you were confusing cells in an inflammatory response with surviving graft cells, but asking if you scored any local inflammatory response to the grafted material as a separate exercise.

BORLAND: No, but in a similar series of experiments in guinea-pigs we did score the inflammatory response, and it was inversely related. The work involved an allogeneic system using inbred strains (unpublished results associated with Borland, Loke & Oldershaw, 1970).

REFERENCES

BAGSHAWE, K. D. & GOLDING, P. R. (1970). Proceedings of a conference on *Immunity and Tolerance in Oncogenesis* (ed. L. Severi), p. 273. Division of Cancer Research: University of Perugia.

BORLAND, R., LOKE, Y. W. & OLDERSHAW, J. (1970). Sex difference in trophoblast behaviour on transplantation. *Nature, Lond.*, **228**, 572.

CHANNING, C. P. & KAMMERMAN, S. (1973). Effects of HCG, sialo-hCG and the subunits of hCG upon luteinization of monkey granulosa cell cultures. *Endocrinology*, **93**, 1035.

JACOBS, B. B. & HUSEBY, R. A. (1967). Growth of tumours in allogeneic hosts following organ culture explanation. *Transplantation*, **5**, 410.

LIPPMAN, M. (1968). Transplantation and cytotoxicity changes induced by acid mucopolysaccharides. *Nature, Lond.*, **219**, 33.

LOKE, Y. W. & PEPYS, M. B. (1974). Effects of human chorionic gonadotropin (HCG) preparations on complement *in vitro*. *Am. J. Obstet. Gynec.*, (in press).

MORELL, A. G., GREGORIADIS, G., SCHEINBERG, I. H., HICKMAN, J. & ASHWELL, G. (1971). The role of sialic acid in determining the survival of glycoproteins in the circulation. *J. biol. Chem.*, **246**, 1461.

TAYLOR, P. C. & HANCOCK, K. W. (1973). Viability of human trophoblast *in vitro*. *J. Obstet. Gynaec. Br. Commonw.*, **80**, 834.

THE IMMUNOGENICITY OF THE PLACENTA AND TROPHOBLAST

By KENNETH BAGSHAWE
AND
SYLVIA LAWLER

It seems probable that the evolution of an interface between mother and foetus endowed with appropriate immunological properties has been essential for the success of viviparity. Two salient characteristics of foeto–maternal relationships suggest that the interface may have both afferent and efferent immunological functions. The first is that although normal pregnancy initiates various immunological processes it does not, with certain notable exceptions, result in conventional hypersensitivity of the mother towards foetal antigens. The second is that if sensitization of the maternal host to foetal antigens is induced artificially, the foetus generally escapes harm.

The normal foeto–maternal confrontation is complex both at the morphological and molecular levels. Although the foetal interface is predominantly trophoblastic and vast in area it is not exclusively trophoblastic. Lymphocytes as well as red cells are known to be transferred from foetus to mother (Schroder & de la Chapelle, 1972) and the possibility that placental stroma is exposed to direct maternal contact through disruptions of placental integrity cannot be excluded. In an area of such complexity the contributions of the various elements need to be studied, so far as possible, separately. Only in the final analysis can reconstruction of the whole picture be attempted. In the case of foeto–maternal relationships we are still at the dissection stage.

In this paper we shall concentrate on what we think is one of the central issues, namely, the role of the molecular topography of normal trophoblast in the induction of the pattern of immunological response characteristic of pregnancy. Unfortunately, the intrinsic difficulties which attend a proper assessment of the immunogenicity of any tissue are increased in the case of trophoblast.

One difficulty is that trophoblast changes its morphological and endocrinological characteristics during embryogenesis and corresponding changes in immunogenicity can be anticipated. Another is the fact that trophoblast does not occur naturally, free of other foetal tissues, except as choriocarcinoma. Trophoblast isolated artificially is liable to be contaminated with other tissue elements; the fact that trophoblastic cells can be readily demonstrated in a preparation by virtue of their endocrine activity, does not preclude contamination. Moreover, the diversity of morphological forms which appear when trophoblast is grown in culture raises some doubts about their relevance to the normal in-vivo situation.

Although the concept of immunogenicity implies antigen recognition by the maternal immunological system, the direct demonstration of allo-antibody or xenoantibody fixation on a cell surface is not informative about its immunogenicity. Immunogenicity can be recognized by primary exposure of an individual to antigen, and comparison of the response with that of unprimed individuals before and after appropriate challenge. But what sort of response are we to look for in the case of foeto–maternal relationships? We know that, at least in the mouse, inter-strain mated multiparous females show delayed rejection of skin grafts from the male strain (Breyere 1961). So in looking for evidence of the immunogenicity of trophoblast should we look for conventional hypersensitivity and accelerated graft rejection or unresponsiveness and decelerated rejection? A null response may indicate a balance between opposing effects.

Finally, there may be a high rate of turnover of the trophoblastic cell surface producing a state of trophoblastic antigen excess in maternal body fluids. This possibility was first discussed by Hulka & Brinton (1963) and it has lately been transplanted to the field of tumour–host immunity.

OBSERVATIONS

Whole placenta as an immunogen

It may be observed here that morphological changes in the thymic–lymphatic system during pregnancy provide prima facie evidence of foetal immunogenicity (Nelson & Hall, 1964; Maroni & de Sousa 1973), but they tell us nothing about its source and nature. The possibility of placental organ-specific antigens has been discussed throughout this century. Much of the interest has centred on their possible role in the aetiology of eclampsia and the nephrotoxicity of heterologous anti-placental sera suggested the presence of antigens common to both placenta and kidney. Such antisera have also been known to have abortifacient actions. Recent studies in this area (Koren, Srinnaboon & Behrmann, 1970), indicate that some, though not all, of the renal damage is attributable to the deposition of antigen–antibody complex in glomerular capillary walls.

Haskova (1962) found that relatively large subcutaneous or intra-peritoneal inoculations of F_1 hybrid mouse placenta were relatively in-effective in inducing hypersensitivity to paternal strain alloantigens. Subsequent studies have indicated, however, that under experimental conditions whole placenta induces hypersensitivity to paternal strain alloantigens and have also shown that placental tissue grafts are susceptible to rejection (Uhr & Anderson, 1962; Schlesinger, 1964; Haskova, 1963;

Billingham, 1971). Whole placenta, it must be concluded, contains cells which express paternally derived antigens in a form capable of eliciting hypersensitivity phenomena, though it is not known whether these are expressed by foetal leucocytes, fibroblasts, trophoblast or other elements. Under physiological conditions the placento–maternal interface is formed largely by the trophoblast which we now consider.

Trophoblast as an immunogen

Park (1971) has pointed out that in the human, with the exception of the first two weeks of gestation, a leucocytic and predominantly a lymphocytic infiltration is seen in varying degree in close association with trophoblast throughout pregnancy. This association has also been observed with choriocarcinoma and a positive correlation observed between the extent of lymphocytic infiltration and the response to chemotherapy (Elston, 1969; Elston & Bagshawe, 1973). The lymphocytic infiltration occurring early in pregnancy is consistent with immunogenicity expressed during the blastocyst or trophoblastic phase, but it provides no indication of whether the antigens involved are tissue-specific, individual-specific or associated with the neoplastic state.

Hulka & Mohr (1968) inserted ectoplacental cone trophoblast into the kidneys of allogeneic recipients and challenged these mice and untreated controls with genetically similar trophoblast. They reported that the success rate, size of grafts and host containment indicated immunization by the primary graft of trophoblast. The observations, as the authors point out, were somewhat subjective; also the unprimed controls underwent no sham operation and no group of immunized mice received syngeneic trophoblast so that the nature of the antigens involved, whether tissue specific or allogeneic, could not be inferred.

Despite the obstacles to devising clean experiments which might define the immunogenicity of trophoblast we can consider first the evidence suggesting the presence of tissue-specific antigens for trophoblast and then the evidence for individual-specific or transplantation antigens.

Tissue-specific immunogenicity of trophoblast

Evidence which has sometimes been interpreted as indicating the immunogenicity of trophoblast came from the study by Kirby (1968). He found that single isogeneic blastocysts grew with diminished vigour and showed reduced trophoblastic outgrowth when transplanted to the testes of C57 males which had received, on four previous occasions, C3H blastocysts (see also paper by Johnson, p. 89). Also, when C3H skin grafts were applied following blastocyst grafts, rejection was delayed. This evidence is open to several interpretations. Unless one accepts that the reaction is directed at trophoblastic precursors, the impaired outgrowth of trophoblast is difficult to reconcile with the resistance of normal trophoblastic and foetal development in hyperimmunized females. The susceptibility of the blastocyst to rejection is not in itself proof for the immunogenicity of trophoblast but the observation of delayed skin-graft rejection by the blastocyst recipients is clearly significant.

Recent studies by Beer, Billingham & Yang (1972) provide strong evidence for a unique trophoblastic antigen in the rat. Rabbit antisera were raised against preparations of trophoblast obtained from placenta at 12–14 days of gestation. After absorption of haemagglutinins the antisera still had some toxicity which could be removed with rat lymphoid cells without impairing its antitrophoblastic activity. The antisera which were highly species-specific, were abortifacient at an intravenous dose of 1 ml at all stages of post-implantation pregnancy irrespective of the genetic constitution of the pregnant rat and foetus. The abortifacient activity was removed only by absorption with trophoblastic cells.

Heterologous antisera raised against human chorionic gonadotrophin (HCG), an endocrine product of trophoblast, are also cytotoxic to trophoblast *in vitro* (Bagshawe, 1970). Recently HCG has been shown to consist of an α subunit homologous with the α subunits of some pituitary polypeptides and a β subunit which is distinguished by an apparently unique amino acid sequence (Bahl, 1968; Morgan & Canfield, 1971). These observations have aroused considerable interest for if β-HCG were only experienced in pregnancy the possibility of sensitization to HCG may exist. In our experience heterologous antisera raised against β-HCG show only a partial discrimination between HCG and luteinizing hormone. Exposure to luteinizing hormone may still afford protection to spontaneous autoimmunization against β-HCG.

Antigens which, though tissue-specific and species-specific, are detectable only in xenogeneic recipients may, of course, have practical and even therapeutic potential but their relevance to the immunogenicity of trophoblast in the physiological sense has still to be established. The xenogeneic host may distinguish antigenic features not discernible to the allogeneic host and it is therefore inappropriate to describe such antigens as autoantigens even though they may have tissue specificity.

Evidence for species-specific antigens on trophoblast

In view of its role as the foeto–maternal interface it would have been surprising if studies had revealed that trophoblast antigenicity were equivalent to that of other tissues composed of nucleated cells. Indeed, it is clear that, up to a point, at any rate, trophoblast can resist interspecies histocompatibility barriers as is evidenced by the variety of mammalian inter-species hybrids that have been recorded (Gray, 1954). Kirby (1962) found that murine trophoblast could flourish in rat hosts at a time when the embryo proper elicited a destructive mononuclear cell reaction. Prior sensitization of rat hosts inhibited the growth of mouse trophoblast indicating that it expressed species-specific antigens and other successful species-combinations were reported by Billington (1966). However, the interest here lies in whether trophoblast expresses alloantigens at all and, if it does, the form of their expression.

Evidence for alloantigens on trophoblast

The starting point of many studies in the past decade were the observations by Simmons & Russell (1962) that embryonic grafts were totally destroyed when transplanted beneath the renal capsules of maternal-strain hosts which had been pre-sensitized to paternal-strain antigens. By contrast, trophoblast allografts proliferated and invaded with little mononuclear cell reactivity on the part of the host and these grafts survived as long as trophoblast grafts in syngeneic hosts.

The possibility that placental fibrinoid, an amorphous mucopolysaccharide rich in sialic acid, might represent the product of immunological interaction had been discussed during the 1950s. The idea that it might constitute an immunological barrier was developed by Kirby and his colleagues (Kirby, Billington, Bradbury & Goldstein, 1964), and Bradbury, Billington & Kirby (1965).

Their observations and the known presence of sialomucin in the periphery or glycocalyx of many types of malignant cells led to the hypothesis (Currie & Bagshawe, 1967*a*) that sialated amino sugars in the glycocalyx might mask or somehow interfere with antigen recognition by the receptors in host cells, a possibility which was heightened by the known net negative electrostatic charge on lymphocytes. The original proposition of electrostatic repulsion between antigen and receptor was later widened to include possible effects from steric hindrance (Currie & Bagshawe, 1968). Experimentally, it was found that desialation of some but not all experimental tumours by in-vitro treatment with neuraminidase significantly increased their immunogenicity (Currie, 1967; Bagshawe & Currie, 1968; Sanford, 1967; Simmons, Rios, Lundgren, Ray, McKhann & Haywood, 1971).

When neuraminidase treated A_2G ectoplacental cone trophoblast was injected intraperitoneally into CBA mice they subsequently showed accelerated rejection of A_2G skin grafts when examined ten days after application, whereas CBA mice receiving untreated A_2G trophoblast did not (Currie, van Doorninck & Bagshawe, 1968). Using different inbred strains of mice and macroscopic assessment of graft rejection, Simmons, Lipschultz, Rios & Ray (1971) were, on the contrary, unable to find any evidence of alloantigens revealed by neuraminidase treatment. It is clear that this conflict of results can only be resolved by further study and perhaps with other techniques. In designing new experiments it should be borne in mind that neuraminidase might act as an immunological adjuvant rather than through its cleavage of sialic acid or might in some other way modify surface antigens. (See also paper by Johnson and Discussion after it.)

The cytotoxic effect of maternal and allogeneic lymphocytes has been observed on normal trophoblast and choriocarcinoma cells (Currie & Bagshawe, 1967*b*) and on trypsinized human placental monolayers (Douthwaite & Urbach, 1971). Such studies must be interpreted with caution partly because syngeneic lymphocytes are not usually available as controls in studies on human trophoblast and partly because the prolonged periods of lymphocyte contact are comparable to the time now known to induce in-vitro T cell sensitization to syngeneic cell lines (Feldman, Cohen & Wekerle, 1972) (see also paper by Howe, p. 136).

In man, a wide array of alloantigens have been defined and these can be studied in relation to trophoblast. Not all the known alloantigens are expressed on all cells; some, such as the Rhesus and Kell antigens are found only on erythrocytes and are thus tissue-specific alloantigens. On the other hand, the ABO antigens are widely distributed throughout the tissues and in individuals with the appropriate genes they are secreted into body fluids. In clinical transplantation the ABO antigens are treated with respect and kidney donors are selected for compatibility in the sense that the donor can give a blood transfusion to the recipient.

Antibodies to the A and B antigens are always found in normal subjects who lack the corresponding antigen. Exceptionally, an ABO-incompatible foetus can be at risk as a result of a placental leak of foetal erythrocytes, if the mother is group O and the foetus group A. The maternal antibody may change in character after such a stimulus and cross the placenta causing foetal haemolysis.

A good deal of effort has been expended on trying to correlate foeto-maternal ABO compatibility with fertility (Boettcher & Hay, 1968), birth weight and placental weight, but effects are only revealed, if at all, in large samples and the ABO system does not appear to exert a strong selective effect.

The concept of the placenta as an inert immunological barrier gained support from the observation of Witebsky & Reich (1932) that placental villi are deficient in blood group antigens. Although Gross (1966) found group A substance on early trophoblast, Theide, Choate, Gardner & Santay (1965) and Szulman (1972) carried out extensive studies with negative results. Szulman examined trophoblast from the fifth week of pregnancy to term using immunofluorescent techniques and found it negative for A, B and H antigens. Using a mixed anti-globulin technique with group A red cells, Loke & Ballard (1973) found evidence for sparsely distributed group A antigen on human trophoblast and supported this with further evidence from electron micrographs of immuno-ferritin-labelled sections.

In contrast to the weak effects of ABO factors on fertility a strong selective effect is exerted in the development of choriocarcinoma. Thus, when compared with group O women, the risk for group A women is less if the husband is also group A, but greatly increased when he is group O (Bagshawe, Rawlins, Pick & Lawler, 1971). The risk is not related to the ABO group of the foetus and the effect is not explicable in any known immunological sense. Metastatic choriocarcinomas arise from ABO-incompatible conceptions; the prognosis of patients with ABO-incompatible tumours treated with chemotherapy is probably better than that for compatible tumours but we do not yet know whether this is immunologically mediated.

Loke, Joysey & Borland (1971) have reported the presence of HL-A antigens on trophoblast grown in culture. One antigen (HL-A2) was demonstrated on one out of three samples tested. Clearly further direct evidence of this nature is required to show whether this is a general or an exceptional phenomenon but maternal responses to HL-A antigens expressed on the foetus can also be sought. HL-A antigens have been detected on foetal tissues in the early weeks of gestation (Siegler & Metzgar, 1970).

Most of the antibodies used for HL-A typing come from women who have been pregnant. In contrast to Rhesus immunization, HL-A antibodies are frequently formed by primigravida and 27% of such women were recently found to have antibodies by the thirty-second week of gestation (Bodmer, 1973 personal communication).

Pregnancies in the human species may terminate in the evacuation of a hydatidiform mole, a hydropic placental product of conception that generally lacks an intact foetus. The incidence varies markedly between different races, being rarer amongst Caucasians, with an incidence of about 1 in 1000 pregnancies, than amongst Asians in whom the incidence is about 1 in 200 (Bagshawe, 1969). Such a conceptus consists of trophoblastic elements, hydrated chorionic villi together with mesenchymal core; capillaries can be present at an early stage and red cells are rarely seen in histological preparations but there is no record of lymphocytes being identified in the capillaries of moles.

The mole as evacuated is benign, but trophoblast may persist for up to 9–10 months post-conception before regressing. Progression to a malignant choriocarcinoma occurs in about 3% of cases and about another 3% have invasive moles which require treatment. Such molar pregnancies are now registered in the United Kingdom and the persistence of trophoblastic tissue is subsequently monitored by regular determinations of human chorionic gonadotrophin levels (Bagshawe *et al.*, 1973).

Patients under treatment for invasive mole or choriocarcinoma, their husbands and families where relevant, have been tested for the HL-A antigens (Lawler, Klouda & Bagshawe, 1971). The serum of the patients has been tested for HL-A antibodies (Klouda, Lawler & Bagshawe, 1973; Lawler, *et al.*, 1973) and since 1970 the serum has been tested at weekly intervals. Amongst this series of 161 patients (Lawler, *et al.*, 1974) are 31 who, in the first pregnancy, had a hydatidiform mole. A careful enquiry was made concerning blood transfusions and 11 of the 31 patients giving the history of a single molar pregnancy had not been transfused before the first serum sample was tested for HL-A antibody. Four of these patients had made specific antibody against HL-A antigens present on the husbands' lymphocytes. Thus in these cases there is no doubt that the molar pregnancy was immunogenic as far as the HL-A system is concerned. Furthermore, it is likely that the trophoblastic elements were the carriers of the antigenic stimulus.

Amongst the 20 patients who had been transfused during the first molar pregnancy, eight had made HL-A antibody against the husbands' cells. It is reasonable to suppose that the patients were primed by the molar pregnancies and immunization was re-inforced by the transfusions.

Another point is that in both the transfused and non-transfused patients the incidence of formation of HL-A antibodies against husbands' antigens, about 40%, was higher than expected in a normal pregnancy at 32 weeks. Thus the molar pregnancies are more likely to provoke the formation of HL-A antibodies than a normal conceptus.

Amongst the series as a whole the HL-A antibodies usually persist whilst the patient is under treatment and tumour is still present. The effect of transfusions is in general to broaden the specificities of the antibodies and in time, as such antibodies disappear, there is often a progressive narrowing of specificity. Many of the patients under constant HL-A antibody screening have choriocarcinoma. It is possible to assume that the stimulus for antibody production continues to be present whilst the tumour is active. The difficulty in assessing the significance of the duration of the HL-A antibody levels is that the consequences of normal pregnancy can be so variable. Furthermore, adequate controls for the patients who have had multiple transfusions as well as pregnancies are not available in the normal population. Another point of interest is that, as in other situations in which patients are exposed to adequate doses of incompatible HL-A antigens, there are responders and non-responders.

Since choriocarcinoma is foreign to the maternal host through the hemizygous set of paternal genes why does the tumour persist? Attempts to find pecularities in the distribution of mating types of husbands and wives on the HL-A system in trophoblastic neoplasia have failed (Klouda *et al.*, 1972). In some instances choriocarcinoma is preceded by a normal-term birth. In these cases the assumption may be made that the antecedent

TABLE 1 *HL-A and choriocarcinoma following live term birth (24 antecedent children)*

Incompatible with mother at both loci	16
Compatible with mother only at 1st locus	5
Compatible with mother only at 2nd locus	1
Compatible with mother at both loci	2

child carries the same genetic information as the malignancy. During the last few years, 24 such families have been tested. Any child has the chance of being incompatible with the mother for one antigen at the first and one antigen at the second HL-A locus. The compatibility situation at the first and second loci is shown in Table 1. The incompatibilities range over the whole spectrum of HL-A antigens. There are two antigenically compatible infants in the series; both had the phenotype *HL-A2*, 12 and both these children were ABO incompatible. In normal single child families in the United Kingdom, 24% would be expected to be compatible for one HL-A antigen, and 2% for two HL-A antigens. However, we have yet to find an antecedent child who is antigenically compatible with the mother for both the HL-A and ABO systems. Thus, the conclusion is that even if these histocompatible antigens are present on the trophoblast, they cannot preclude the development of malignancy, although the ABO system substantially alters the risk factor between different mating pairs.

Other genetic determinants are carried in the HL-A region of the chromosome and are known to influence immune responses and it remains to be seen what role they have to play in foeto–maternal relationships and in the malignancies of trophoblastic tissue.

Conclusions

Evidence for the immunogenicity of the placenta and, in particular, of the trophoblast is inadequate in many respects. Nevertheless, it suggests that under physiological condition it is not immunogenically inert. The pattern of immunological responses characteristic of pregnancy may be determined in part by the form in which antigens are presented on the surface of the blastocyst and trophoblast.

SUMMARY

Under experimental conditions mammalian placenta has been shown previously to be immunogenic with respect to alloantigens. Study of the immunogenicity of trophoblast presents intrinsic difficulties mainly because it is liable to be contaminated with other elements when isolated artificially. Between species, trophoblast is undoubtedly immunogenic although the production of some inter-species hybrids is not precluded.

Tissue-specific antigens have been demonstrated on trophoblast in a limited sense but only by employing xenogeneic recognition systems. Evidence for the presence of alloantigens on the trophoblast of laboratory animals remains controversial. In man, both ABO and HL-A antigens are probably represented on trophoblast and in the case of HL-A antigens may even be immunogenic.

REFERENCES

BAGSHAWE, K. D. (1969). *Choriocarcinoma. The clinical biology of the trophoblast and its tumours.* Arnold: London.

BAGSHAWE, K. D. (1970). Some immunological aspects of choriocarcinoma. In *The Biology and Surgery of Tissue Transplantation* (ed. E. J. Maxwell Anderson), pp. 139–52. Blackwell: Oxford.

BAGSHAWE, K. D. & CURRIE, G. A. (1968). Immunogenicity of L1210 murine leukaemia cells after treatment with neuraminidase. *Surgery*, **218**, 1254.

BAGSHAWE, K. D., RAWLINS, G., PICK, M. C. & LAWLER, S. D. (1971). The ABO blood groups in trophoblastic neoplasia. *Lancet*, **i**, 553.

BAGSHAWE, K. D., WILSON, H., DUBLON, P., SMITH, A., BALDWIN, M. & KARDANA, A. (1973). Follow-up after hydatidiform mole: Studies using radio-immunoassay for urinary human chorionic gonadotrophin (hCG). *J. Obstet. Gynaec. Br. Commonw.*, **80**, 461.

BAHL, O. P. (1968). On the nature of polypeptide and carbohydrate chains in human chorionic gonadotrophin. *Fedn. Proc.*, **27**, 530.

BEER, A. E., BILLINGHAM, R. E. & YANG, S. L. (1972). Further evidence concerning the autoantigenic status of the trophoblast. *J. exp. Med.*, **135**, 1177.

BILLINGHAM, R. E. (1971). The passenger cell concept in transplantation immunology. *Cell Immunol.*, **2**, 1.

BILLINGTON, W. D. (1966). Vascular migration of transplanted trophoblast in the golden hamster. *Nature, Lond.*, **211**, 988.

BOETTCHER, B. & HAY, J. (1968). An investigation of a proposed mechanism for infertility based on ABO blood group incompatibility. *Am. J. Obstet. Gynec.*, **100**, 437.

BRADBURY, S., BILLINGTON, W. D. & KIRBY, D. R. S. (1965). A histochemical and electron microscopical study of the fibrinoid of the mouse placenta. *J. R. Microsc. Soc.*, **84**, 199.

BREYERE, E. J. (1961). Tolerance induced by parity in mice incompatible at the H-2 locus. *J. natn. Cancer Inst.*, **27**, 409.

CURRIE, G. A. (1967). The masking of antigens on the Landschutz ascites tumour. *Lancet*, **ii**, 1336.

CURRIE, G. A. & BAGSHAWE, K. D. (1967*a*). The masking of antigens on trophoblast and cancer cells. *Lancet*, **i**, 708.

CURRIE, G. A. & BAGSHAWE, K. D. (1967*b*). The antigenicity of normal and malignant human trophoblast – some implications. *Advance in Transplantation*, **34**, 523.

CURRIE, G. A. & BAGSHAWE, K. D. (1968). The role of sialic acid in antigenic expression: further studies of the Landschutz ascites tumour. *Brit. J. Cancer*, **22**, 843.

CURRIE, G. A., VAN DOORNINCK, W. & BAGSHAWE, K. D. (1968). Effect of neuraminidase on the immunogenicity of early mouse trophoblast. *Nature, Lond.*, **219**, 191.

DOUTHWAITE, R. M. & URBACH, G. I. (1971). In vitro antigenicity of trophoblast. *Am. J. Obstet. Gynec.*, **109**, 1023.

ELSTON, C. W. (1969). Cellular reaction to choriocarcinoma. *J. Path. Bact.*, **97**, 261.

ELSTON, C. W. & BAGSHAWE, K. D. (1973). Cellular reaction in trophoblastic tumours. *Brit. J. Cancer*, **28**, 245.

FELDMAN, M., COHEN, I. R. & WEKERLE, H. (1972). T-cell mediated immunity in vitro: and analysis of antigen recognition and target cell lysis. *Transplant. Rev.*, **12**, 57.

GRAY, A. P. (1954). *Mammalian hybrids*: checklist with bibliography. *Tech. Commun, no.* **10**. Commonw. Bur. Anim. Breed. Genet: Edinburgh.

GROSS, S. J. (1966). Human blood group A substance in human endometrium and trophoblast localized by chromatographed rabbit antiserum. *Am. J. Obstet. Gynec.*, **95**, 1149.

HASKOVA, V. (1962). Transplantation non-antigenicity of the fetal placenta. *Nature, Lond.*, **193**, 278.

HASKOVA, V. (1963) Differences in the antigenic effectiveness of the foetal part of mouse placenta depending on the strain combination employed. *Folia Biol., Praha*, **9**, 99.

HULKA, J. F. & BRINTON, V. (1963). Antibody to trophoblast during early postpartum period in toxemic pregnancies. *Am. J. Obstet. Gynec.*, **86**, 130.

HULKA, J. F. & MOHR, K. (1968). Trophoblast antigenicity demonstrated by altered challenge graft survival. *Science*, **161**, 696.

KIRBY, D. R. S. (1962). Reciprocal transplantation of blastocysts between rats and mice. *Nature, Lond.*, **194**, 785.

KIRBY, D. R. S. (1968). The immunological consequences of extra-uterine development of allogenic mouse blastocysts. *Transplantation*, **6**, 1005.

KIRBY, D. R. S., BILLINGTON, W. D., BRADBURY, S. & GOLDSTEIN, D. (1964). Antigen barrier of the mouse placenta. *Nature, Lond.*, **204**, 548.

KLOUDA, P. T., LAWLER, S. D. & BAGSHAWE, K. D. (1972). HL-A matings in trophoblastic neoplasia. *Tissue Antigens*, **2**, 280.

KLOUDA, P. T., LAWLER, S. D & BAGSHAWE, K. D. (1973). HL-A antibodies in patients with trophoblastic neoplasia. I General Survey. *Symp. Series Immunobiol.*, **18**, 268, Karger: Basel.

KOREN, Z., SRINNABOON, S. & BEHRMANN, S. J. (1970). Purification and fractionation of anti-mouse placental serum. *Am. J. Obstet. Gynec.*, **108**, 666.

LAWLER, S. D., KLOUDA, P. T. & BAGSHAWE, K. D. (1971). The HL-A system in trophoblastic neoplasia. *Lancet*, **ii**, 834.

LAWLER, S. D., KLOUDA, P. T. & BAGSHAWE, K. D. (1973). HL-A antibodies in patients with trophoblastic neoplasia. II After molar pregnancies. *Symp. Series Immunobiol.*, **18**, 272, Karger: Basel.

LAWLER, S. D., KLOUDA, P. T. & BAGSHAWE, K. D. (1974). The immunogenicity of molar pregnancies in the HL-A system. *Am. J. Obstet. Gynec.*, (in press).

LOKE, Y. W. & BALLARD, A. C. (1973). Blood group A antigens on human trophoblast cells. *Nature, Lond.*, **245**, 329.

LOKE, Y. W., JOYSEY, V. C. & BORLAND, R. (1971) HL-A antigens in human trophoblast cells. *Nature, Lond.*, **232**, 403.

MARONI, E. S. & DE SOUSA, MARIA A. B. (1973) The lymphoid organs during pregnancy in the mouse – a comparison between a syngeneic and an allogeneic mating. *Clin. exp. Immunol.*, **13**, 107.

MORGAN, F. J. & CANFIELD, R. E. (1971). Nature of the subunits of human chorionic gonadotrophin. *Endocrinology*, **88**, 1045.

NELSON, J. H. & HALL, J. E. (1964). Studies on the thrombolymphatic system in humans. I. Morphologic changes in lymph nodes in pregnancy at term. *Am. J. Obstet. Gynec.*, **90**, 482.

PARK, W. V. (1971). *Choriocarcinoma – a study of its pathology*. Heinemann: London.

SANFORD, B. H. (1967). An alteration in tumor histocompatibility induced by neuraminidase. *Transplantation*, **5**, 1273.

SCHLESINGER, M. (1964). Serologic studies of embryonic and trophoblastic tissues of the mouse. *J. Immunol.*, **93**, 255.

SCHRODER, J. & DE LA CHAPELLE, A. (1972). Fetal lymphocytes in the maternal blood. *Blood*, **39**, 153.

SIEGLER, H. F. & METZGAR, R. S. (1970). Embryonic development of human transplantation antigens. *Transplantation*, **9**, 478.

SIMMONS, R. L., LIPSCHULTZ, M. L., RIOS, A. & RAY, P. K. (1971). Failure of neuraminidase to unmask histocompatibility antigens on trophoblast. *Nature New Biol.*, **231**, 111.

SIMMONS, R. L., RIOS, A., LUNDGREN, G., RAY, P. K., MCKHANN, C. F. & HAYWOOD, G. R. (1971). Immunospecific regression of methyl-cholanthene fibrosarcoma with neuraminidase. *Surgery*, **70**, 38.

SIMMONS, R. L. & RUSSELL, P. S. (1962). The immunologic problem of pregnancy. *Ann. N. Y. Acad. Sci.*, **99**, 717.

SZULMAN, A. E. (1972). The A, B and H blood group antigens in human placenta. *New Eng. J. Med.*, **286**, 1028.

THEIDE, H. A., CHOATE, J. W., GARDNER, H. H. & SANTAY, H. J. (1965). Immunofluorescent examination of the human chorionic villus for blood group A and B substance. *J. exp. Med.*, **121**, 1039.

UHR. J. W. & ANDERSON, S. G. (1962). The placenta as a homotransplant. *Nature, Lond.*, **194**, 1292.

WITEBSKY, E. & REICH, H. (1932). Zur gruppenspezifischen differenzierung der placentarorgane. *Klin. Wochenschr.*, **11**, 1960.

DISCUSSION

LAWLER: The observation on molar pregnancies raises the possibility that trophoblastic tissue can immunise the mother in a normal pregnancy. I am indebted to Julia Bodmer for an estimate of 20% for the frequency of HL-A antibody formation in primiparous women tested at 32 weeks of gestation. The instance of maternal immunization against HL-A serological determinants in patients having a single pregnancy of 16–24 weeks with a hydatidiform mole, in the absence of blood transfusions, is about 40%. Transfusions can re-inforce any immunization that has occurred during a pregnancy, or provoke antibody formation against donor antigens not present in the foetus, so that a higher incidence of antibody formation would be expected amongst transfused than non-transfused patients. Given the same degree of stimulation, patients may or may not respond by forming HL-A antibodies. Another point is that if the Ir system is involved, as mentioned earlier by Davies, (see p. 105 in Discussion after Johnson's paper) as far as foeto–maternal relationships are concerned, it may be necessary to consider this as a hemizygous system!

EDWARDS: Could you indicate the difference among the various patients in your series?

LAWLER: The patients are usually first tested for HL-A antibodies about six months after the beginning of the pregnancy, before treatment starts.

BAGSHAWE: The moles are of course evacuated at various times after conception, usually between 16 and 24 weeks, and the patients are admitted to our unit at intervals ranging from one month to six months after that. So the range would be from five months to 11 months post-conception.

DAVIES: If the incompatibility that matters is actually at the Ir locus, then this is the worst possible situation for using HL-A antisera because they tend to be unrelated to the Ir system. The people in your studies would be unrelated, and only in related people can the correlation between the HL-A and Ir be seen.

BODMER: The question of whether a relationship between serologically-detected determinants and another product of that region, probably not Ir but something in that region, can actually be found will depend on what is called 'linkage disequilibrium'.

I would like to mention some unpublished data that my wife, Bonner, myself and others have obtained on severe toxaemia of pregnancy. We are interested in any possible relationship between HL-A and toxaemia. We have two ideas in mind; one, to discover if the presence of maternal HL-A antibodies has any effect on the pregnancy and as mentioned before a lot of evidence now indicates that there is no effect of such antibodies. A second point is whether there might be some relationship with incompatability. A group of 24 husband–wife pairs, in which there were pregnancies with very severe toxaemia, have been typed for HL-A and antibodies, and to our surprise something unusual was found. In a normal population, approximately 75% of people have four different HL-A antigens, about 23–24% have three antigens – presumably because they are homozygous at one of the two loci, and less than 2% have only two antigens and are presumably homozygous at both loci. The toxaemia parents were apparently homozygous at both loci in roughly 20% of the cases, which is a very high incidence, and obviously these data must be extended. The distribution lies equally in the males and the females – an observation to be puzzled over further.

EDWARDS: Do you mean that the husband is homozygous for the opposite antigens to those of the wife?

BODMER: No, there is no evidence from our data of either an increased or reduced incidence of incompatibility. Also there is no evidence in these patients of an increased incidence of HL-A antibody.

JENKINS: We have also been looking at this problem. No evidence of increased incompatibility was found in our patients; some evidence in our retrospective studies (Jenkins, Good & Good, 1973) revealed some incompatibility factor but not as yet in any of our prospective studies. Antigen incidence may be altered in the babies.

DAVIES: The prolonged survival of skin grafts from F_1s has been attributed to enhancement. It is very difficult to enhance skin grafts, and enhancement for even a few days is very significant. If the enhancing system is good, for example for a tumour or for a heart or kidney, the same effect shows up with skin as only of very limited duration. The small amount of prolongation of skin grafts could thus be an indicator of a strong enhancing effect.

BAGSHAWE: All the evidence we have suggests that the prolonged survival of skin grafts on patients with choriocarcinoma is restricted to skin grafts from the husband.

EDWARDS: How would this specificity relate to the work described by Kirby (1968) and by Johnson earlier in this Symposium (p. 89)? There were no indications of either enhancement or more rapid rejection of skin grafts of allogeneic animals after the repeated transfer of blastocysts or other sources of trophoblast. The evidence is evidently not fully decisive on this point at present.

BILLINGTON: Kirby repeatedly transferred batches of four blastocysts to ectopic sites in the mouse, and showed a progressive decrease in blastocyst proliferation together with a prolonged survival of allogeneic skin grafts. He did not transfer pure trophoblast although I have since found that the serial transfer of ectoplacental cones leads neither to an inhibition of the growth of trophoblast nor to an alteration in the survival of allogeneic skin grafts (Billington, 1973).

McLAREN: Some of the results of Kirby's work could have been due to a non-immunological inhibition of the growth of blastocysts. Could a non-immunological inhibition occur during choriocarcinoma?

JENKINS: Females in some species have been shown to develop a greater unresponsiveness to paternal antigens with increasing parity (Jenkins & Hancock, 1972). Four cases described by Bagshawe needed prolonged treatment, and high doses of chemotherapy. Is there more difficulty in controlling the tumour in women of high parity, because of possible enhancement with successive births?

BAGSHAWE: There is no correlation between parity and sensitivity to drugs.

ALLISON: A strong immunological response would arise against choriocarcinoma, as judged from the effectiveness of chemotherapy. Perhaps this tumour is effectively controlled by drug treatment, because immunity can cope with the small number of residual cells.

BAGSHAWE: This is a very complex problem for the reaction against trophoblastic tumours is highly variable. Generally speaking, there is probably a strong reaction against trophoblastic tumours but a small proportion of the patients are immunosuppressed when they are admitted. We do not know the mechanism, but so far we have seen four patients who were suppressed as judged by BCG and DNCB sensitization and other tests. These four patients were unique in failing to respond to their initial chemotherapy. The high rate of response found in chemotherapy may therefore be partly due to the high rate of host killing of tumour cells. Chemotherapeutic agents for cancer may be much more effective when a high rate of cell turnover in the target cell population is accompanied by cell death from any other mechanism. If immune mechanisms contribute to the high rate of cell death, a reduction of this form of killing by any form of immunosuppression would result in drug resistance.

DAVIES: But did you not show that in most cases there was a cellular infiltration around the tumours?

BAGSHAWE: I showed here one example of a choriocarcinoma with a marked cellular reaction. Others have none. The majority of tumours show something between these two extremes.

JOHNSON: The lymphocytic infiltration in what is clearly choriocarcinoma tissue is interesting. Davies suggested that enhancement might be involved but are there lots of lymphocytes in tissues that are enhanced?

DAVIES: I believe that there are few.

HOWE: Abbas *et al.* (1974) have contrary evidence, for enhancing antibodies delayed the lymphocyte infiltration, but ultimately the histology of normal and enhanced kidneys were similar.

MUNRO: In the cases where antibodies arise in the mother against the husband's HL-A antigens, were the sera taken before or after the chemotherapy had begun?

BAGSHAWE: They were taken before, during and after chemotherapy, although the serum sample used mostly was usually that before chemotherapy began.

DAVIES: The ease with which a tumour can be removed is quite fantastic, when the host and tumour differ at the major H locus. A mouse can remove a tumour of half its own weight over a weekend. The only precedent one has for growth of tumour in a histoincompatible environment is again, enhancement, as in the data on skin grafts quoted earlier.

May I ask a further question concerning the trophoblast antigens? We have immunized rabbits with mouse placentae and obtained sera which would induce abortion in mice, and of course these sera were highly toxic for mouse cells. Nevertheless, when absorbed to zero cytotoxicity for mouse lymphocytes, the sera still reacted with a cell suspension made from placentae. The cells were extracted from placentae and were heterogeneous with regard to cell type. We could get a specific cytotoxic effect on those cells with absorbed sera (unpublished data).

BAGSHAWE: Beer & Billingham (1971) showed that the abortifacient activity of such sera could be absorbed by trophoblast but not by other cells.

DAVIES: Our situation was different to Billingham's in that we removed the ability of these sera to induce abortion without absorbing with trophoblast, and while retaining cytotoxicity for trophoblast cells.

BILLINGTON: Beer & Billingham used rat placentae as their homogenate, and many contaminating cells must have been included. They reported that the abortifacient activity was absorbed by lymphoid cells but not by trophoblast. The antisera were totally ineffective in mice and hamsters, and effective only in rat – remember that they were using a rabbit anti-rat serum. They also tried to prepare an anti-placental antibody in rat against rat placenta, but were unable to show any abortifacient property. The situation is obviously quite complicated.

BAGSHAWE: Yes, this situation involves xenogeneic antisera to examined differences involving tissue-specific antigens, and its importance in physiological circumstances may be negligible even though interesting.

ALLISON: In the work of Beer & Billingham, is a direct effect of antibody or complement involved and is there any collaboration with K cells? This could be determined by complement fixation; cobra venom could be used and the K cell activity could be depleted by using methotrexate, anti-lymphocytic serum or other methods.

JOHNSON: With respect to K cells and the enhancement of these cells, are there any systems which have shown either *in vitro* or *in vivo* that enhancement can exist when K cells are present?

ALLISON: Immune complexes will block killing by K cells very efficiently and will also block killing by T cells. Antibody by itself will block killing by T cells but not by K cells.

LOKE: I have certain empirical findings of relevance to the immunogenicity of trophoblast (unpublished data). Sera were taken from patients with trophoblast tumours and from multiparous women, and trophoblast used as target cells in mixed antiglobulin tests. All seven sera from patients with tumours had antibodies which were localized on the trophoblast, but none of the sera from the multiparous women had this property. This result could, or course, be due to sampling, in that we happened to choose patients with moles who had developed antisera from previous pregnancies, but it could indicate that patients with tumours are sensitized most of the time. When these sera were tested against a panel of lymphocytes, only two out of the seven appeared to have HL-A antibodies. Is the mixed antiglobulin reaction therefore more sensitive in the detection of anti-HL-A antibodies or are we testing something other than HL-A antibodies, such as antibodies against trophoblast, in the serum of patients with hydatidiform moles?

BAGSHAWE: There may be many other antigens than those of the HL-A system, some common to trophoblast, others to trophoblast that has undergone some form of malignant transformation, and you may be detecting any of these.

LAWLER: The seven patients with tumours are more likely to have been transfused than those who have not had tumours. A transfusion on top of pregnancy increases the chances of antibody formation and these antibodies could be HL-A or other antibodies; we, of course, detect HL-A antibodies in our system. Also, it is just possible that trophoblast can cause immunization during the normal pregnancy.

BODMER: As far as I am aware, the mixed agglutination or mixed anti-globulin test is decidedly more sensitive than the cytotoxic assay, so more activity is found using mixed anti-globulin tests when screening a range of sera from women who are normally pregnant. A decision about a trophoblast-specific antigen could be made by examining trophoblast from many pregnancies for, if all were reactive, the antibody would be directed against a trophoblastic antigen rather than to an HL-A antigen.

LOKE: So far we have only had three cell lines for testing, and so we have not covered this point.

MCLAREN: Could you comment on the claim some years ago by Hulka, Brinton, Schaaf & Baney (1963) on the appearance of circulating antibody against placenta soon after parturition of women. The antibody was found before parturition only in cases of toxaemia. Could you also comment on the fragments of trophoblast that break off from the placenta and are found in the lungs of women; do they provoke a cellular response, is there any evidence that they provoke antibody response, do they persist after parturition, and what happens if pregnancy is terminated early – do they persist then?

BAGSHAWE: Concerning antibodies against trophoblast in the post-partum woman, Hulka stated that the antibodies were found only after parturition. Of course, these antibodies could be absorbed out during pregnancy. Moreover, the specificity of the antibodies was not determined – all that can be said is that they reacted with placenta and I believe that there has been no further follow-up on this work.

CURZEN: With reference to Hulka's work, they removed gammaglobulin, labelled it with fluorescein and then used it to stain cryostat sections of the placenta. If their results are correct, full-term human placenta should be stained with a labelled anti-human globulin, and if some anti-trophoblast antibody is attached, the trophoblast should also be stained. In his presentation, Faulk demonstrated fluorescence of subtrophoblast deposits in the placenta. A great deal of work with labelled anti-human gammaglobulin has indicated a virtual impossibility to stain the trophoblast.

JENKINS: We labelled maternal IgG and have been unable to find it on trophoblast, which confirms Curzen's point (Hancock & Taylor, 1973).

BAGSHAWE: Concerning the fate of trophoblastic fragments in the circulation raised by Anne McLaren, there has been a report by Iklé (1964) that 100 000 trophoblast cells enter the maternal circulation daily, and these are assumed to be filtered out in the pulmonary circulation because they are not found in the peripheral circulation. The mechanism whereby they are killed in the pulmonary system is not known, but they do die out soon after parturition. Any form of vascular obstruction in the lungs during pregnancy is rare, although there have been two cases recorded of a fatal trophoblastic embolism in apparently normal pregnancies. This condition is much more frequent in patients who have hydatidiform mole, and until the mole is evacuated large amounts of trophoblast and sometimes intact villae can enter the circulation and cause vascular obstruction of the lungs, although these elements appear to die out quite quickly. In the case of choriocarcinoma there is a tendency for a progressive obliteration of the pulmonary blood vessels.

BODMER: Is it true that pieces of trophoblast are released into the circulation in cases of severe toxaemia?

CURZEN: I am not aware of this and I would like to see much more evidence of this. I know that Arthur Williams (personal communication) was investigating this phenomenon, especially by using blood returning from the uterus to find trophoblast cells in it, although I have not seen any publication.

ALLEN: You mentioned that a powerful anti-HCG serum would kill trophoblast cells in culture. Is this a complement-dependent system and would it be a feasible method for the treatment of hydatidiform mole?

BAGSHAWE: In culture, complement is necessary. We have given large amounts of rabbit anti-HCG to a small group of patients with drug-resistant choriocarcinoma, even up to 150 ml of antiserum by slow intravenous infusion. One patient had some renal pain without any observable effect on renal function. These patients had quite low levels of circulating gonadotrophins prior to this treatment even though we could not eliminate the cancer by drugs. In general, the HCG levels fell to very low levels and the titres remained very low or undetectable for a few days but they then rose again. We have no good evidence that administering this antibody was in any way cytolytic although of course we may have problems with getting the right amount of antibody into the patients. It is encouraging that there was no morbidity and it is of note that one patient was a young man with a teratocarcinoma, and the man evidently went into remission after treatment, but he was the first one we treated and had further chemotherapy. He died a few months later from an unknown illness, with profound hypoglycaemia of unknown cause. At autopsy we found no evidence of residual tumour, and this is very unusual in teratocarcinoma. Before chemotherapy he had had extensive disease including brain metastases and therefore there may have been some cytotoxic action in this patient but we have not been able to demonstrate this convincingly.

EDWARDS: May I raise a question about the β sub-unit of HCG, which is now being purified. Is there any evidence that this can be used in your system, or that immunization against this subunit may arise naturally? The only evidence I know is that antibodies against HCG were present in two pregnant women, whereas none were found in non-pregnant women (Ružić *et al.*, 1973).

BAGSHAWE: We have no evidence in our patients of any antibodies against HCG during pregnancy, nor do I know of any work on autoantibodies against any of the polypeptide hormones. These ideas were prevalent in the 1930s, and Stevens & Crystle (1973) recently obtained evidence suggesting that hapten-coupled HCG is immunogenic in human females. As far as the β subunit is concerned all that we can say is that xenogeneic antisera to β-HCG do discriminate more decisively between HCG and luteinizing hormone than do antisera raised against whole HCG but individual anti-β HCG vary considerably in their power to discriminate. I know of no evidence that the β subunit is autoantigenic.

ALLISON: Immunotherapy could be possible in the choriocarcinoma situation, and the K cell is interesting because its possibilities for immune therapy are greater than with T cells. Has anyone tried to use maternal sera with antibodies directed against the paternal antigen of the HL-A under conditions where the activity of the K cells is monitored? If these cells are depleted by chemotherapy they would then return in numbers and a large amount of antibody directed against the paternal haplotypes could be given.

BAGSHAWE: No-one has used adoptive transfer of antibody. We have actively immunized one of these unresponsive patients using her husband's leukocytes and BCG, hoping to augment her reactions against paternal antigens in this way. We have collected lymphocytes from a sibling which was HL-A identical to the patient and incubated these *in vitro* with the cells of the patient's husband, and these were then transfused after a five-day incubation back into the patient. This one patient, having been unresponsive to chemotherapy previously, now showed a distinct response. Her tumour cell population as judged by HCG excretion was reduced by roughly 99% which is incomparably better than we had seen before in that patient. She was not given both antibody and sensitized lymphocytes at the same time (Bagshawe and others, unpublished observations).

DAVIES: Is this difference about chemotherapy in relation to K cells so distinct? A xenogeneic, fully absorbed, tumour-specific antibody will protect mice from tumour growth up to a point, and it is possible to arrange for an amount of drug that will give the same amount of protection. If both are used together, there is no improvement – and if antibody is given first followed by drug there is no special effect. The other way around, however, i.e. drug first followed by antibody, gives a great enhancement of the anti-tumour effect. I do not understand this; it would be easier to understand if the other way around. In this case, we are neither dealing with K cells nor knocking them out.

HEAP: Could I raise a question about the normal delivery of the placenta at full term? Concerning antibodies against trophoblast, is there any prospect that the delivery of the placenta could be related to some component of the normal immunological mechanisms? Comparative studies of the endocrinology of parturition have so far failed to identify a unitary mechanism for the initiation of parturition in various species.

BAGSHAWE: The immediate answer is, of course, that in the intra-strain matings, where there can presumably be no strong immunological effect, parturition occurs quite normally.

REFERENCES

ABBAS, A. K., CORSON, J. K., CARPENTER, C. B., MERRILL, J. P. & DAMMIN, G. J. (1974). Immunological enhancement of rat renal allografts. I. Comparitive morphology of acutely rejecting and passively enhanced grafts. *Am. J. Path.*, **75**, 255.

BEER, A. E. & BILLINGHAM, R. E. (1971). Immunology of mammalian reproduction. *Adv. Immunol.*, **14**, 46.

BEER, A. E., BILLINGHAM, R. E. & YANG, S. L. (1972). Further evidence concerning the autoantigenic status of the trophoblast. *J. exp. Med.*, **135**, 1177.

BILLINGTON, W. D. (1973) Does trophoblast express tissue-specific antigens– In *Immunology of Reproduction* (ed. K. Bratanov), p. 475. Bulg. Acad. Sci.: Sofia.

FAULK, W. PAGE, CONOCHIE, L. B., TRENCHER, P. & DORLING, J. (1975). An immunohistological study of the distribution of immunoglobulins, collagen, and contractile proteins in human placentae. *Protides of the Biological Fluids* (XX11 Colloquium). Pergamon Press (in press).

HANCOCK, K. W. & TAYLOR, P. V. (1973). The significance of maternal lymphocytotoxic antibodies in human pregnancy. In *Immunology of Reproduction* (ed. K. Bratanov), p. 426. Bulg. Acad. Sci.: Sofia.

HULKA, J. F., BRINTON, V., SCHAAF, J. & BANEY, C. (1963). Appearance of antibodies to trophoblast during the postpartum period in normal human pregnancies. *Nature, Lond.*, **198**, 501.

IKLÉ, F. A. (1964). Dissemination von Syncytiotrophoblastzellen im mütterlichen Blut während der Gravidität. *Bull. Schweiz Akad. med. Wiss.*, **20**, 62.

JENKINS, D. M., GOOD, W. & GOOD, SHEILA, (1973). Serum seromucoid and the materno–paternal mixed leucocyte reaction following previous severe pre-eclampsia. *J. Obstet. Gynaec. Br. Commonw.*, **80**, 19.

JENKINS, D. M. & HANCOCK, K. W. (1972). Maternal unresponsiveness to paternal histocompatibility antigens in human pregnancy. *Transplantation*, **13**, 618.

KIRBY, D. R. S. (1968). The immunological consequences of extra-uterine development of allogeneic mouse blastocysts. *Transplantation*, **6**, 1005.

RUZIC, A., RADOVIC, P., POPOVIC, D., VUJUCIC, V., ZDRAVKOVIC, G. & MITROVIĆ, A. (1973). Anti-chorionic gonadotrophin antibodies in the sera of pregnant women. *1st. Int. Cong. on Immunology in Obstetrics and Gynaecology*, Padua. Excerpta Medica (in press).

STEVENS, V. C. & CRYSTLE, C. D. (1973). Effect of immunization with hapten-coupled HCG on human menstrual cycle. *Obstet. Gynec.*, **42**, 485.

VOISIN, G. (1974). Proceedings of the Second International Planned Parenthood Federation Cambridge Workshop. *J. Reprod. Fert.* (in press).

THE HUMAN PLACENTA AS A FILTER FOR CELLS AND PLASMA PROTEINS

By MATTEO ADINOLFI

I have been asked to discuss the mechanisms of the transfer of cells and proteins through the placenta; however there are recent and excellent reviews of this subject (Mollison, 1967; Benirschke & Sullivan, 1969; Brambell, 1970; Billingham & Silvers, 1971; Wild, 1973). Discretion dictates, therefore, that I should limit my paper mainly to some aspects of the transfer of cells and proteins across the human placenta. I shall restrict the discussion of this subject in other mammals to the framework necessary to interpret the studies carried out in man.

TRANSFER OF RED CELLS AND LYMPHOCYTES

Red cells

It is now believed that the transfer of red cells from the maternal into the foetal circulation occurs frequently in man both during pregnancy and at delivery although there is still a good deal of disagreement about details (Lewi, *et al.*, 1961; Finn, Harper, Stallings & Krevans, 1963; Zipursky, *et al.*, 1963; Cohen, Zuelzer, Gustafson & Evans, 1964; Woodrow & Finn, 1966; Mollison, 1967; McConnell, 1969; Brambell, 1970). The presence of foetal red cells in maternal blood has been demonstrated by detecting cells with a high concentration of Hb_f using the acid-elution test or by immunofluorescence using antibodies against antigens possessed by the foetus and absent in the mother. The frequency with which foetal red cells are found in the blood of pregnant women seems to be affected by the stage of the pregnancy and most authors agree that the proportion of women in whom foetal red cells can be detected is greater towards the end of gestation than in the early stages. (Lewi *et al.*, 1961; Cohen, *et al.*, 1964; Woodrow & Finn, 1966; McConnell, 1969). ABO incompatibility appears to affect the frequency with which foetal red cells are detected in maternal blood (Lewi *et al.*, 1961; Finn *et al.*, 1963). The total amount of foetal red cells in normal pregnant women ranged between 0.1 and 3 ml, but only near 1% of women have 3 ml or more foetal red cells in their circulation.

In man, like in some other mammals notably chinchillas and hamsters, trophoblastic cells also invade the maternal blood vessels (Billington, 1969; Billingham & Silvers, 1971); in pregnant women cellular fragments of the syncytio trophoblast frequently gain access to the maternal circulation after the eighteenth week of gestation.

On the other hand, the transplacental passage of foetal red cells into the maternal circulation seems to be far less common than the traffic in the reverse direction (Cohen & Zuelzer, 1964; Mollison, 1967). Using fluorescein labelled anti-A and anti-B, maternal red cells were identified only in the circulation of three out of 82 newborn infants (Cohen & Zuelzer, 1964).

Lymphocytes

Studies of the transplacental passage of lymphocytes in man and other mammals have produced contradictory results (Adinolfi, 1970, Barnes & Tuffrey, 1970; Billingham & Silvers, 1971). The importance of the problem is obvious, in view of the biological properties of the cells involved (Fig. 1). If maternal lymphocytes gain access into the foetus, the expected consequences would be either (*a*) the immunological rejection of the foetus, (*b*) the insurgence of a form of runt disease during perinatal life or (*c*) some degree of immunological tolerance of the newborn towards the maternal tissue antigens. The outcome of the transfer of maternal lymphocytes would depend upon the time of the transfer and the number and type of cells transferred. If foetal lymphocytes cross the placenta, the analysis of maternal blood, after stimulation of the lymphocytes with

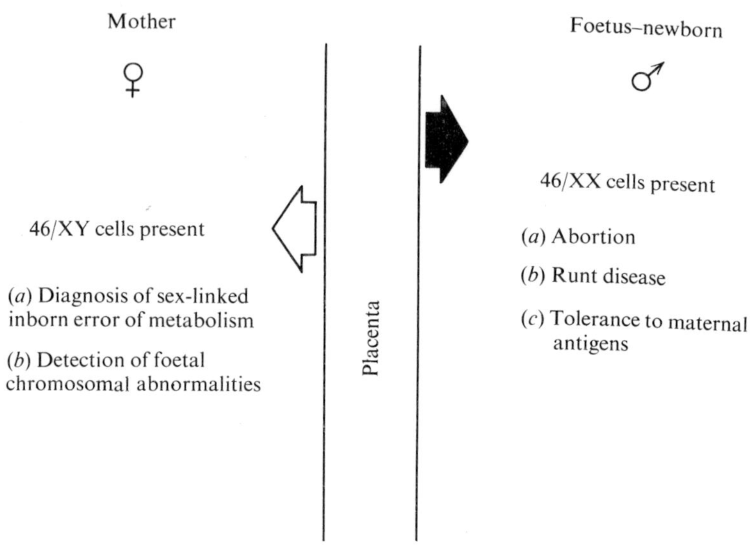

Fig. 1. Expected consequences of the transfer of maternal lymphocytes into the foetal circulation and vice versa.

PHA and chromosome analysis, should provide useful information about the sex of the foetus (in cases of sex-linked inborn errors of metabolism) and about the presence of foetal chromosome anomalies.

Recent studies of the transfer of human lymphocytes have been carried out by analysing the metaphase chromosomes of PHA-stimulated lymphocytes from pregnant women and male infants. The presence of 46/XY cells in the maternal circulation and 46/XX cells in male newborn blood was evaluated either by counting the number of the acrocentric chromosomes (five in 46/XY and four in 46/XX cells) or by the quinacrine method; the latter technique is less susceptible to errors since it provides a clear demonstration of the Y chromosome and offers the possibility to identify specific autosomes.

Many factors may influence the results of these investigations; the most important are: (1) Mode and time of the transplacental passage of lymphocytes. If the mechanism of the transfer of lymphocytes is similar to that of red cells, the total number of foetal lymphocytes in the maternal circulation should correspond to the volume (0.1–3 ml) of foetal blood present in the mother in normal pregnancies. If the transfer is selective, a higher percentage of lymphocytes may be expected. (2) Survival of foetal lymphocytes in the maternal circulation and vice versa. Since most mothers produce anti-HL-A antibodies, the incompatible foetal lymphocytes should survive only for a short period of time in the maternal circulation. The foetus is equally capable of mounting a cellular immune response against foreign antigens and one would expect that maternal lymphocytes should be lysed by the immune response of the foetus. (3) Proportion of a minor population of histoincompatible lymphocytes which survive in a 72 hour PHA culture. Although artificial mixtures of 46/XX and 46/XY cells can be prepared and used as controls, it is difficult to mimic a situation in which a minor population of foetal (or maternal) cells has been in contact *in vivo* for an unknown period of time with maternal (or foetal) lymphocytes. Other technical difficulties arise when the identification of 46/XY cells is based on counting the number of acrocentric chromosomes since breakages are frequent in normal lymphocyte cultures. (4) The lymphocytes stimulated by PHA represent only a subpopulation of these cells present in circulation; furthermore, at the present time, the number of cells which can be analysed for each subject is limited; both factors may reduce the chance of detecting a minor population of transferred lymphocytes.

The chromosome complement of PHA-stimulated lymphocytes has been analysed in at least six groups of pregnant women (Tables 1 and 2). In four series, 46/XY cells were recognized by counting the acrocentric chromosomes (Walkanowska, Conte & Grumbach, 1969; De Grouchy & Trébuchet, 1971; Schindler, Graf & Martin-du-Pan, 1972; Takahara, Kadotani, Kusumi & Makiko, 1972). In the remaining two groups, foetal

TABLE 1 *Evidence of 46/XY cells in maternal blood*

Sex of newborn	Series* A Yes	A No	B Yes	B No	C Yes	C No	D Yes	D No
Male	13	3	8	3	4	5	3	7
Female	2	6	4	6	5	2	1	12

* 46/XY cells in maternal blood were detected using a cytogenetic method by Walknowska *et al.* (1969), series A; De Grouchy & Trébuchet (1971), series B; Schindler *et al.* (1972), and series C and D, Takahara *et al.* (1972).

46/XY cells were detected using the quinacrine method (Schröder & de la Chapelle, 1972; Adinolfi & Gorvette, 1973). As shown in Table 1 and Fig. 2, 46/XY cells were detected in many women whose pregnancy terminated with the delivery of either a male or a female infant. The percentage of 46/XY cells in maternal blood was found to vary between 0.1 and 2% (Fig. 2).

The most perplexing observation reported in these studies was the detection of 46/XY cells in a high percentage of pregnant women, who delivered a female infant (Fig. 2 and Table 1). The presence of 46/XY cells in the circulation of mothers whose pregnancies resulted in the delivery of female infants has been attributed to the survival of foetal cells derived from previous male foetuses (Walkanowska *et al.*, 1969; De Grouchy & Trébuchet 1971). However, in the series studied by Schindler *et al.* (1972) 46/XY cells were detected in five primigravidae with female infants. Furthermore, the hypothesis that the 46/XY cells detected in these mothers were lymphocytes which survived a previous 'male' pregnancy is not supported by the observation, in one of these studies, that cells with male chromosome complement could not be detected in mothers re-tested only a few months after delivery.

The persistence of foetal lymphocytes would also be surprising in view of the high incidence of maternal cytotoxic antibodies in many pregnant

TABLE 2 *Presumptive Y body in interphase lymphocytes from pregnant women**

Sex of newborn	Y body: Present	Absent
Male	7	2
Female	4	8

* From Schröder & de la Chapelle, 1972. The Y body was detected in interphase cells using the quinacrine staining method.

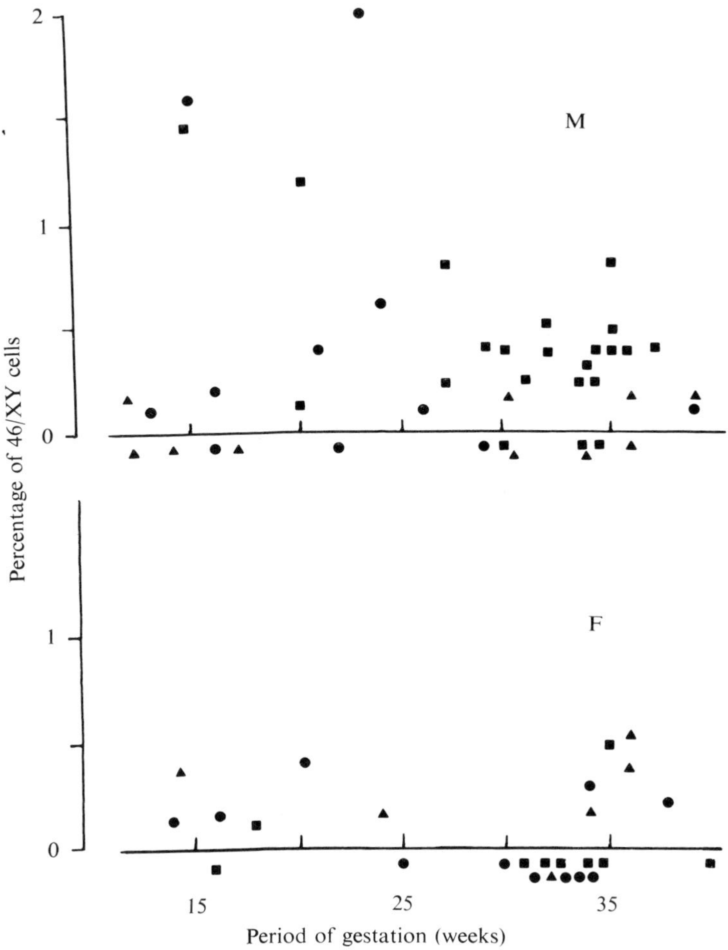

Fig. 2. Percentage of 46/XY cells in the circulation of pregnant mothers whose pregnancies resulted in the delivery of male (M) and female newborns (F). From Walkanowska *et al.* (1969). ■, De Grouchy & Trébuchet (1971), ▲, and Schindler *et al.*, (1972), ●. The karyotype of the cells was determined by counting the number of small acrocentric chromosomes. See also Table 1.

women; these antibodies should be capable of lysing the foetal cells. There is also a marked discrepancy between the estimated volume of foetal red cells known to cross the placenta into the maternal circulation (0.1 to 3 ml of blood during gestation) and the number of foetal PHA-stimulated lymphocytes detected in the maternal blood in these studies (mean values: 0.3–0.4%, corresponding to about 14–18 ml blood).

The fluorescence of the Y chromosome in interphase cells, after staining with quinacrine derivatives, has been used in the fourth series (Schindler *et al.*, 1972) in an attempt to detect foetal lymphocytes in maternal circulation. Maternal lymphocytes were stimulated with PHA and the presumptive presence of Y body was analysed in interphase cells. As shown in Table 2, Y bodies were detected in lymphocytes from seven out of nine women who delivered a male infant; Y bodies were also detected in four women who delivered female infants; two of these women were primigravidae; one had a previous female newborn and one an abortion. No Y bodies were observed in the buccal cells of the four female infants. Retrospective studies of the metaphase chromosomes of the four mothers with presumptive Y bodies showed unusual bright fluorescent spots of various autosomes. In a short note Zimmerman & Schmickel (1971) mentioned that an attempt to demonstrate Y bodies in unstimulated lymphocytes from 32 primiparae failed to detect, with accuracy, the sex of the foetus.

In the fifth series (Adinolfi & Gorvette, 1973) the analysis of metaphase chromosomes of PHA-stimulated lymphocytes from maternal blood did not demonstrate the presence of 46/XY cells derived from male foetuses. In this series two groups of eight women were studied, one at mid-gestation and the other at delivery (Table 3).

In conclusion, it appears that foetal lymphocytes cannot be detected with certainty in maternal blood either before or after stimulation with PHA. Too often, in fact, presumptive foetal cells 46/XY have been detected in mothers whose pregnancies terminated with the delivery of a female infant. It is not possible to exclude, however, that in normal pregnancies a limited number of foetal lymphocytes may cross the placenta, especially at delivery when leaks may occur. It should be emphasized that, if as much as 3 ml of foetal blood is present in the maternal circulation and if it is assumed that lymphocytes and red cells cross the placenta at a similar rate, the expected ratio of foetal to maternal lymphocytes should be 1:1500. Using the present techniques it would be difficult to detect foetal lymphocytes in maternal blood at this dilution. The high incidence of anti-HL-A cytotoxic antibodies in pregnant women does not necessarily imply that transfer of lymphocytes takes place during an early stage of gestation. In fact, recent studies have shown that HL-A antigens, in addition to being constituents of cell membranes, are also

TABLE 3 *PHA-stimulated lymphocytes from male foetuses (46/XY) in the maternal blood*

Number of mothers	Gestation	Number of cells examined 46/XX	46/XY
8 (Series L)	14–34 weeks	5 564	0
8 (Series D)	At delivery	6 120	0
Total 16		11 684	0

present in a soluble form (Charlton & Zmijewski, 1970; Van Rood, Van Leeuwen & Van Santen, 1970; Aster, Miskovich & Rodey, 1973).

The transfer of maternal lymphocytes into the foetal circulation has also been the object of several investigations both in man and experimental animals (Angell & Adinolfi, cited by Adinolfi & Wood, 1969; Desai & Creger, 1963; Turner, Wald & Quinlivan, 1966; Benirschke & Sullivan, 1969; Billington, 1969; Adinolfi & Wood, 1969; Tuffrey, Bishun & Barnes, 1969; Billington, *et al.*, 1969; Brambell, 1970; Beer & Billingham, 1972; Adinolfi & Gorvette, 1973). In 1964, Billingham suggested that runt disease could occur as a consequence of the passage of a sufficient number of maternal lymphocytes across the placenta. Soon after, maternal lymphocytes were detected in three male foetuses or infants and in at least one of these, symptoms highly suggestive of runt disease were described (Adinolfi & Wood, 1969; Billingham & Silvers, 1971).

Since then, transfer of maternal lymphocytes into the foetal circulation has been studied in at least three series of newborns using a cytogenetic approach (Table 4). In 1966 Turner and collaborators analysed about 5500 lymphocytes obtained from 183 male newborns; 46/XX and 46/XY cells were distinguished by counting the number of acrocentric chromosomes. All but two infants were found to have cells containing 46/XY chromosomes; in one infant four of the thirty karyotypes examined were 46/XX; this baby died a few hours after delivery and the autopsy revealed polycystic disease of liver, kidney, lung and pancreas. The second infant showed ten 46/XX and two 46/XY cells out of thirty examined; this baby died one month after birth and the autopsy revealed several malformations.

An investigation using a similar cytogenetic approach was carried out in another group of 18 *normal* male infants (Angell & Adinolfi cited in Adinolfi & Wood, 1969). In this series, chromosome analysis was performed on 300–400 cells for each child and a total of about 6500 cells were examined; maternal 46/XX cells were not detected.

In the third series the analysis of the chromosomes was performed using the quinacrine method (Adinolfi & Gorvette, 1973). Maternal 46/XX

TABLE 4 *Evidence of 46/XX cells in the circulation of male newborns*

References	No. newborns tested	No. newborns with 46/XY cells only	No. newborns with 46/XY and 46/XX cells	Total cells tested
1. Turner *et al.*, (1966)*	183	181	2	5 500
2. Angell & Adinolfi (1969)*	18	18	0	6 482
3. Present observation†	16	15	1	12 126

*The analysis of 46/XY cells was carried out by counting the number of acrocentric chromosomes.

†The analysis of the chromosome was carried out using the quinacrine staining method.

cells were detected in one infant; in this baby, out of 1600 cells examined four 46/XX cells were observed. These cells were characterized by the absence of the typical fluorescent Y, the presence of two X chromosomes, four chromosomes belonging to the G group and the presence of strongly fluorescent satellites on one of the G chromosomes which was absent in the baby's cells. The chromosome banding and satellite pattern exactly matched those of PHA-stimulated maternal lymphocytes. In the second sample of blood taken from this baby at six weeks of age, one maternal 46/XX cell was observed out of 750 metaphase chromosomes examined.

These results suggest that in only a few pregnancies maternal lymphocytes may be detected in foetal blood after stimulation with PHA. If a high number of maternal lymphocytes gain access into the foetal circulation, the foetus or the newborn will be severely affected as in the cases described by Turner and collaborators. On the other hand, if a few maternal lymphocytes cross the placenta they are probably lysed by the cellular immune response of the foetus. In this regard it is of interest that the transfer of lymphocytes from mothers with leukaemia has rarely been described out of many cases investigated.

TRANSFER OF PLASMA PROTEINS

Selective transfer of proteins

The route and the degree to which plasma proteins are transferred through the placenta vary in different species of mammals (Vahlquist, 1958; Freda, 1962; Schultze & Heremans, 1966; Mollison, 1967; Adinolfi & Wood, 1969; Brambell, 1970; Brambell & Halliday, 1956; Gitlin & Morphis, 1969). Selective transfer of maternal proteins before birth occurs by way of the uterine lumen and the foetal yolk sac in the rabbit, guinea pig and rat; this appears to be the only route in the rabbit and the main route in guinea pig in which transfer also occurs through the chorio-allantoic placenta (Hemmings, 1956; King & Enders, 1970; Brambell, 1970; King & Enders, 1971; Wild, 1973). Transmission of proteins in man and monkeys, on the evidence available, seems to be mainly by way of the chorio-allantoic placenta.

Selection is perhaps one of the most remarkable features of the transfer of proteins; thus for example, whereas IgM molecules are readily transmitted from the mother to the foetus in the rabbit (Hemmings & Jones, 1962), this class of immunoglobulins does not cross the placenta in man (Adinolfi & Wood, 1969). There is evidence that the reason why some maternal proteins are, or are not, transferred into the foetal circulation is not related to the number or the thickness of the placenta membranes, but depends upon the capacity of the cells forming such membranes to allow a percentage of the endocytosed proteins to be transferred across without being degraded (Brambell, 1970; Wild, 1973).

The first indirect evidence that selectivity is an intracellular event and does not occur at the surface of the yolk sac endoderm was obtained in rabbits by Hemmings (1956). Direct evidence that selection indeed occurs within the endothelial cells has been produced by Wild (1970, 1973) and Slade (1970) in rabbits, and King & Enders (1970, 1971) in guinea pigs.

Transfer of homologous and heterologous proteins into the foetus

Experimental work has also shown that the transmission of homologous and heterologous proteins is clearly independent of the molecular size and that for example, whereas IgG molecules cross the placenta, their light chains are not readily transferred from the maternal into the foetal circulation (Brambell, 1970). In man, maternal IgG molecules, and to a lesser extent albumin, have been shown to be transferred into the foetal blood (Dancis *et al.*, 1961; Gitlin, Kumate, Urrusti & Morales, 1964). Maternal IgG antibodies start to cross the human placenta at about three months of gestation; at this stage of development foetal Rh(+) red cells may be agglutinated by an anti-globulin test due to the presence of maternal anti-Rh incomplete antibodies (Mollison, 1967). At the end of normal gestation, the mean levels of IgG in newborn sera is slightly higher than that in the corresponding maternal samples (Schultze & Heremans, 1966). Variations in the levels of IgG in the mother are usually reflected in the concentration of these proteins in cord blood; in fact, a deficiency of IgG in maternal serum is associated with a similar deficiency in the newborn blood; on the other hand, high levels of IgG are present in infants born to mothers with hypergammaglobulinemia (Adinolfi & Wood, 1969).

The discovery of the genetic markers of human IgG, the Gm system (Grubb & Laurell, 1956; Grubb, 1970), has made it possible to investigate both the selective transfer of the four subclasses of this group of proteins and their synthesis during life *in utero*.

The major observations which have emerged from these studies are the following:

(a) Most of the IgG molecules present in newborn sera carry Gm specificities present in the corresponding maternal samples, irrespective of the genotype of the newborn (Grubb & Laurell, 1956; Grubb, 1970). Slowly, in about two to three months, the maternal IgG molecules are replaced by those produced by the newborns.

(b) Sera from normal newborns contain traces of IgG molecules with Gm specificities absent in the maternal blood; these immunoglobulins are produced by the foetus under the control of genes inherited from the father (Fudenberg & Fudenberg, 1964; Adinolfi & Wood, 1969).

(c) The foetal IgG molecules carrying Gm specificities absent in the corresponding maternal blood cross the placenta and induce the synthesis of anti-Gm in the mother. Antibodies directed against the genetic markers of the chains (anti-Gm) and of light chains (anti-Inv) have been detected in about 4.1% of normal pregnant women (Fudenberg & Fudenberg, 1964; Adinolfi & Wood, 1969.)

(d) Over 75% of children between 14 months and 10 years of age produce anti-Gm against the maternal Gm specificities (Adinolfi & Wood, 1969). The synthesis in normal children of antibodies against the maternal Gm factors is surprising, since one would expect them to become tolerant to the antigenic determinants of the maternal IgG molecules.

(e) Experimental work in mice and rabbits has shown that maternal antibodies directed against allotypic specificities of the foetal Ig molecules induce a suppression of the synthesis of the subclasses of IgG displaying the genetic markers (Dray, 1972). This phenomenon of allotypic suppression does not induce hypogammaglobulinemia, since the deficiency of one subclass is fully compensated by the increased synthesis of the other subclasses of immunoglobulins. A similar mechanism of allotypic suppression by maternal anti-Gm has not been observed in man. This is probably due to the fact that the maternal anti-Gm and anti-Inv are usually associated with IgM – and therefore do not cross the placenta – or are present at low levels.

Estimation of the levels of the four subclasses of IgG using monospecific antibodies has confirmed that these proteins are present in cord sera in concentrations similar to those present in the corresponding maternal blood with the exception of IgG2; it has been suggested that the mean level of this subclass of IgG is lower in samples from newborns than that observed in maternal sera (Wang, Faulk, Stuckey & Fudenberg, 1970). In contrast to the evidence that IgG molecules readily cross the human placenta, the molecules of the other four subclasses of human Ig either are not transferred from the maternal into the foetal circulation or cross the placenta only in small quantities.

IgA molecules are present in low concentrations in the serum of normal neonates; high levels have been detected in infants with congenital infections or in infants previously transfused during life *in utero*. High levels of IgA have also been detected in neonate blood in Nigeria during the wet season when malaria, parasitic and viral infections are most frequent (McFarlane & Udeozo, 1968; Adinolfi & Wood, 1969). Evidence that IgA is synthesized during foetal life and that these molecules cross the placenta has been obtained by detecting maternal antibodies against genetic specificities of IgA (Am system); thus mothers who are *Am(a-)* have been found to produce anti-Am directed against the IgA of the corresponding infants.

IgM molecules are present in newborn sera in concentrations ranging between 10 to 20% of the levels present in maternal blood. Specific maternal IgM antibodies (anti-A, anti-I) do not seem to cross the placenta.

IgD molecules are not present in normal cord sera (Rowe, Crabbe & Turner, 1968), however the synthesis of this class of Ig starts during foetal life: in fact, between 10 and 14% of foetal lymphocytes react with anti-IgD when tested by immunofluorescence but apparently IgD molecules are not released into the circulation (Rowe *et al.*, 1973).

IgE molecules are present in sera from normal newborns in concentrations lower than those detected in normal adults. There is good evidence that the levels of IgE in cord sera are not correlated with their levels in maternal blood; this holds true also when the amounts of IgE are compared between pairs of allergic mothers and their respective newborns (Johansson & Bennick, 1967).

The transmission of the other human maternal plasma proteins has been investigated either by injecting into the circulation of pregnant women ^{131}I-labelled proteins (Dancis *et al.*, 1961; Gitlin *et al.*, 1964; Schultze & Heremans, 1966), or by comparing the phenotypic patterns of the genetic variants of single plasma proteins in pairs of maternal and newborn sera (Adinolfi, 1970). The results of these studies have shown that albumin is transferred across the placenta although less readily than IgG. So far, genetic variants of maternal plasma proteins showing a genetic polymorphism, such as transferrins (Tf), haptoglobins (Hp), β-lipoproteins (Ag) and α-glycoproteins (Gc) have not been detected in the foetus or newborn lacking the corresponding gene. However, the usual techniques employed cannot detect proteins present at nanogram levels and traces of maternal protein variants may have escaped detection. Antibodies against Ag (β-lipoprotein) variants have been observed in a pregnant woman.

Transfer of foetal proteins into the mother

Studies of the transfer of plasma proteins from the foetal into the maternal circulation have shown that some proteins cross the placenta in this direction. The evidence of the transmission of foetal IgG and IgA molecules is indirect and is based on the observation that antibodies against genetic markers of these two groups of plasma proteins are occasionally detected in maternal blood. The foetal α-globulin (AFP) is another protein which seems to be transferred into the maternal circulation in small quantities. AFP is present in human foetuses in concentrations reaching 300 mg/100 ml between the twelfth and eighteenth week of gestation; later the levels decline and at delivery the amount of AFP in normal infants ranges between 2 and 20 mg/100 ml. The protein is present in low amount in amniotic fluid, and fetal urine and CSF (Bergstrand & Czar, 1956; Halbrech & Klibanski, 1956; Gitlin & Boesman, 1966; Van Furth & Adinolfi, 1969; Abelev, 1971; Adinolfi, Adinolfi & Lessoff, 1974). Using the radioimmunoassay technique, only traces (10–20 ng/ml) of AFP can be measured in sera from normal adults; however during gestation the concentration of AFP increases progressively from the twelfth week (Sëppala & Ruoslahti, 1972; Adinolfi *et al.*, 1974).

No correlation has been observed between the levels of AFP in pairs of maternal and foetal sera; in fact the amount of AFP slightly increases in maternal blood while it declines in the corresponding foetuses (Adinolfi *et al.*, 1974). In a group of pregnant women samples of sera were collected just before delivery, at delivery and within a few days after delivery (Adinolfi & Kohn, 1973).

In some women the levels of AFP were found to increase sharply at the time of delivery; this suggests that occasionally, at delivery, a variable number of leaks may develop in the placental barrier through which AFP and other blood components are transferred from the foetal into the maternal circulation.

Mechanisms of protein transfer across the placenta

Only a few attempts have been made to study the mechanisms of the transport of plasma proteins across the human placenta. Using fluorescent antibodies, little accumulation of homologous albumin and immunoglobulin could be seen in the trophoblast although the two proteins were dispersed throughout the villus structure (Bardawill, Toy & Hertig, 1958).

IgG molecules can be eluted from human placentae that have been washed free of blood and plasma proteins (McCormick, *et al.*, 1971). These eluates, labelled with fluorescein, tend to return to the trophoblastic basement membrane of cryostat sections of normal human placenta. So far, the eluates have not been found to contain specific antibodies against red cells, HL-A antigens, platelets, and normal serum proteins (Faulk, Carbonara & Jeannet, 1973).

The only study by electron microscopy of the transfer of proteins through a chorion-allantoic placenta, similar to that present in man, has been carried out in the guinea pig (King & Enders, 1971). Two proteins, horseradish peroxidase and ferritin, and thorotrast (a colloidal thorium dioxide in dextrin) were injected into the circulation of pregnant animals. The two proteins were found to be taken up by vesicles formed at the surface of the trophoblast, which then migrates deeper into the cytoplasm. Within 10–20 minutes after the injection, peroxidase and ferritin particles were observed in the basement membranes. In the areas where peroxidase was demonstrable in the basement membrane, the protein was observed in micropynocytic vesicles at the albuminal site of the capillary endothelium, in the intracellular spaces between the endothelial cells and in the lumina of foetal capillaries. Macrophages in the foetal connective tissue also contained peroxidase ten minutes after the injection. In spite of its concentration near the surface of the trophoblast and in multivesicular bodies, thorotrast was not found to cross the trophoblast barrier during the time of observation.

The cytological basis for protein absorption and transport by the visceral yolk sac placenta has been investigated in the guinea pig (King & Enders, 1971) and the rabbit (Slade, 1970; Wild, 1970; Wild, Stanber & Slade, 1972; Wild, 1973). In the guinea pig the distribution of horseradish peroxidase, ferritin or colloidal thorium dioxide was studied by electron microscopy at 2, 10, 20, 40 and 160 minutes after the injection into the uterine lumen (King & Enders, 1971). All three types of molecules were rapidly absorbed by endoderm cells. Two minutes after the injection, both horseradish peroxidase and ferritin were found to be present in saccular invaginations of the apical cell surface and in vesicles and dense tubules in the underlying cytoplasm. By contrast, particles of thorotrast were seldom observed in the apical invaginations and vesicules.

Ten minutes after the injection, peroxidase was found in numerous vesicles and tubules in the apical cytoplasm and in large vacuoles; ferritin, after the same interval had penetrated well into the endoderm and it was detected in the apical cytoplasm as well as in large vacuoles. The distribution of ferritin did not change after 20 minutes; horseradish peroxidase instead, at this stage, was observed also in a small number of vesicles and tubules in the basal cytoplasm of the endoderm cells; peroxidase activity was also found in macrophages in the stroma of the yolk sac indicating that some material had left the endoderm cells.

Only 160 minutes after the injection, ferritin particles were found in dense droplets in the basement membrane beneath the endoderm cells and within the lumen of the vitelline blood vessels; ferritin was also present in macrophages in the stroma of the yolk sac. At this stage, increased numbers of droplets containing peroxidase were found in the basal cytoplasm.

Particles of thorotrast were detected only in small vesicles and vacuoles in the apical cytoplasm until 40 minutes after the injection. A small number of thorotrast particles was found in the basal cytoplasm 160 minutes after injection; none of the vacuoles in this area seem to concentrate these particles in a manner similar to peroxidase or ferritin.

These results clearly show that horseradish peroxidase (molecular weight: 40 000) was rapidly transmitted across the yolk sac epithelium in guinea pig; ferritin (molecular weight 480 000) was transmitted across the endoderm cells more slowly; in contrast, thorotrast was not transmitted within the period of observation. From the rapid build-up of the large absorption droplets it is clear that most of the absorbed proteins were retained in the endoderm cells and gradually degraded by lysosomal activity. Once peroxidase and ferritin had left the endoderm cells, the proteins could follow three pathways: (a) they could enter the vitelline vessels through the fenestrations of these capillaries. (b) they could cross the splanchnic mesoderm and mesothelium and enter the exocelom or

(c) after entering the stromal compartment, peroxidase and ferritin could be sequestered by macrophages.

When a variety of proteins were injected into the uterine lumen of pregnant rabbits and their localization followed by fluorescent tracing, only those proteins readily transmitted into the foetal blood, such as rabbit and human IgG, were evident in and below the basement membrane of the yolk sac endoderm, although both transmitted and non-transmitted proteins could be detected in the absorptive vesicles (McCormick *et al.*, 1971; Wild, 1973.) Ferritin-like molecules, which appeared to be discharged from small vesicles into the basement membranes but did not retain their antigenicity (Slade, 1970) were sequestered by macrophages as were fluorescein-labelled rabbit and human immunoglobulins.

In a recent paper Wild and collaborators (1972) have also clearly demonstrated that both readily-transmitted and poorly- or non-transmitted proteins were taken up in the same vesicles. The precise mechanism of transport across the endothelial cells of the capillaries and vessels in rabbits is not known; micropinocytosis, however, seems to be a characteristic feature of such cells in the rabbit where fenestration has not been observed.

In conclusion the results of these studies suggest that in guinea pigs and rabbits the trophoblast is the major barrier to the transfer of proteins across the placenta and that this function is exercised by a low rate of absorption and the presence of multivesicular bodies and lysosomes which sequester and degrade part of the absorbed proteins. Evidence for the presence in the yolk sac endoderm of various lysosomal enzymes within the vesicles has been obtained (Wild *et al.*, 1972) and three cathepsins and three exopeptidases capable of hydrolysing bovine albumin and immunoglobulin have been detected in the rabbit yolk sac splanchnopleur (Wild, 1973).

Probably, most of the proteins taken up in the vesicles are degraded within the cells during the transfer. Once the proteins have reached the basal membranes they have access to the capillaries; in the case of the guinea pig this may occur through the fenestrations of the endothelial cells. Both ferritin and peroxidase have been shown to leave the capillaries through the fenestrae (Clementi & Palade, 1969). Some proteins are taken up by macrophages; their fate is unknown but probably they are degraded. Further it appears that some proteins can cross the splanchnic mesoderm and mesothelium in small vesicles which seem to empty their contents in the exocelom.

SUMMARY

In man, IgG molecules are the only ones which readily cross the placenta; the transmission of this group of proteins from the maternal into the foetal circulation and vice versa induces synthesis of antibodies directed against *Gm* and *Inv* genetic factors in both mothers and infants. Although, at present, the maternal anti-Gm and anti-Inv antibodies seem to have little or no effect during pregnancy and early life, their possible role in relation to diseases in adulthood remains to be investigated. The selective transfer of IgG starts at about three months and increases progressively during gestation. Recent work in experimental animals suggests that transmission of these proteins is an intracellular event and is mediated by specific receptors present in the molecules. Particles which are not transferred are taken up equally by the trophoblastic cells but do not reach the basement membranes.

Other plasma proteins may occasionally be transferred in small quantities and, if they express genetic determinants absent in the maternal blood, may induce the synthesis of antibodies directed against the allotypic variants. In man, this has been observed for IgA and Ag (β-lipoprotein) molecules.

Studies of the transfer of AFP have shown that a small amount of this protein is transferred during gestation into the maternal blood. The estimation of the levels of AFP in maternal sera has been shown to be a valuable indication of neural-tube defects of the foetus and foetal or placental distress. At time of delivery a rapid increase of the level of this foetal protein has been observed in some mothers, suggesting that a variable number of leaks may develop in the placenta barrier. It is probably through these leaks that foetal red cells gain access into the maternal blood at time of delivery.

In normal neonates, maternal lymphocytes have been detected in only one case out of 16 newborns investigated; the incidence of maternal cells was 1 in 750. A massive transmission of maternal immunocompetent cells has been shown to impair the survival of the foetus and the newborn. There are no conclusive data as yet about the possible transfer of maternal lymphocytes during early foetal life.

A careful cytogenetic examination of infants failing to thrive should be carried out in order to exclude the possibility that the syndrome may be due to the presence of immunocompetent cells derived from the mother during gestation.

The work carried out in the Paediatric Research Unit was supported by the Spastics Society and the Medical Research Council.

REFERENCES

ABELEV, G. I. (1971). Alpha-fetoprotein in ontogenesis and its association with malignant tumours. In *Advances in Cancer Research*, **14**, p. 295.

ADINOLFI, A., ADINOLFI, M. & LESSOFF, M. (1974). Human alphafoetoprotein in ontogeny and disease. *J. med. Genet.*, (in press).

ADINOLFI, M. (1970). Genetic polymorphism and in vitro cultures in studies of the ontogenesis of human fetal proteins. In *Biochemistry of development* (ed. Ph. Benson), p. 224. Spastics International Medical Publication: London.

ADINOLFI, M. C. & GORVETTE, D. P. (1974). The transfer of lymphocytes through the human placenta. *Proceeding of first International Congress of immunology in obstetrics and gynaecology*, (Padua) Exerpta Medica: Foundation: Amsterdam. (In press).

ADINOLFI, M. & KOHN, J. (1973). Transfer of foetal α_1-globulin into the maternal circulation. In *Immunology of Reproduction* (ed. K. Bratanov) p. 558, Bulg. Acad. Sci.: Sofia.

ADINOLFI, M. & WOOD, C. (1969). Ontogenesis of immunoglobulins and components of complement in man. In *Immunology and Development* (ed. M. Adinolfi), p. 27. Spastic International Medical Publication: London.

ASTER, R. H., MISKOVICH, B. H. & RODEY, G. E. (1973), Histocompatibility antigens of human plasma. Localization to the HLD-3 lipoprotein fraction. *Transplantation*, **16**, 205.

BARDAWILL, W. A., TOY, B. L. & HERTIG, A. T. (1958). Localization of homologous plasma proteins in the human placenta by fluorescent antibody. *Am. J. Obstet. Gynec.*, **75**, 708.

BARNES, R. D. & TUFFREY, M. (1970). Maternal cells in the newborn. In *Advances in the Biosciences*, **6**, 457.

BEER, A. & BILLINGHAM, R. E. (1972). Concerning the uterus as a graft site and the foetus as a natural parabiotic organismic homograft. In *Ontogeny of acquired immunity*, Ciba Found. Symposium, p. 149.

BENIRSCHKE, K. & SULLIVAN, M. M. (1969). The human placenta in relationship to the development of chimerism. In *The foeto-placental unit*, p. 37. Excerpta Medica Foundation: Amsterdam.

BERGSTRAND, C. G. & CZAR, B. (1956). Demonstration of a new protein fraction in serum from the human fetus. *Scand. J. Clin. Lab. Invest.*, **8**, 174.

BILLINGHAM, R. E. (1964). Transplantation immunity and the maternal–fetal relationship. *Trans. Col. Physicians*, **31**, 187.

BILLINGHAM, R. & SILVERS, W. (1971). *The Immunobiology of Transplantation*. Prentice-Hall; Inc.: Englewood Cliffs, New Jersey.

BILLINGTON, W. D. (1969). Immunological processes in mammalian reproduction. In *Immunology and Development* (ed. M. Adinolfi), p. 89. Spastic International Medical Publication: London.

BILLINGTON, W. D., KIRBY, D. R. S., OWEN, J. J. T., RITTER, M. A., BURTONSHAW, M. D., EVANS, E. P., FORD, C. E., GOULD, I. K. & MCLAREN, A. (1969). Placenta barrier to maternal cells. *Nature, Lond.*, **224**, 704.

BRAMBELL, F. W. R. (1970). *The transmission of passive immunity from the mother to young. Frontiers of Biology*, **18**, North-Holland Publ. Co.: Amsterdam.

BRAMBELL, F. W. R. & HALLIDAY, R. (1956). The route by which passive immunity is transmitted from mother to foetus in the rat. *Proc. R. Soc. B*, **145**, 170.

CHARLTON, R. K. & ZMIJEWSKI, C. M. (1970). Soluble HL-A7 antigen: localization in β-lipoprotein fraction of human serum. *Science*, **170**, 636.

CLEMENTI, F. & PALADE, G. E. (1969). Intestinal capillaries. I. Permeability to peroxidase and ferritin. *J. Cell Biol.*, **41**, 33.

COHEN, F., & ZUELZER, W. W. (1964). Identification of blood group antigens by immunofluorescence and its application to the detection of the transplacental passage of erythrocytes in mother and child. *Vox. Sang, Basel*, **9**, 75.

COHEN, F., ZUELZER, W. W., GUSTAFSON, D. C. & EVANS, M. M. (1964). Mechanisms of iso-immunization. I. The transplacental passage of fetal erythrocytes in homospecific pregnancies. *Blood*, **23**, 621.

DANCIS, J., LIND, J., ORATZ, M., SMOLENS, J. & VARA, P. (1961). Placental transfer of proteins in human gestation. *Am. J. Obstet. Gynec.*, **82**, 167.

De Grouchy, J. & Trébuchet, C. (1971). Transfusion foeto-maternelle de lymphocytes sanguins et detection du sexe du foetus. *Ann. Genet.*, **14**, 133.

Desai, R. G. & Creger, W. P. (1963). Maternal-fetal passage of leucocytes and platelets in man. *Blood*, **21**, 665.

Dray, S. (1972). Allotype suppression. In *Ontogeny of acquired immunity*, Ciba Found. Symposium, p. 87.

Faulk, W. P., Carbonara, A. & Jeannet, M. (1973). Immunobiological studies of the human placenta. In *Immunology of Reproduction*, p. 405. Bulg. Acad. Sci.: Sofia.

Finn, R., Harper, D. T., Stallings, S. A. & Krevans, J. R. (1963). Transplacental hemorrhage. *Transfusion, Philad.*, **3**, 114.

Freda, V. J. (1962). Placental transfer of antibodies in man. *Am. J. Obstet. Gynec.*, **84**, 1756.

Fudenberg, H. H. & Fudenberg, B. R. (1964). Antibody to hereditary human gamma-globulins (Gm) factors resulting from maternal – fetal incompatibility. *Science*, **145**, 170.

Gitlin, D. & Boesman, M. (1966). Serum alpha-feto-protein, albumin and γ globulin in the human conceptus. *J. Clin. Invest.*, **45**, 1826.

Gitlin, D. & Morphis, L. G. (1969). Systems of maternal–foetal transport of γG immunoglobulin in the mouse. *Nature, Lond.*, **223**, 195.

Gitlin, D., Kumate, J., Urrusti, J. & Morales, C. (1964). The selectivity of the human placenta in the transfer of plasma proteins from mother to fetus. *J. Clin. Invest.*, **43**, 1938.

Grubb, R. (1970). *The genetic markers of human immunoglobulins* Springer-Verlag: Berlin, New York.

Grubb, R. & Laurell, A. B. (1956). Hereditary serological human serum groups, *Acta Path. Microbiol. Scand.*, **39**, 390.

Halbrech, I. & Klibanski, C. (1956). Identification of a new normal embryonic haemoglobin. *Nature, Lond.*, **178**, 794.

Hemmings, W. A. (1956). Protein selection in the yolk sac splanchnopleur of the rabbit – the distribution of isotope following injection of ^{131}I-labelled serum globulin into the uterine cavity. *Proc. R. Soc. B*, **145**, 186.

Hemmings, W. A. & Jones, R. E. (1962). The occurrence of macroglobulin antibodies in maternal and foetal sera of rabbits as determined by gradient centrifugation. *Proc. R. Soc. B*, **157**, 27.

Johansson, S. G. & Bennick, H. (1967). Studies on a new class of human immunoglobulin. In *Gamma globulins Nobel symposium* no. **3**. (ed. J. Killander), p. 193.

King, B. F. & Enders, A. C. (1970). Proteins absorption by the guinea pig visceral yolk sac placenta. *Am. J. Anat.* **129**, 261.

King, B. F. & Enders, A. C. (1971): Protein absorption by the guinea pig chorio-allantoic placenta. *Am. J. Anat.*, **130**, 409.

Lewi, S., Clarke, T. K., Guentat, P., Walter, P., Mayer, M. (1961). Les erythrocytes foetaux dans la circulation maternelle. *Bull. Fed. Soc. Gynec. Obstet. franc.*, **13**, 535–45.

McCormick, J. N., Faulk, W. P., Fox, H. & Fudenberg, H. H. (1971). Immunohistological and elution studies of the human placenta. *J. exp. Med.*, **133**, 1.

McConnell, R. B. (1969). Immunological relationship between mother and fetus. In *Immunology and Development* (ed. M. Adinolfi), p. 159. Spastics International Medical Publication: London.

McFarlane, H. & Udeozo, I. O. K. (1968). Immunochemical estimation of some proteins in Nigerian paired maternal and fetal blood. *Arch. Dis. Child.*, **43**, 42.

MOLLISON, P. L. (1967). *Blood transfusion in clinical medicine*, 4th edn. Blackwell: Oxford.

ROWE, D. S., CRABBE, P. A. & TURNER, M. W. (1968). Immunoglobulin D in serum, body fluids and lymphoid tissues. *Clin. exp. Immunol.*, **3**, 477.

ROWE, D. S., HUG, K., FAULK, W. P., MCCORMICK, J. N. & GERBER, H. (1973). IgD on the surface of periferal blood lymphocytes of the human newborn. *Nature New Biol.*, 242, 155.

SCHINDLER, A-M., GRAF, E. & MARTIN-DU-PAN, R. (1972). Prenatal diagnosis of fetal lymphocytes in the maternal blood. *Obstet. Gynec.*, **40**, 340.

SCHRÖDER, J. & DE LA CHAPELLE, A. (1972). Fetal lymphocytes in the maternal blood. *Blood*, **39**, 153.

SCHULTZE, H. E. & HEREMANS, J. F. (1966). *Molecular biology of human proteins*, vol. I. Elsevier Publ. Co.: Amsterdam.

SËPPÄLA, M. & RUOSLAHTI, E. (1972). Radioimmunoassay of maternal serum alpha fetoprotein during pregnancy and delivery. *Am. J. Obst. Gynec.*, **112**, 208.

SLADE, B. S. (1970). An attempt to visualise protein transmission across the rabbit yolk sac splanchnopleur. *J. Anat.*, **107**, 531.

TAKAHARA, H., KADOTANI, T., KUSUMI, I. & MAKIKO, S. (1972). Some critical aspects on prenatal diagnosis of sex in leucocyte cultures from pregnant women. *Proc. Japan. Acad.*, **48**, 603.

TUFFREY, M., BISHUN, N. P. & BARNES, R. D. (1969). Porosity of the mouse placenta to maternal cells. *Nature, Lond.*, **221**, 1029.

TURNER, J. H., WALD, N. & QUINLIVAN, W. L. G. (1966). Cytogenetic evidence concerning possible transplacental transfer of leukocytes in pregnant women. *Am. J. Obstet. Gynec.*, **95**, 851.

VAHLQUIST, B. (1958). The transfer of antibodies from mother to offspring. *Adv. Pediat.*, **10**, 305.

VAN FURTH, R. & ADINOLFI, M. (1969). In vitro synthesis of the foetal α-globulin in man. *Nature, Lond.*, **222**, 1296.

VAN ROOD J. J., VAN LEEUWEN, A. & VAN SANTEN, M. C. T. (1970). Anti HL-A2 inhibitor in normal serum. *Nature, Lond.*, **226**, 366.

WALKANOWSKA, J., CONTE, F. A. & GRUMBACH, M. M. (1969). Practical and theoretical implications of fetal/maternal lymphocyte transfer. *Lancet*, **i**, 1119.

WANG, A. C., FAULK, W. P., STUCKEY, M. A. & FUDENBERG, H. H. (1970). Chemical differences of adult, fetal and hypo-gamma-globulinemic IgG immunoglobulins. *Immunochemistry*, **7**, 703.

WILD, A. E. (1970). Protein transmission across the rabbit foetal membrane. *J. Embryol. exp. Morphol.*, **24**, 313.

WILD, A. E. (1973). Transport of immunoglobulins and other proteins from mother to young. In *Lysosomes in Biology and Pathology* (ed. J. T. Dingle), p. 169. North-Holland Publ. Co.: Amsterdam.

WILD, A. E., STANBER, V. V. & SLADE, B. S. (1972). Simultaneous localization of human γ-globulin I^{125} and ferritin during transport across the rabbit yolk sac. *Z. Zellforsh.*, **123**, 168.

WOODROW, J. C. & FINN, R. (1966). Transplacental haemorrhage. *Brit. J. Haemat.*, **12**, 297.

ZIMMERMAN, A. & SCHMICKEL, R. (1971). Fluorescent bodies in maternal circulation. *Lancet*, **i**, 1305.

ZIPURSKY, A., POLLOCK, J., NEELANDS, P., CHOWN, B. & ISRAELS, L. G. (1963). The transplancental passage of fetal red blood-cells and the pathogenesis of Rh immunisation during pregnancy. *Lancet*, **ii**, 489.

DISCUSSION

LAWLER: Only those lymphocytes that respond to mitogens would be detected in your work, and although some of the foetal lymphocytes do respond, others do not. Could some of the lymphocytes home to somewhere in the maternal system rather than remaining in the peripheral circulation?

ADINOLFI: I agree that only lymphocytes which respond to PHA stimulation could be detected. This has probably reduced our chance of observing foetal cells in maternal blood. I do not know if a percentage of transfused lymphocytes is usually sequestered in the spleen, lung or liver.

BODMER: Given up to 3 ml of foetal blood in the maternal circulation, the proportion of foetal cells in the maternal blood must be less than one in 8000. Could this low number of cells be detected in your testing system?

ADINOLFI: With the techniques available at present, it would be very difficult to detect foetal lymphocytes in the maternal circulation if less than 3 ml of foetal blood are present at any time in the mother. This is one reason why I think that the claimed incidence of 0.3–2.0% of foetal lymphocytes in maternal blood does not reflect a real situation.

BODMER: HL-A could be used to select against the maternal antigens and concentrate cells carrying the paternal antigen.

COOMBS: Some years ago in our experiments on lymphocyte markers, we found distinct evidence of maternal lymphocytes in foetuses (Wolf *et al.*, 1971). Altogether there were five pregnancies, and in one of them a single foetus amongst ten showed maternal cells in the blood ten days after parturition. These declined considerably in number by five weeks.

GARDNER: After the transfusion *in utero* of females with rhesus foetuses, maternal lymphoid cells persist in the foetus. Tuffrey, Bishun & Barnes (1969) used the chromosome marker T6, and found in contrast to others (Billington *et al.*, 1969) that a reasonable percentage of foetuses had been colonized, often to a considerable extent. The one obvious difference between these experiments by the two groups was the use of multiparous females and not virgins, as recipients by Tuffrey *et al.*

EDWARDS: Did you study women who had had several children, or women having their firstborn? Increasing parity could influence lymphocyte migration.

ADINOLFI: Some were primigravidae; others had had several children.

EDWARDS: Can we form a general conclusion that a large number of maternal lymphocytes are found in some foetuses, although few or none are transferred to most foetuses?

ADINOLFI: In a normal pregnancy perhaps a few maternal lymphocytes are transferred across the placenta into the foetal circulation; some of these cells are destroyed by the immune response of the foetus and the remaining lymphocytes are too few in number to be detected after stimulation of foetal cells with PHA. Occasionally, the number c maternal lymphocytes is high and can be detected in foetal blood; th foetus and the newborn do not seem to be affected. In rare cases the pe centage of maternal lymphocytes is very high and they may be responsibl for the described 'runt' disease observed in some newborns, or for th death of the foetus.

EDWARDS: I still have doubts about possible selection of the childre and foetuses used in your study. The group of babies had reached full terr quite naturally and I would like to know if the same result would b obtained with foetuses early in pregnancy.

ADINOLFI: The time of the transfer of maternal lymphocytes may b of great importance. The only male foetuses which have been studied fo the presence of maternal 46/XX cells were those investigated for th presence of chromosome abnormalities. The number of cells carefull analysed for each foetus has always been limited and the data are no conclusive.

BILLINGTON: I agree with the comments and the data by Adinolfi, i contrast with other studies which claim to show positive results.

EDWARDS: Yes, I would not question Adinolfi's data, but my concerr is with the selective nature of the data in relation to the time of pregnancy We have to relate these results to the observations by Beer & Billinghar (1971, 1973) that lymphocyte transfer might occur early in implantation and their type of observation could be missed under the conditions o the present work. Beer and Billingham actually induced runting ir foetuses by sensitizing the mother at a critical time before mating. Wha is your opinion on the nature of this runting in foetuses?

BILLINGTON: These authors have now reported an extensive series They have used Ag-B incompatible rats, and worked with several other species. In syngeneic pregnancies, the adoptive transfer of allogeneic lymphoid cells obtained from a donor sensitized against the antigens of a syngeneic type of pregnancy led to a high proportion of runted offspring. Even more surprising, if the female was then left unmated for several months and then remated to a syngeneic male, a number of offspring in the second litter were also affected by the runting syndrome. In allogeneic pregnancies, the same effect was obtained by active immunization before mating. These authors believe that the placenta is porous to sensitized cells at a particular stage, and these cells then cause runting.

EDWARDS: One of the interesting points in their work was the necessity to immunize at a specific time before mating. Nevertheless, I think it is fair to say that they have not yet identified any maternal cells in the foetus. The cause of runting is thus open to debate.

ADINOLFI: Yes, in the results so far published by Beer & Billingham, there was no direct evidence of maternal lymphocytes in the newborn rats.

ALLISON: One cause of runting is virus infection and this could be one cause of runting in their system.

JOHNSON: Beer & Billingham's results could indicate a specific sensitivity to allogeneic interactions between lymphocytes and the target antigen in the region of the deciduo-trophoblast interface. Their observations on adoptive transfer can be explained if lymphocytes homing in on decidual tissue exerted some form of allogeneic effects which led to a greater porosity of tissue in the area. Likewise, if lymphocytes from the critically-timed grafting experiments homed in to the decidual tissue to give a local allogeneic effect, the same general increase in placental porosity might occur increasing the chances of runting. Thus, there would be in fact no selective transfer of sensitized cells but a generalized increase in placental porosity. These alternatives could be distinguished by using a second, non-lymphocytic cell marker for placental porosity.

HOWE: I have recently used an immunoincompetent leukemia cell that preferentially homes to the spleen. When pregnant animals received these tumour cells in the presence of sensitized allogeneic lymphocytes, the leukemic cells could be recovered from the spleens of the offspring. This is direct evidence for transplacental cell traffic (Howe, 1974).

EDWARDS: We are going to hear more, from Allen, later in this meeting about decidual cells. Lymphocytes may be present in the uterine lumen at implantation and there are reports that lymphocytes associate closely with decidual cells during early human implantation (see Discussion after Howe's paper p. 149). This is just the time when Beer & Billingham find the period of maximum sensitivity for the foetus.

Foetal damage could be missed in Adinolfi's study for another reason, since the foetuses may die before they can be typed, just as with many chromosomally-imbalanced foetuses. For this and other reasons, I think that the question about transfer at these early stages is still open to debate.

BILLINGTON: I disagree slightly because Beer & Billingham found that the period of sensitivity occurred before or after pregnancy.

EDWARDS: But in their work the lymphocytes would become maximally sensitized at implantation, and the time of effective immunization before or during mating was critical.

HOWE: In their earlier experiments, Beer & Billingham also got runting when non-sensitized cells were transferred into mothers.

ALLISON: I am interested in the mechanism of protein transfer and how it compares with that found in other situations. In capillaries of muscles a micropinocytic cell exists in which there is diffusion from one side to the other in very small vesicles. According to the Brambell hypothesis, material traverses the cell by diffusion without coming into contact with the lysosomal system. This mechanism would be rather specific for homologous IgG, but not with heterologous IgG because the vesicles would be modified in some way and so attract the attention of the lysosomal system. This would be a reasonable suggestion to account for the transfer of proteins across the placenta.

EDWARDS: Does the distribution of the two types of immunoglobulin in the vesicles help with this point?

ADINOLFI: The distribution was identical in the vesicles.

ALLISON: Some points about basement membranes might prove interesting. In different situations, these membranes display differing degrees of permeability. In the kidney, the membranes of the renal glomeruli cut off transfer at a particular molecular size which is well defined, but in the lung this is not the case. Some of the pictures showed by Adinolfi suggested that the basement membrane is permeable to peroxidase. It would be interesting to know if this is true in relation to proteins of other sizes.

ADINOLFI: The distribution of ferritin and peroxidase looks different in each vesicle.

AMOROSO: At what age or stage in pregnancy were these experiments by Enders conducted?

ADINOLFI: Two weeks before delivery.

AMOROSO: This is an important point because my impression was that your evidence controverted the Brambell theories. I believe the guinea-pig is amongst the group of animals in which transfer occurs via both the yolk sac and the chorio-allantoic placenta. Early in pregnancy, transfer takes place across the yolk sac placenta and later in pregnancy across the allantoic placenta. In the rabbit, of course, transfer is across the yolk sac placenta.

BILLINGTON: In the guinea-pig, the transplacental passage of immunoglobulin is greater (Wild, 1973).

AMOROSO: There were three categories, including those in which transfer was post-natal. Kingsley & Enders have the impression that protein was not transmitted via the allanto-chorionic placenta but did transfer via the yolk sac.

BODMER: Is it fair to say that in the ungulates there is no evidence of either cellular or protein transfer via the placenta from foetus to mother? If so, the fact that cattle make anti-HL-A-like antibodies due to maternal–foetal stimulation (Iha *et al.*, 1973) must be explained through stimulation by trophoblast material. Alternatively, it could be due to solubilized antigen.

EDWARDS: How many examples are there of foetal protein other than immunoglobulin initiating a response in the mother? I believe there is a report about a lipoprotein (Durwald, Leopold & Krämer 1965). Could you describe any other examples of this phenomenon?

ADINOLFI: An anti-Ag (β-lipoprotein variant) has been observed in maternal blood and the antibody was supposed to be produced following foetal stimulation.

ALLISON: There is also sensitization against Gm (Fudenberg & Fudenberg, 1964).

AMOROSO: This may be the appropriate place to remind this audience of the aphorism of Thomas Henry Huxley who once said that advances are more surely made from a correct exposition of errors than from the nebulous expression of truth.

REFERENCES

BEER, A. E. & BILLINGHAM, R. E. (1971). Immunobiology of mammalian reproduction. *Adv. Immunol.*, **14**, 2.

BEER, A. E. & BILLINGHAM, R. E. (1973). Maternally acquired runt disease. *Science*, **179**, 240.

BILLINGHAM, R. E. (1971). The transplantation biology of mammalian gestation. *Am. J. Obstet. Gynec.*, **110**, 469.

BILLINGTON, W. D., KIRBY, D. R. S., OWEN, J. J. T., RITTER, M. A., BURTONSHAW, M. D., EVANS, E. P., FORD, C. E., GOULD, I. K. & MCLAREN, A. (1969). Placental barrier to maternal cells. *Nature, Lond.*, **224**, 704.

DÜRWALD, W., LEOPOLD, D. & KRÄMER, K. H. (1965). The formation of precipitating antibodies after multiple pregnancies. *Vox Sang.*, **10**, 94.

FUDENBERG, H. H. & FUDENBERG, B. R. (1964). Antibody to hereditary human gamma-globulins (Gm) factors resulting from maternal–fetal incompatibility. *Science*, **145**, 170.

HOWE, C. W. S. (1974). Facilitated transplacental transfer of tumour cells. *Nature, Lond.*, (in press).

IHA, T. H., GERBRANDT, G., BODMER, W. F., MCGARY, D. & STONE, W. H. (1973). Cross-reactions of cattle lymphocytotoxic sera with HL-A and other human antigens. *Tissue Antigens*, **3** , 291.

TUFFREY, M., BISHUN, N. P. & BARNES, R. D. (1969). Porosity of the placenta to maternal cells in normally derived mice. *Nature, Lond.*, **224**, 701.

WILD, A. E. (1973). Transport of immunoglobulins and other proteins from mother to young. In *Lysosomes in Biology and Pathology* (ed. J. T. Dingle), *Frontiers in Biology*, **29**, p. 129. North-Holland: Amsterdam.

WOLF, B., JANEWAY, C. A., COOMBS, R. R. A., CATTY, D., GELL, P. G. H. & KELUS, A. S. (1971). Immunoglobulin determinants on the lymphocytes of normal rabbits. III. As4 and As6 determinants on individual lymphocytes and the concept of allelic exclusion. *Immunology*, **20**, 931.

IMMUNOLOGICAL ASPECTS OF THE EQUINE ENDOMETRIAL CUP REACTION

By W. R. ALLEN

The sudden and rapid differentiation of spindle-shaped connective tissue cells of the uterine stroma into large, polyhedral, epithelioid, decidual cells, is a striking feature of pregnancy in man, rodents and some other mammals. The process of decidualization commences at the time of implantation and the characteristic decidual cells rapidly accumulate to form a solid mass of cells, the decidua, around the developing blastocyst. Although a variety of non-specific traumatic stimuli will induce a full decidual reaction in the uterus of non-pregnant rodents (deciduoma), the exact nature of the stimulus from the blastocyst which induces decidualization during pregnancy is not known. Decidual cells contain large quantities of glycogen and the decidua probably functions as an important source of nutrition for the developing embryo. The decidua also functions to limit the invasion of trophoblast into maternal tissues and in general, the degree of decidualization of the uterus is directly proportional to the invasiveness of the placenta. The reaction is particularly marked in the haemochorial placenta of rodents, where extensive erosion of endometrial tissues by the invading trophoblast brings the foetal chorion in direct contact with maternal blood.

Recent comprehensive reviews of decidualization by De Feo (1967) and Finn (1971) obviate the necessity for any detailed description of the process here. Rather, it is the purpose of this paper to describe some of the interesting aspects of the lesser known and much less well understood decidual-like reaction which occurs in the placenta of the mare and other female equidae during early pregnancy.

OBSERVATIONS

Gonadotrophin and progesterone production during pregnancy in the mare

High concentrations of gonadotrophin (PMSG) are present in the blood of mares during the first half of gestation. PMSG first becomes detectable in the serum between 37 and 40 days after ovulation and levels rise rapidly to a peak of 50–250 i.u./ml between days 55 and 75 (Fig. 1). The concentration declines steadily thereafter and PMSG has usually disappeared again from the blood by day 150 (Cole & Hart, 1930; Day & Rowlands, 1940; Allen, 1969*b*). PMSG is a high molecular weight glycoprotein which exhibits both follicle stimulating (FSH) and luteinizing

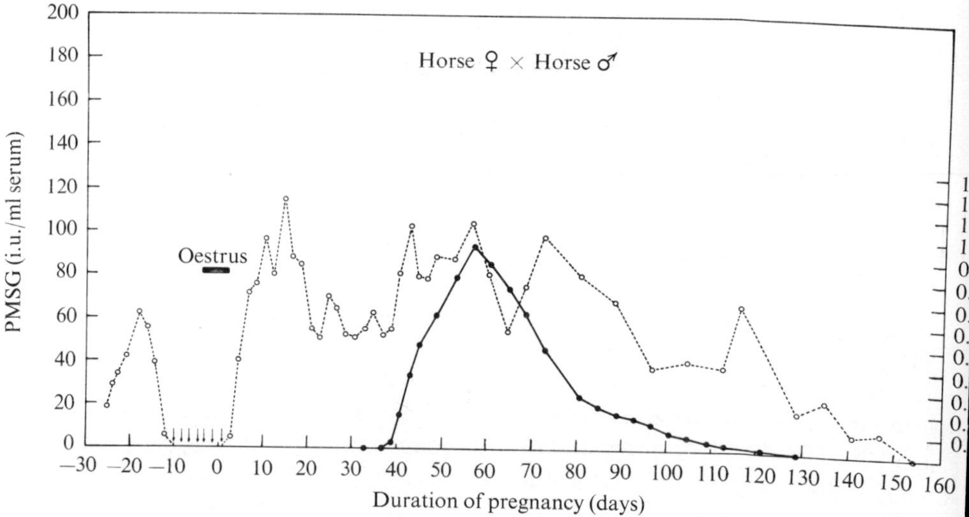

Fig. 1. The concentrations of PMSG (solid line) in the serum and progesterone (dashed line) in the peripheral plasma of a Welsh Pony mare during an oestrous cycle and the first half of pregnancy. (Allen, W. R., Smith, Janet, G. & Short, R. V. (1969)–unpublished.)

(LH) hormone activities. The molecule contains a high proportion of carbohydrate and readily dissociates into two subunits (Morris, 1966; Gospodarowicz & Papkoff, 1967; Gospodarowicz, 1972). PMSG is used widely in agriculture, veterinary medicine and biological research, to stimulate follicular development, ovulation and superovulation in laboratory and domestic animals.

The primary corpus luteum of pregnancy in the mare begins to regress from as early as the twentieth day after ovulation thereby causing a decline in peripheral plasma progesterone levels (Fig. 1). A number of secondary follicles begin to enlarge between 17 and 23 days after ovulation and one or more of these ovulate around day 40. Thereafter, a variable number of secondary corpora lutea develop in the ovaries, either as a result of normal ovulations or by luteinization of unruptured follicles (Amoroso, Hancock & Rowlands, 1948; van Niekerk, 1965, 1973; van Rensburg & van Niekerk, 1968; Allen, 1970*b*; Allen, W. E., 1971). The secondary corpora lutea also regress around day 150 of pregnancy and the mare's ovaries then remain inactive until term. Progesterone required for maintenance of gestation is secreted by the placenta from day 90 onwards (Short, 1959) and bilateral ovariectomy performed after the fifth month of gestation does not interrupt the pregnancy (Amoroso & Finn, 1962).

Endometrial cups – the source of PMSG

Soon after the discovery of PMSG by Cole & Hart (1930), Catchpole & Lyons (1934) made an exhaustive study of PMSG concentrations in various foetal and maternal tissues and fluids throughout pregnancy. They concluded from their findings that PMSG was probably produced by the foetal chorion and selectively stored in the maternal endometrium. A few years later, however, Cole & Goss (1943) and Clegg, Boda & Cole (1954) produced convincing evidence that PMSG is secreted by the so-called endometrial cups. They measured very high concentrations of PMSG in endometrial cup tissue and even higher concentrations in the exocrine secretion liberated by the cups.

Endometrial cups were first described by Schauder (1912) and consist of a series of small, ulcer-like, endometrial outgrowths which develop

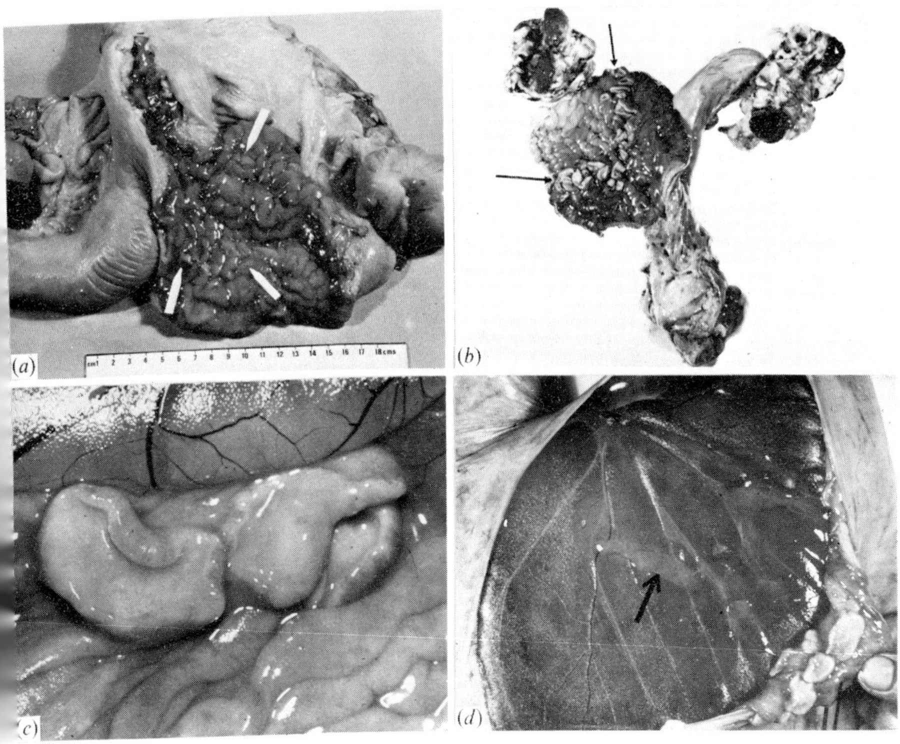

Fig. 2. (*a*) Mare's uterus at 39 days of gestation with the pregnant horn opened and the conceptus removed. The young endometrial cups (arrowed) are visible as small, discrete, pale areas in the endometrium.
(*b*) Mare's uterus at 63 days of gestation with the pregnant horn opened and the conceptus removed. The mature endometrial cups (arrowed) are seen as cup-shaped, ulcer-like endometrial outgrowths arranged in a circle around the lumen of the uterus. A secondary corpus luteum is present in each ovary.
(*c*) Close-up view of the pregnant horn of a mare's uterus at 73 days of gestation. The foetal membranes have been pulled back to expose two well-developed endometrial cups.
(*d*) Mare's uterus at 85 days of gestation with the pregnant horn opened to show endometrial cups in the early stages of degeneration. The characteristic, honey-coloured, endometrial cup exocrine secretion (arrowed) is adhered to the overlying foetal membranes.

in the uteri of pregnant equidae between 40 and 150 days of pregnancy. They are never found in non-pregnant animals and develop only in the pregnant horn of the uterus in a circle around the enlarging foetal sac (Allen, 1970*b*). The cups are first seen macroscopically between 38 and 40 days of gestation as a series of small, discrete, pale areas in the endometrium (Fig. 2*a*). They grow rapidly during the next three weeks and reach a maximum size (2–10 cm in length, 1–3 cm in breadth) between days 60 and 70 (Fig. 2*b*). During development, the cups become increasingly concave or saucer-shaped due to an overgrowth of the peripheral regions and consequent regression in the central area of the structure (Fig. 2*c*). After about day 70 of gestation, the cups begin to undergo a steady degeneration and necrosis, becoming increasingly pale and cheesy in appearance. At the same time there commences an outpouring of a sticky honey-coloured exocrine secretion which accumulates on the surface of the overlying chorio-allantois (Fig. 2*d*). This material is extremely rich in PMSG and may contain up to 1 000 000 i.u. PMSG/gram fresh tissue (Clegg *et al.*, 1954; Rowlands, 1963).

The necrotic endometrial cup tissue and admixed, inspissated exocrine secretion is eventually dehisced from the surface of the endometrium between 130 and 180 days of gestation, leaving a pale scar in the endometrium which has normally disappeared by term. The sloughed material sits upon the surface of the overlying foetal membranes where, by virtue of gravity and other physical forces, it sinks towards the centre of the conceptus and so gives rise to the pedunculated chorio-allantoic pouches described by Clegg *et al.* (1954). A curious finding is that although very high concentrations of PMSG are present in endometrial cup secretion and are therefore in direct contact with the chorio-allantois, PMSG cannot be detected in any foetal fluids or tissues throughout pregnancy (Catchpole & Lyons, 1934; Allen, 1970*b*).

Histologically, the fully developed endometrial cup is composed of a discrete and densely packed mass of very large, epithelioid decidual-like cells (Fig. 3*a*). These cells are irregular in outline and possess pale-staining, frothy cytoplasm with a prominent densely staining area around the nucleus (Fig. 3*b*). The cytoplasm is filled with short profiles of endoplasmic reticulum which resemble granules with uniform density. Lipid droplets are present and a few mitochondria are scattered throughout the cells. The Golgi apparatus is significant and many smooth-surfaced, Golgi-associated vesicles are present in the cytoplasm. Many of the cup cells are binucleated. The nuclei are ovoid, contain little or no heterochromatin and possess one or more large, dense nucleoli (Hamilton, Allen & Moor, 1973).

The endometrial epithelium overlying the endometrial cups becomes obliterated during initial development of the cups between 37 and 40 days of gestation but it rapidly becomes re-established over the next two or three

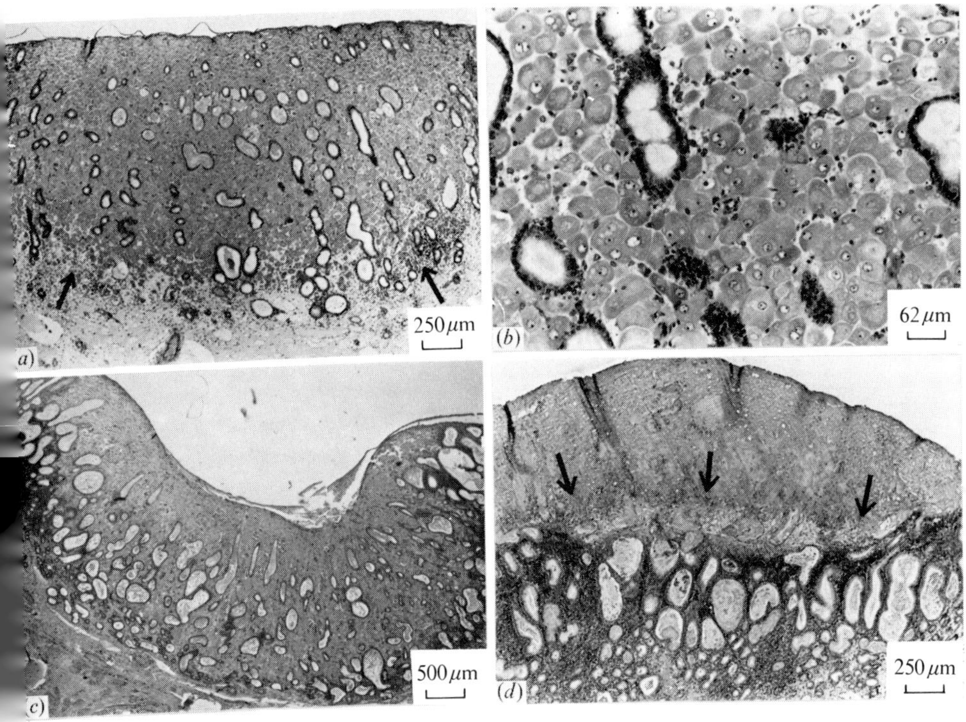

Fig. 3 (*a*) Low-power section of a horse endometrial cup at 53 days of gestation. The large endometrial cup cells are densely packed together and the outlets of many endometrial glands in the cup have been obliterated so that the basal or fundic portions of the glands are becoming distended with accumulated exocrine secretions. Leucocytes (arrowed) have already begun to accumulate in the endometrial stroma surrounding the cup.
(*b*) Higher magnification near the base of (*a*) showing the large, epithelioid decidual-like endometrial cup cells, many of which are binucleated.
(*c*) Low power section of a horse endometrial cup at 65 days of gestation. The cup has now become concave or saucer-shaped due to overgrowth of the peripheral regions and commencing degeneration of the cup cells in the uppermost central area. The fundic portions of the endometrial glands are grossly distended with secretion and lymphocytes, plasma cells and eosinophils are densely accumulated in the endometrial stroma at the periphery of the cup.
(*d*) Low power section of a completely degenerate horse endometrial cup at 132 days of gestation. The necrotic cup tissue and admixed exocrine secretion is in the process of being dehisced from the surface of the endometrium (arrowed). The uterine stroma beneath the cup is densely packed with invading leucocytes.

days. It begins to degenerate again around day 70 when regression and necrosis of the cups commences. The apical portions of the endometrial glands within the cup area are also obliterated during growth and development of the cups. The fundic portions at the base of the cup tissue remain intact however and become increasingly distended due to accumulation of their secretory products (Fig. 3*c*).

A striking feature of the endometrial cup reaction and certainly the most pertinent from the point of view of the present paper is the accumulation of leucocytes in the uterine stroma at the periphery of the endometrial cup tissue. This aggregation of cells commences immediately after the cups begin to develop and it increases steadily during the next 60–80 days, giving rise to a band of cells which appear to wall off the cup tissue from the surrounding normal endometrium. At first, the invading cells consist almost exclusively of small lymphocytes but later on, increasing numbers of plasma cells and eosinophils become evident.

The cups begin to regress from about day 70 and the large cup cells at the luminal surface and in the centre of the cup are first to undergo necrosis. As this tissue degenerates and sloughs off the surface, outlets for the swollen endometrial glands at the base of the cup are re-established and so exocrine release of the characteristic endometrial cup secretion commences. As degeneration proceeds, the accumulated leucocytes begin to invade the cup tissue and phagocytose the dying cells and eventually, the whole necrotic structure is sloughed from the endometrium between days 130 and 150 of gestation (Fig. 3*d*).

The origin of the endometrial cups

Placentation in the mare is diffuse, non-invasive and epithelio-chorial in nature (Amoroso, 1952). Development of branching chorio-allantoic villi which interdigitate with corresponding endometrial crypts or sulci does not commence until after the sixth week of gestation (Amoroso, 1952; Samuel, Allen & Steven, unpublished). Until this time, the foetal sac is only tenuously attached to the endometrium by a small area of bilaminar omphalopleur remaining at the abembryonic pole and is largely held in place by increased myometrial tone (Ewart, 1897; Amoroso, 1952; Allen, 1970*b*).

Endometrial cups develop entirely within the maternal endometrium and there exists no physical connection between the cups and the overlying foetal membranes; indeed, those areas of the chorio-allantois which are directly opposed to the cups are devoid of chorio-allantoic villi and are seen as a series of white scars in the foetal placenta at term. For this reason, it was generally considered for many years that the endometrial cups were entirely maternal in origin (Clegg *et al.*, 1954; Amoroso, 1955; Rowlands, 1963; Allen, 1970*b*). However, recent experiments have demonstrated that the large decidual-like endometrial cup cells are of foetal origin, being derived from a specialized area of the trophoblast known as the chorionic girdle (Allen & Moor, 1972; Allen, Hamilton & Moor, 1973).

Ewart (1897, 1915), gave classical accounts of the early development of horse foetal membranes, the basic features of which are summarized

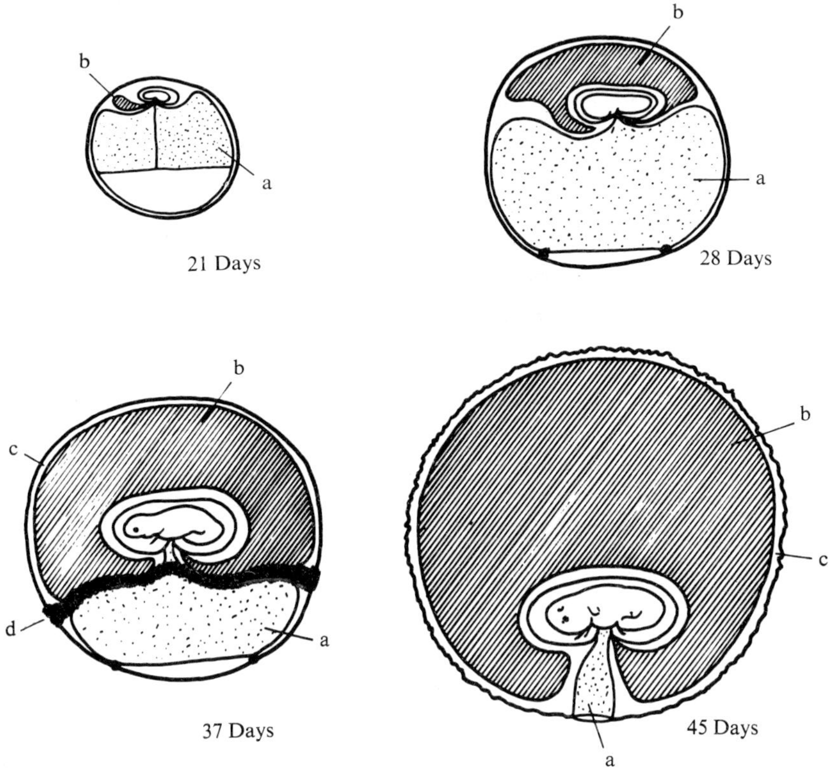

Fig. 4. Diagrammatic representation of the development of horse foetal membranes between 21 and 45 days of gestation. a, yolk sac; b, allantois; c, chorio-allantois; d, chorionic girdle.

diagrammatically in Fig. 4. At 21 days of pregnancy, the foetal sac is spherical in shape and measures 6 cm in diameter. The embryo is already completely surrounded by the amnion and the whole is enclosed in the chorio-vitelline membrane. Vascular mesoderm is developing between chorion and yolk sac and the sinus terminalis is situated in the equatorial region of the foetal sac. The allantois is just beginning to develop as an outpouching of the hind gut of the embryo. By 25 days of gestation, vascular mesoderm has further advanced so that only a small area of bilaminar omphalopleur remains at the abembryonic pole of the vesicle; tenuous attachment between trophoblast and endometrium exists in this region. The allantois is growing rapidly and by day 28 it completely surrounds the embryo and amnion. As the allantois enlarges, the yolk sac regresses and the embryo, attached by the yolk stalk, is withdrawn towards the abembryonic pole of the conceptus. The chorion is composed of a single layer of tall columnar trophoblast cells and fusion of the allantois with the chorion gives rise to the chorio-allantois. At the point where the develop-

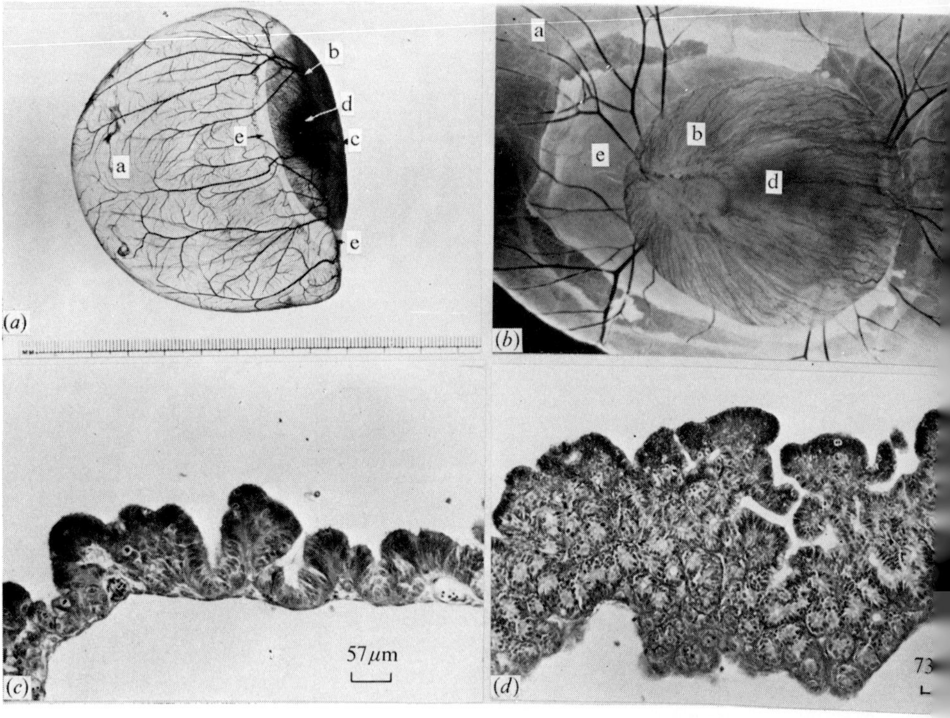

Fig. 5. (*a*) Intact horse conceptus at 36 days of pregnancy showing the prominent chorionic girdle (e) at the junction of the enlarging allantois (a) and regressing yolk sac (b). The embryo (d) has been drawn towards the abembryonic pole of the foetal sac where a small area of bilaminar omphalopleur (c) still persists.
(*b*) Intact horse conceptus at 38 days of gestation viewed from the abembryonic pole and showing the chorionic girdle (e) in the process of sloughing off the surface; a, normal chorio-allantois; b, regressing yolk sac; d, shadow of the embryo.
(*c*) Section of horse chorionic girdle at 28 days of gestation. The single layer of elongated columnar trophoblast cells is thrown into a series of shallow ridges.
(*d*) Section of horse chorionic girdle at 35 days of gestation. The ridges of trophoblast cells have now developed into elongated, branching, villous folds which project from the surface of the conceptus. Many of the villi have been sectioned transversely. The apical regions of the cells covering the luminal surfaces of the villi are filled with darkly staining material.

ing allantoic and regressing yolk sac membranes are adjacent to one another, a discrete area of the chorion becomes thrown into a series of simple, shallow corrugations (Fig. 5*c*). As development continues during the next ten days, folding of the chorion at the front of the enlarging allantois increases markedly, giving rise by day 35, to a series of closely-packed, tall, villous-like projections of trophoblast separated by deep crypts (Fig. 5*d*). This area, the chorionic girdle, is clearly visible as a discrete, pale, annulate band of tissue around the entire circumference of the conceptus (Fig. 5*a*).

The allantois is not carried into the villous projections of the chorionic girdle and each villous is covered by a layer of highly modified trophoblast cells. Intervillous spaces are filled with an amorphous, dark-staining material which extends over the free surface of each villous and into the lumina of adjacent endometrial glands. This material gives the impression of an adhesive which binds foetal and maternal surfaces together. The girdle cells have elongated supra-nuclear regions and are

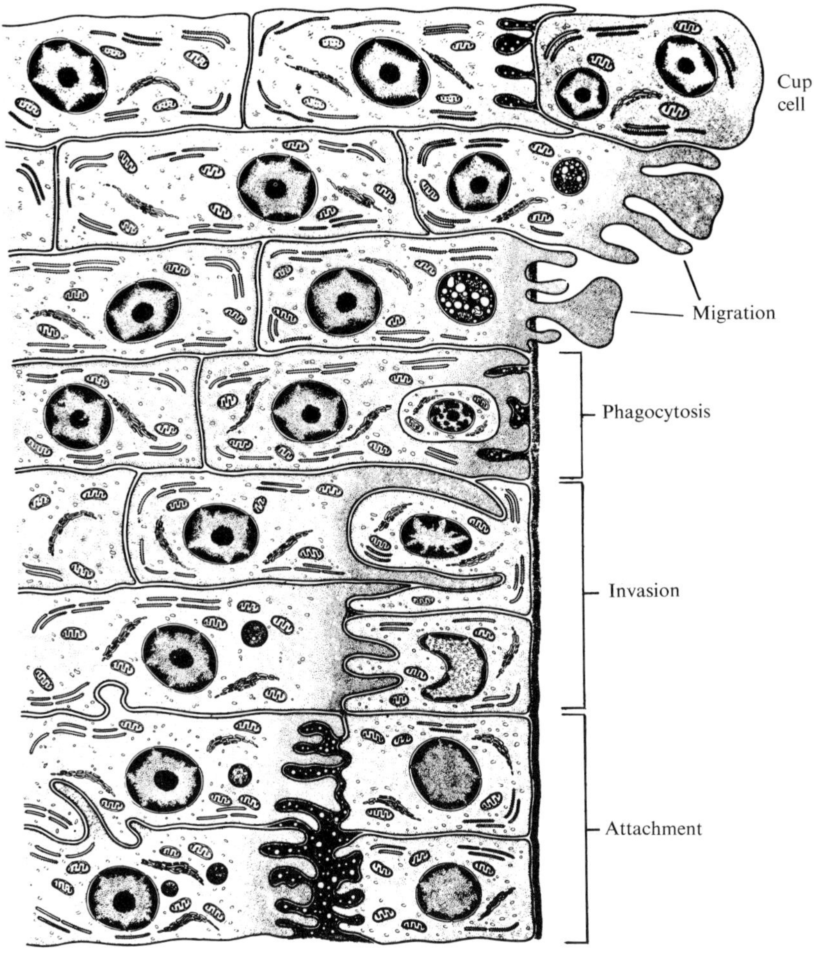

Fig. 6. Diagrammatic representation of the histogenesis of an endometrial cup. The chorionic girdle becomes firmly attached to the endometrium and girdle cells invade and phagocytose the endometrial epithelium. The girdle cells then migrate through the basal lamina into the endometrial stroma where they form endometrial cup cells. (From Allen, Hamilton & Moor, 1973.)

often binucleated. Those on the uppermost surface of the villi exhibit large pseudopodia-like extensions which project towards the surface of the endometrium through the extracellular matrix (Allen *et al.*, 1973).

Between 36 and 38 days of gestation, the whole of the now highly developed chorionic girdle separates from the foetal membranes and becomes firmly attached to the underlying endometrium (Fig. 5*b*). The specialized girdle cells then rapidly invade, destroy and phagocytose the uterine epithelium, pass through the basal lamina and migrate out into the endometrial stroma (Fig. 6). The process of invasion is very rapid and is accomplished by means of the large pseudopodic processes which project from the apical surface of the cells. These force their way both between and through the epithelial cells and also through the basal lamina (Fig. 6). Once the invading girdle cells reach the uterine stroma, they quickly lose their pseudopodia and round up to form the characteristic large decidual-like endometrial cup cells (Allen *et al.*, 1973).

Thus, the endometrial cups are now known to be composed primarily of a discrete colony of highly specialized, gonadotrophin-secreting foetal trophoblast cells which retain no physical connection with the remainder of the conceptus.

The influence of foetal genotype upon PMSG and progestagen levels in vitro

Yet another noteable and experimentally useful feature of equidae is their ability to conceive and carry to term, a great variety of interspecific hybrid offspring (King, Short, Mutton & Hamerton, 1966; Gray, 1972). The mule, resulting from the crossing of a female horse (*Equus caballus*) to a male donkey (*Equus assinus*) and the hinny, resulting from the mating of a male horse to a female donkey, are the most common equine hybrids. Mules and hinnys are similar in appearance, have a diploid chromosome complement of 63 which is half the sum of the diploid number of the parents, and are infertile (Chandley *et al.*, 1974).

Using a relatively insensitive frog spermiation bioassay, Bielanski, Ewy & Pigoniowa (1956) were unable to detect any PMSG in the blood of mares carrying mule conceptuses. However, a few years later, Clegg, Cole, Howard & Pigon (1962) showed that PMSG is present in the blood of mares carring mules, but only in concentrations of one tenth of the normal levels found in mares carrying normal horse conceptuses. Allen (1969*c*) confirmed these observations with an immunoassay, finding maximum PMSG levels of only 5–20 i.u./ml between 45 and 56 days of gestation in mares carrying mule pregnancies and a disappearance of gonadotrophin from the serum as early as days 80–90. Similarly in donkeys carrying hinny conceptuses, Allen (1969*c*) measured maximum serum PMSG levels (200–300 i.u./ml) which were six to eight times those found in donkeys carrying normal donkey foetuses. Furthermore, PMSG persisted in the blood as late as day 140 of gestation (Fig. 7).

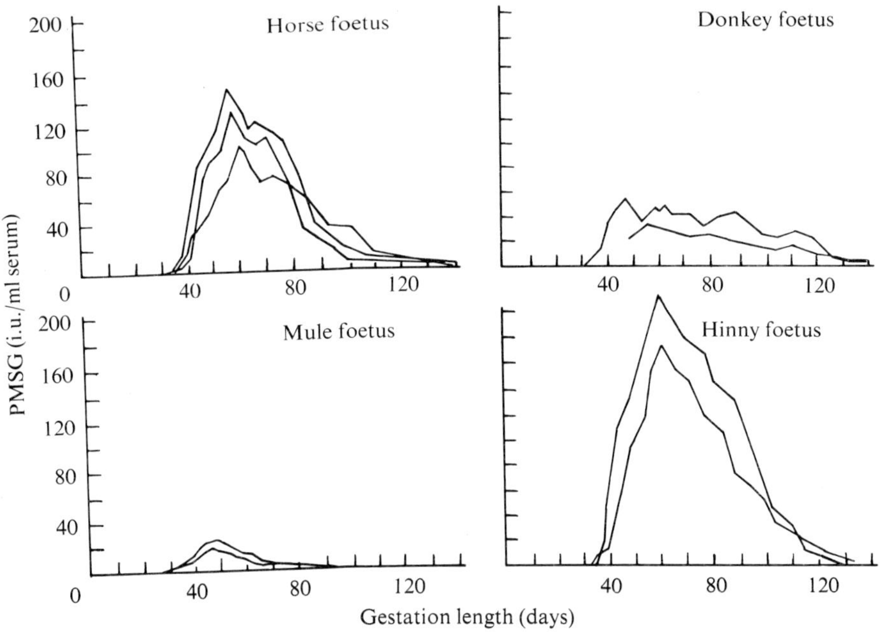

Fig. 7. Comparison of PMSG concentrations in the serum of 3 mares carrying normal horse conceptuses, two donkeys carrying normal donkey conceptuses, two mares carrying mule conceptuses and two donkeys carrying hinny conceptuses. (From Allen W. R., 1969*c*.)

Figure 1 shows the progesterone concentrations measured by the method of Schomberg, Coudert & Short (1967) in the peripheral plasma of a pony mare during an oestrous cycle and the first half of pregnancy. As mentioned previously, the primary corpus luteum of pregnancy in the mare begins to regress from the twentieth day of pregnancy, leading to a decline in the concentration of progesterone in the peripheral plasma. Levels rise again round day 40 with the first of the secondary ovulations and then fluctuate for the next 60–80 days. Thereafter, a second decline occurs until around day 150 when the secondary corpora lutea also regress. Although individual mares may vary greatly in peripheral plasma progestagen levels between 40 and 120 days of gestation (4–68 ng progestagen/ml), the great majority of mares show fluctuations in the range 5–15 ng/ml during this period (Allen, Stewart & Crowhurst, unpublished observations).

Although foetal genotype has a marked influence upon PMSG levels in the mare (*Equus caballus*), it does not appear to influence progestagen production. Figure 8 shows the levels of PMSG and progestagen measured by haemagglutination-inhibition and competitive protein binding assays respectively (see Allen, 1969*a*; Allen & Rossdale, 1973) in two pony mares, one carrying a normal horse conceptus and the other a mule.

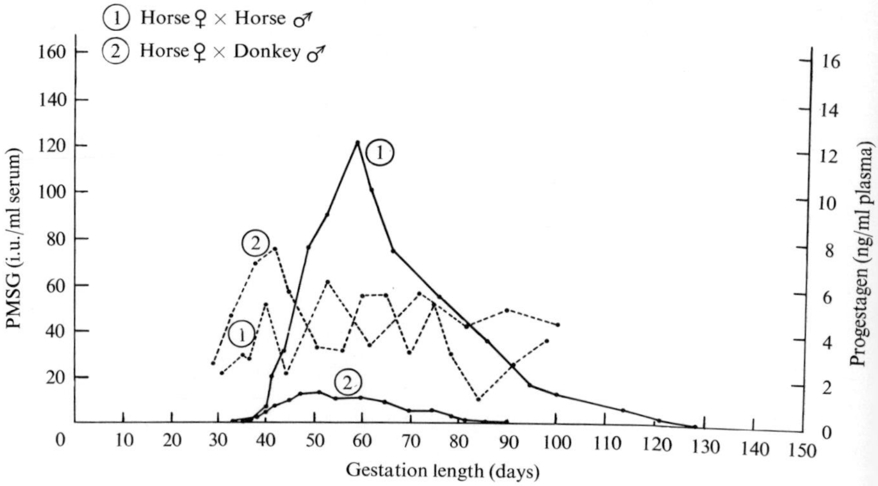

Fig. 8. PMSG concentrations (solid line) in the serum and progestagen concentrations (dashed line) in the peripheral plasma of two mares, one carrying a normal horse conceptus (No. 1) and the other a mule conceptus (No. 2).

While PMSG levels are severely depressed in the latter, progestagen levels are essentially similar in both animals.

Foetal genotype does profoundly influence progestagen production in the donkey (*Equus assinus*) however, and Fig. 9 shows the concentrations of PMSG and progestagen measured in the peripheral blood of a donkey carrying a hinny conceptus. Levels of both hormones are very greatly increased and the rise and fall of each closely follows the other. Progestagen concentrations reach as high as 800 ng/ml on day 75 of pregnancy, some 80–100 times the maximum levels found in donkeys carrying normal donkey conceptuses. Peripheral plasma progestagen levels of this magnitude have not been reported in any other species during pregnancy, including women (see Heap, Perry & Challis, 1973). High concentrations of progesterone (up to 280 ng/ml) are found in the plasma of guinea pigs between 15 and 50 days of pregnancy (Heap & Deanesly, 1966) but these elevated levels result from a marked increase in progesterone-binding protein in plasma and do not reflect an increase in steroid production (Challis, Heap & Illingworth, 1971).

Heap & Allen (unpublished observations) could find no evidence of an increase in progesterone binding protein in five plasma samples taken from a donkey carrying a hinny conceptus between 30 and 80 days of pregnancy and examination of one 60-day hinny conceptus at laparotomy indicated that the maternal ovaries are the principal source of the high levels of progestagen found in the peripheral blood of donkeys carrying hinny pregnancies. The ovaries of this particular animal contained five well-developed corpora lutea which had resulted from normal ovulations and

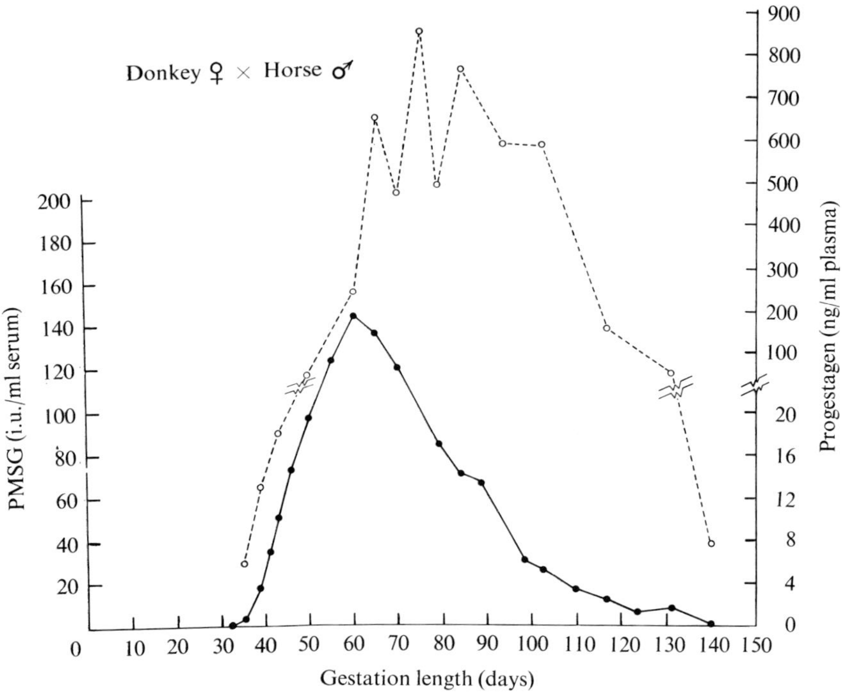

Fig. 9 PMSG concentrations (solid line) in the serum and progestagen concentrations (dashed line) in the peripheral plasma of a female donkey carrying a hinny conceptus.

13 heavily luteinized but unruptured follicles. Progestagen concentrations in ovarian vein plasma from the pregnant (1938 ng/ml) and non-pregnant (3227 ng/ml) sides of the animal were considerably higher than those in uterine vein plasma from the pregnant (129 ng/ml) and non-pregnant (113 ng/ml) horns of the uterus. The uterine vein levels were in turn insignificantly different from the mean of the progestagen concentrations measured in four samples of jugular vein plasma (118 ng/ml) which were recovered simultaneously with the ovarian and uterine vein samples.

Effect of foetal genotype upon PMSG production in vitro

Allen & Moor (1972) demonstrated that chorionic girdle tissue recovered from the horse conceptuses between 33 and 36 days of gestation forms stable monolayers of characteristic, large, epithelioid, binucleate cells in tissue culture. These cells secrete large quantities of PMSG into the culture medium and are morphologically indistinguishable from endometrial cup cells in the mare. Chorionic girdle cells cultivated *in vitro* can remain viable and secrete gonadotrophin for over 180 days (Fig. 10) whereas endometrial cup cells survive for only 80–90 days in the mare

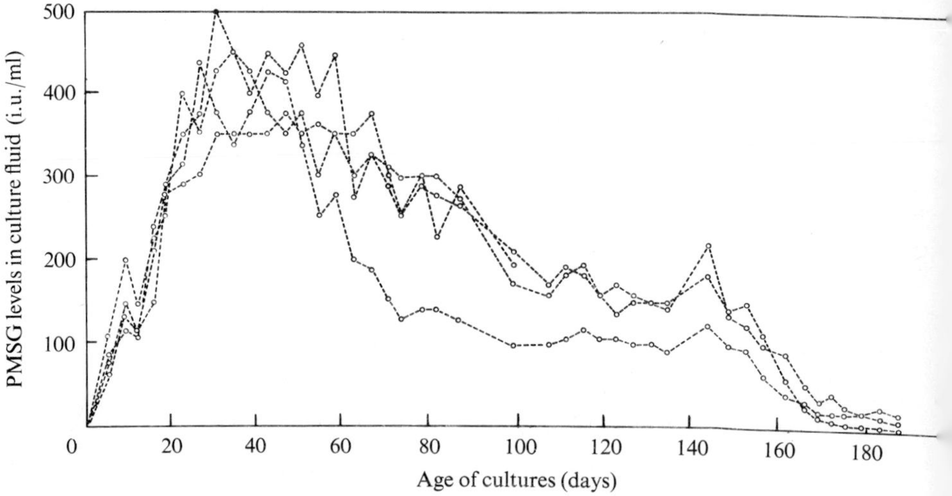

Fig. 10. PMSG concentrations measured in the culture medium recovered at 3–4 day intervals from 4 individual cultures of horse chorionic girdle tissue. (From Allen, W. R. & Moor, R. M., 1972.)

carrying a horse foetus and for less than 20 days in the mare carrying a mule foetus.

More recently, Moor, Allen, Rowson & Hamilton (unpublished) have shown that the pronounced influence of foetal genotype upon PMSG levels *in vivo* is closely mirrored *in vitro*. Chorionic girdle tissue recovered from mule conceptuses secretes only one-tenth of the PMSG in culture that the same tissue taken from horse conceptuses produces, even though the degree of cell growth and the morphological characteristics of the cultured cells are similar in both cases. Conversely, chorionic girdle tissue from hinny conceptuses produces very large quantities of PMSG in culture whereas donkey girdle tissue secretes relatively small amounts. Thus, PMSG production by chorionic girdle tissue is high when the horse (*Equus caballus*) provides the paternal component of the conceptus as in the horse and hinny, but is low when the donkey (*Equus assinus*) is the sire as in the donkey and mule. Although the maternal environment in which the chorionic girdle develops may in some way influence the subsequent capacity of the cells to secrete gonadotrophin, the findings of Moor *et al.* nevertheless suggest that the unusual situation exists in equidae where the paternal genome exerts an overriding control upon the development and secretory capacity of a specialized portion of the conceptus.

Effect of repeated hybrid pregnancies upon PMSG levels in vivo

Fig. 11 shows the PMSG concentrations measured in the serum of two pony mares during two consecutive mule pregnancies. During the first pregnancy in both mares, the entire conceptus was surgically removed

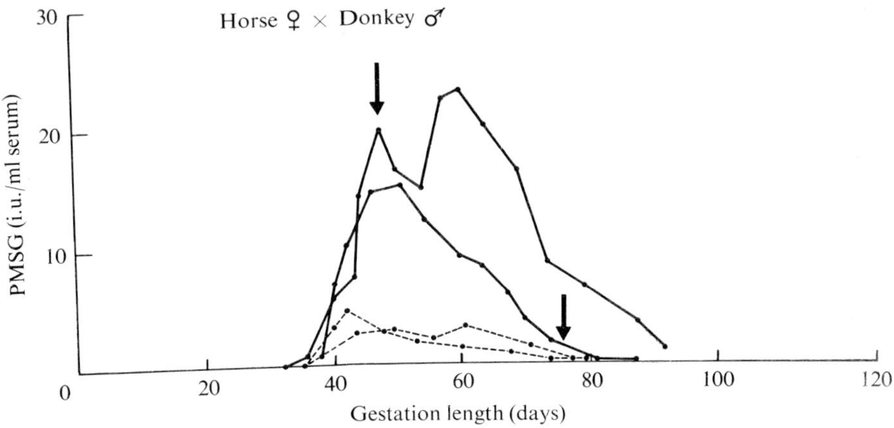

Fig. 11. PMSG concentrations measured in the serum of two Welsh Pony mares during two consecutive mule pregnancies. First pregnancy, solid line; second pregnancy, dashed line. The conceptus was removed (arrows) surgically from the uterus of both mares during the first pregnancy at 45 and 75 days respectively. The second pregnancy was allowed to go to term.

from the uterus at 45 and 75 days of gestation respectively. The mares returned to oestrus 43 and 14 days respectively after the operation and were covered by the same jack donkey that had sired the original foetus. Both mares conceived to this mating and subsequently produced live mule foals after gestation periods of 332 and 325 days respectively.

PMSG levels were normal for mares carrying mule conceptuses in both animals during the first pregnancy, despite the loss of the conceptuses after 45 days of gestation (see p. 226); maximum serum levels of 22 and 16 i.u./ml occurred on days 60 and 52 after ovulation and PMSG had become undetectable by days 93 and 81 respectively. PMSG production was markedly depressed in both mares during the second uninterrupted mule pregnancy, however. Maximum levels of only 6 and 4.5 i.u./ml occurred on days 42 and 45 of gestation respectively and PMSG had disappeared from the blood of both mares by day 78 of gestation. These results suggested that the endometrial cups were destroyed even more rapidly than normal during the second concurrent mule pregnancy as a result of prior sensitization of the mares against donkey transplantation antigens during the first gestation.

The effect of foetal genotype upon the histogenesis of the endometrial cups

Fig. 12*a–f* shows the marked influence of foetal genotype upon the development and regression of endometrial cups in mares and donkeys. Although large numbers of leucocytes begin to accumulate in the vicinity of endometrial cups in normal intraspecific horse and donkey pregnancies from as early as 38 days of gestation, the cells remain aggregated in the endometrial stroma around the cup and only begin to actively invade and destroy the cup tissue after about day 70 (Fig. 2*a–d*). However, in mares carrying mule conceptuses, the whole process of leucocyte invasion is very greatly enhanced in magnitude and speed and as early as day 45, large numbers of lymphocytes and plasma cells have been attracted to the cups.

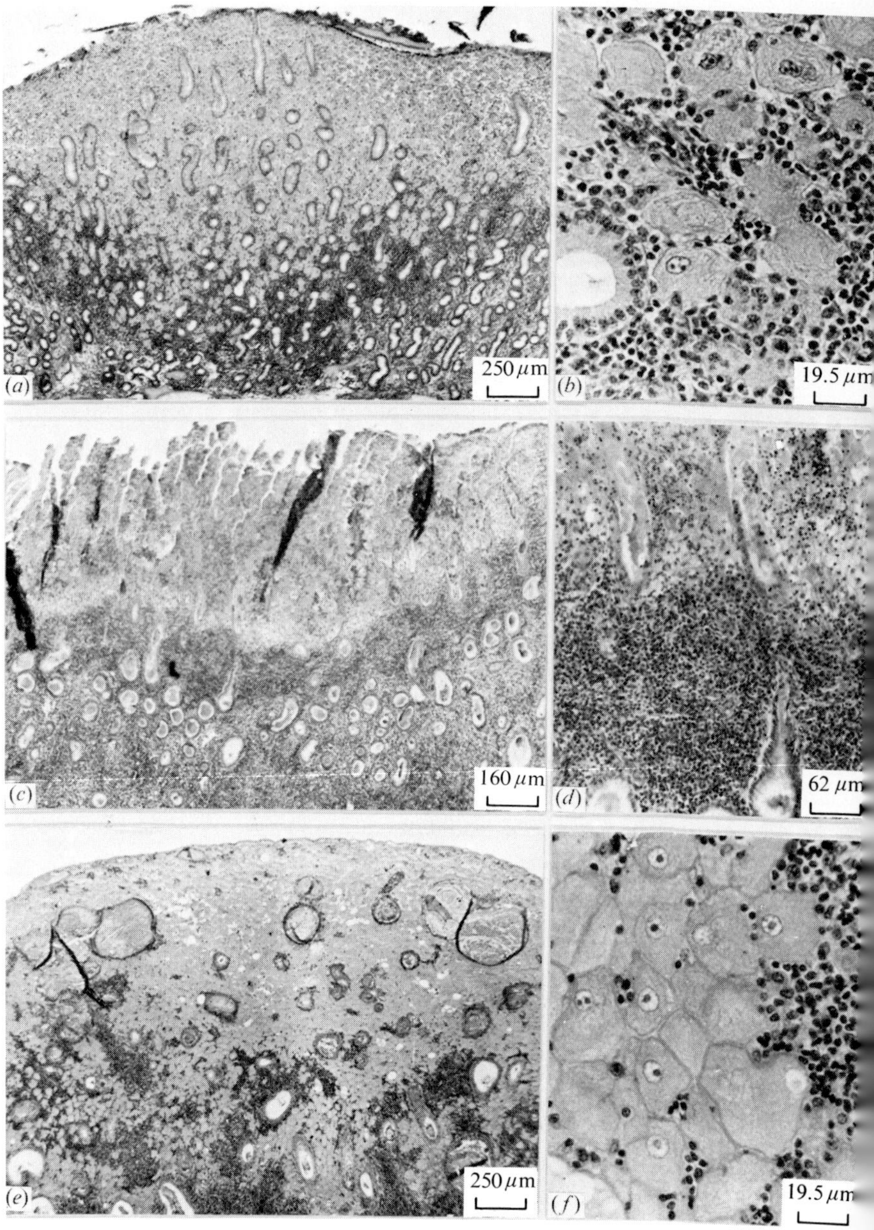

Instead of remaining clustered in the endometrial stroma, they are already invading deep into the cup tissue and destroying and phagocytosing the endometrial cup cells (Fig. 12*a*, *b*). By day 60, the entire endometrial cup has undergone premature necrosis and is in the process of being sloughed from the endometrium. The necrotic cup tissue and a considerable area of the adjacent endometrial stroma is densely packed with invading cells, including large numbers of plasma cells and eosinophils (Fig. 12*c*, *d*).

The leucocyte response to the endometrial cups is similarly greatly enhanced in donkeys carrying hinny conceptuses although is perhaps not quite so intense in terms of cell numbers, as the reaction in mares carrying mule conceptuses. At 60 days of pregnancy, hinny endometrial cup cells are densely packed together and exhibit thick, densely staining cell membranes. Very large numbers of lymphocytes, plasma cells and eosinophils are accumulated in the stromal tissue around the cup and small 'pockets' of these cells are present deep within the cup tissue (Fig. 12*e*). However, in marked contrast to the situation in mule endometrial cups at the same stage of pregnancy, there is no generalized invasion of leucocytes throughout the cup tissue and the majority of the large endometrial cup cells remain intact and morphologically normal, despite being surrounded by large numbers of the invading cells (Fig. 12*f*).

The biological function of PMSG in the mare

The coincidental appearance and disappearance of PMSG from the mare's blood with the development and regression of secondary corpora lutea in her ovaries, has led to the widely-held assumption that PMSG stimulates and maintains luteal function during early pregnancy in the mare. However, Rowlands (1949), Bain (1967), van Rensburg & van

---

Fig. 12. (*a*) Low power section of an endometrial cup from a mare carrying a mule conceptus at 45 days of gestation. Very large numbers of leucocytes have already accumulated in the uterine stroma and these cells are actively invading and destroying the cup tissue.
(*b*) Higher magnification near the base of (*a*) showing the mule endometrial cup cells being destroyed and phagocytosed by the invading leucocytes.
(*c*) Low power section of an endometrial cup from a mare carrying a mule conceptus at 60 days of gestation. The cup tissue has already become completely necrotic and the uterine stroma beneath the cup is packed with lymphocytes, plasma cells and eosinophils.
(*d*) Higher magnification near the base of (*c*).
(*e*) Low power section of an endometrial cup from a donkey carrying a hinny conceptus at 60 days of gestation. Very large numbers of leucocytes have accumulated in the uterine stroma and although isolated 'pockets' of leucocytes have invaded deep into the cup, the bulk of the cup tissue remains intact and apparently normal. Endometrial glands near the surface of the cup have undergone cystic degeneration due to gross overdistension with accumulated secretion.
(*f*) Higher magnification near the base of (*e*). The large endometrial cup cells exhibit a thick and densely-staining cell membrane and appear relatively unharmed by the invading leucocytes which surround them.

Niekerk (1968) and others have observed that renewed follicular growth commences in the mare's ovaries 15–20 days before the development of endometrial cups and this finding, together with the following recently acquired experimental evidence, casts some doubt upon a luteotrophic role for PMSG in the mare.

If the entire conceptus is removed surgically from the uterus of mares after the thirty-eighth day of pregnancy, chorionic girdle cells have already invaded the endometrium and the endometrial cups develop and regress normally as though the mare was still pregnant. Accordingly, high concentrations of PMSG remain in the blood of such mares for many weeks after loss of the pregnancy thereby giving rise to false-positive blood pregnancy tests (Allen, 1970*a*). As judged by peripheral plasma progestagen determinations, luteal function is maintained in some mares while PMSG remains in the blood following removal of the conceptus (Fig. 13), but in others, the corpora lutea regress within 20 days of removal of the pregnancy, despite the continued presence of high concentrations of gonadotrophin in the blood (Fig. 14). The reason for this difference in luteal activity between individual mares following loss of pregnancy is unknown but whether the corpora lutea remain functional or not, mares will

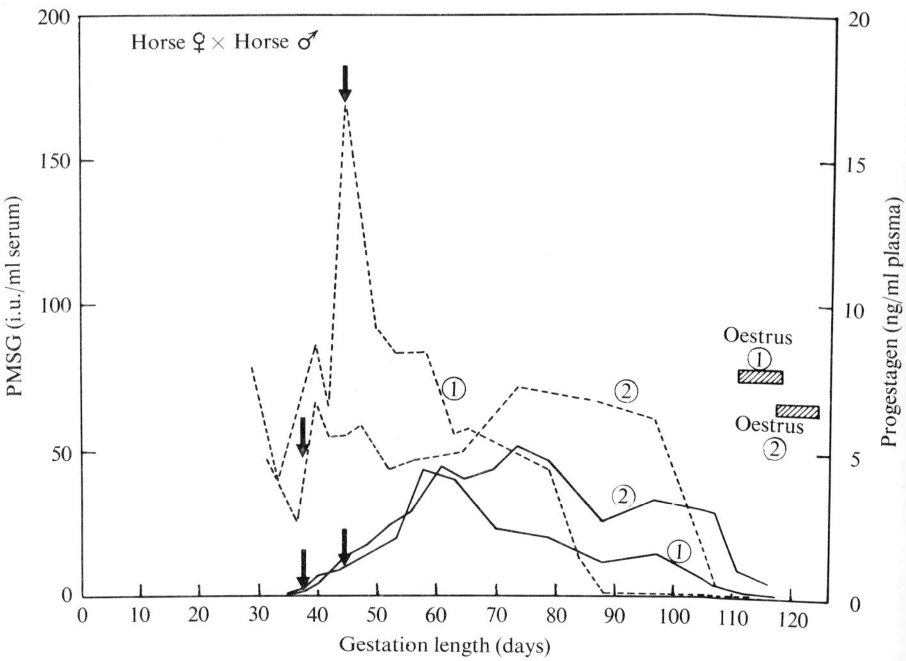

Fig. 13. PMSG concentrations in the serum (solid line) and progestagen concentrations in the peripheral plasma (dashed line) of two Welsh Pony mares following surgical removal of the conceptus (arrows) from the uterus at 38 (No. 2) and 45 (No. 1) days of gestation respectively.

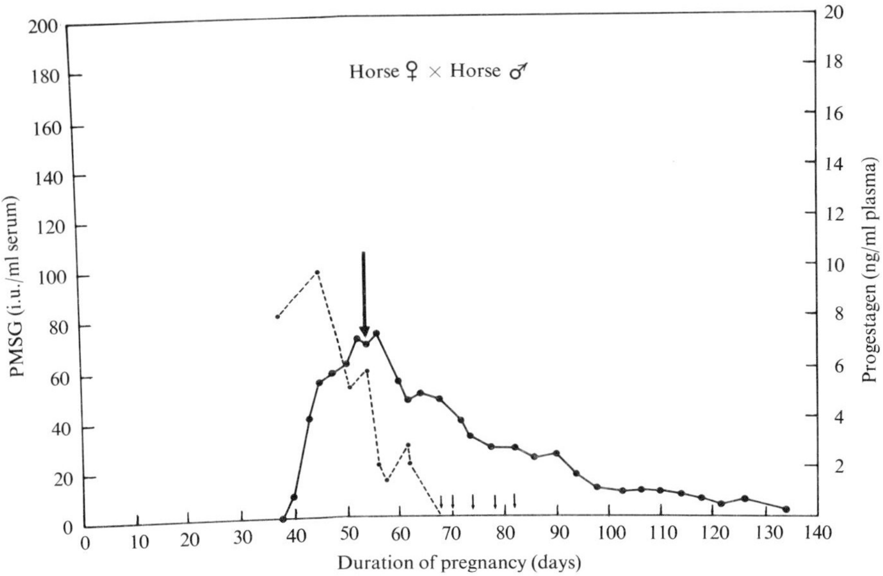

Fig 14. PMSG concentrations in the serum (solid line) and progestagen concentrations in the peripheral plasma (dashed line) of a Welsh Pony mare following surgical removal of the conceptus (arrow) from the uterus at 54 days of gestation.

only return to oestrus and ovulate following surgical removal of the conceptus or spontaneous death and resorbtion or abortion of the pregnancy, *after* the endometrial cups have regressed entirely and PMSG had disappeared from the blood (Allen, 1970*a*; Mitchell, 1971).

Mares carrying twin conceptuses have two sets of endometrial cups and therefore show double the normal serum levels of PMSG found in mares with a single pregnancy (Rowlands, 1949; Allen 1970*b*). Twins are considered undesirable in thoroughbred mares and it is common veterinary practice to remove twin foetal sacs by saline irrigation of the uterus, immediately following their diagnosis by rectal palpation of the uterus at 42–45 days of gestation. The endometrial cups have already begun to develop at this stage and, as mentioned previously, they continue to function normally and secrete high levels of PMSG for the next 80–90 days. Even if luteolysis is induced in these mares by treatment with potent synthetic analogues of prostaglandin $F_2\alpha$ (Allen & Rossdale, 1973), the mares will only return to oestrus and ovulate when PMSG disappears from the blood. Thus, the apparently paradoxical situation can exist in non-pregnant mares where, in the absence of functional luteal tissue in the ovaries and in the presence of very high concentrations of gonadotrophin in the blood, oestrus, follicular development and superovulation does not occur as one might expect. On the contrary, the ovaries become increasingly shrunken and inactive and the mare enters a prolonged period of deep anoestrus. This evidence suggests that although PMSG may play some minor part in the development of secondary corpora lutea in the normal, pregnant mare, it is not the only factor involved in the process. Other hitherto unknown foetal stimuli are apparently required, in the absence of which, PMSG depresses rather than stimulates ovarian function.

DISCUSSION

The nature of the mechanisms by which the mammalian foetus survives as a homograft in the uterus of the mother has received much attention during the past decade and Kirby (1968), Currie (1970), Beer & Billingham (1971) and others have reviewed the subject recently. Kirby (1968) summarized the evidence for and against the four principal hypotheses which have been advanced to account for the protection of the conceptus from the homograft reaction: namely, that the conceptus is antigenically immature, that the mother is rendered immunologically incapable during pregnancy, that the uterus is an immunologically privileged site or that foetal transplantation antigens are either not expressed on the surface of trophoblast cells in direct contact with maternal tissues or are masked from maternal recognition by some form of physical barrier. Kirby (1968) proposed that the bulk of available evidence supports the last of these postulates and argued that, despite the suggestion of Simmons & Russell (1962) and Simmons (1966) that paternal antigens are not expressed upon mouse trophoblast cells, the finding of Kirby, Billington, Bradbury & Goldstein (1964), Bradbury, Billington & Kirby (1965) and Kirby, Billington & James (1966) demonstrated that paternally-derived histocompatibility antigens are present upon mouse trophoblast but are masked by a mucoprotein layer rich in hyaluronic and sialic acids (the 'sialomucin coat') which envelops each trophoblast cell. Currie & Bagshawe (1967) supported this hypothesis and showed that human trophoblast cells from which the sialomucin layer was removed by treatment with neuraminidase were rapidly destroyed *in vitro* by allogeneic lymphocytes whereas control, untreated cells remained unharmed. They confirmed the earlier findings of Kirby *et al.* (1964) that the thickness of the sialomucin coat increases with increasing histoincompatibility between mother and father and proposed that the sialomucin layer carries a strong electronegative charge and functions by repelling similarly charged maternal lymphocytes. Good (1967) challenged this suggestion however, on the grounds that electrostatic repulsion is unlikely to function in an aqueous medium.

The equine endometrial cup reaction involves the rapid organization and development of a discrete area of the foetal chorion followed by active migration and invasion of highly specialized gonadotrophin-secreting trophoblast cells into the maternal endometrium. The whole reaction takes place during the first half of pregnancy and the actual life-span of the endometrial cups is little more than one quarter of the period of gestation. The gross histological changes observed during development and regression of the cups in both intraspecific and interspecific hybrid pregnancies, provide unequivocal evidence of the expression of paternally-derived histocompatibility antigens by endometrial cup cells. Whether such antigens are located upon the surface of the cells or are merely present in the principal secretory product of the cells, PMSG, remains unclear. Whatever the case, the paternal antigens are certainly recognized by the mare's immunological defence mechanisms and she mounts a classical graft-versus-host immune reaction to destroy and reject the foreign tissues.

It is not surprising that the maternal immune reaction against endometrial cup cells is considerably greater in hybrid pregnancies than in normal intraspecific matings. However, the very marked difference between mule and hinny pregnancies in the speed or rate of success of the invoked lymphocytic reaction in destroying and rejecting the cup cells remains puzzling. In mule endometrial cups, the leucocytic reaction is enormous and the cup tissue is totally destroyed by the invading cells within 20 days. In hinny cups on the other hand, the leucocyte response, while perhaps not quite so intense as that in the mule cup, is nonetheless a great deal more severe than in normal donkey endometrial cups. Even so, the individual hinny cup cells appear much better able to withstand the immunological attack directed against them and they remain viable and continue to secrete PMSG for as long as normal, non-hybrid cup cells.

Donkey tissues may be considerably more antigenic to horses than vice versa so that donkey-derived paternal transplantation antigens upon mule cup cells elicit a more intense and purposeful immunological response in the mare, than do horse-derived paternal antigens on hinny cup cells in the donkey. Alternatively, donkey and horse tissues may be equally antigenic to each other but more histocompatibility antigens may be expressed on mule than hinny cup cells. Some evidence in support of this possibility was provided by Hašková (1962, 1963), in studies upon inter-strain mouse pregnancies. She showed that placental tissue homogenates from A♀ × C57♂ conceptuses are not antigenic when injected into A-strain hosts whereas similar tissue homogenates from C57♀ × A♂ conceptuses induce a strong immune reaction in A-strain hosts. Similar results were obtained with CBA- and A-strain combinations. More recently, Baines, Ansell, McLaren & Micklem (1974) have shown by the cluster (rosette) assay of Micklem, Asfi, Staines & Anderson (1970) that following multiple pregnancies in mice, the degree of sensitization of the mother against paternal strain antigens is significantly greater in CBA♀ × C57BL♂ matings than in C57BL♀ × CBA♂ crosses.

Even if a marked disparity does exist in the relative antigenicity of horse and donkey tissues to each other, this alone could not account for the striking differences between hinny and mule cup cells in their ability to withstand the direct attack of the invading lymphocytes and other immune cells (see Fig. 12*c–f*). The in-vitro studies of Moor *et al.* (unpublished) have clearly shown that, apart from any possible influence of maternal environment, hinny cup cells manufacture and secrete very much greater amounts of PMSG than mule cup cells. Furthermore, hinny endometrial cups develop in the presence of considerably higher progesterone concentrations than do mule cups. A striking feature of both horse and hinny endometrial cups, where PMSG production by individual cup cells is high, is the pronounced limitation of the invading leucocytes to the maternal endometrial stroma during early development of the cups. The accumulated cells only begin to invade the cup tissue after the seventieth day of gestation when PMSG production is declining and prior to this time, lymphocytes which do come in contact with cup cells, seem unable to harm them. In mule cups on the other hand, where the individual cup cells are secreting relatively little PMSG, leucocyte invasion proceeds unchecked from the outset and the migrating immune cells rapidly kill and phagocytose cup cells as they encounter them.

The question arises as to whether the increased levels of gonadotrophin, and possibly also progestagen, play any part in diminishing the antigenicity of hinny endometrial cup cells to the mother or, more especially, provide some form of protective barrier which defends the cup cells from attack by sensitized lymphocytes. The recent experiments of Borland and his colleagues (see p. 157–164) with regard to the immunoprotective role of human chorionic gonadotrophin (HCG) in human pregnancy, seem particularly relevant to the situation in equidae. Borland finds that xenogeneic grafts of human trophoblast survive in guinea-pigs for much longer than control tissues such as human foetal skin and that incubation of the control tissues in a medium containing HCG prior to grafting to the xenogeneic host, greatly prolongs their survival time. Furthermore, removal of the sialomucin coat from human trophoblast cells by incubation in neuraminidase prior to explantation, reduces survival time to that of control tissues and subsequent incubation of neuraminidase-treated trophoblast cells in HCG, re-establishes the original survival pattern.

These findings strongly suggest that in the human, HCG is a vitally important integral part, if not the whole, of the immuno-protective fibrinoid layer described earlier by Kirby *et al.* (1964, 1966), Currie & Bagshawe (1967) and others. HCG is a glycoprotein (Bahl, 1969*a,b*, 1973), which possesses many of the chemical characteristics and constituents ascribed to the sialomucin coat of mouse and human trophoblast, especially hyaluronic and sialic acids (Bradbury, Billington & Kirby, 1965; Bradbury, Billington, Kirby & Williams, 1969). PMSG is also a glycoprotein containing a high proportion of sialic acid (Gospodarowicz, 1972) and it is therefore tempting to speculate that PMSG constitutes a similar immunoprotective barrier between foetal and maternal surfaces in equidae, acting to partially, but not completely, mask paternally-derived transplantation antigens from maternal recognition and prevent sensitized lymphocytes from gaining access to the endometrial cup cells. Martinek (1970) has established that physical imperfections and actual gaps exist in the sialomucin layer covering some trophoblast cells in rats and so the original suggestion of Currie & Bagshawe (1967) that the sialomucin coat functions to repel similarly charged maternal lymphocytes electrostatically rather than mask paternally-derived antigens, could perhaps account for the marked difference between mule and hinny cup cells in their response to the presence of sensitized leucocytes.

Apart from the specialized cells of the chorionic girdle which form the endometrial cups, the remainder of the chorio-allantois of the mare is non-invasive (Amoroso, 1952). Nevertheless, intimate contact between foetal and maternal surfaces, including close inter-digitation of microvilli on the surface of endometrial epithelial cells with those on the surface of opposing trophoblast cells, occurs soon after the commencement of cup development, and is well advanced by day 60 (Allen, 1970*b*; Samuel, Allen & Steven, unpublished). Despite the very obvious and pronounced immunization of the mare against paternally-derived foetal transplantation antigens expressed by endometrial cup cells, there occurs no similar maternal recognition of, or immune response to, the trophoblast cells of the normal chorio-allantois. The hypothesis that PMSG forms a physico-chemical barrier which protects endometrial cup cells from immune attack cannot hold true for the remaining chorio-allantois and one must therefore postulate that the trophoblast cells of the chorio-allantois either fail to exhibit paternal histocompatibility antigens or that such antigens, if present, are masked from maternal recognition by some form of physical barrier other than PMSG. In the absence of evidence to the contrary, it seems feasible to suggest that a process of immunological enhancement, as originally proposed by Kaliss (1958), Breyere & Barrett (1960) and others, may play an important role here. Early 'controlled' sensitization of the mare during the initial invasion of chorionic girdle cells into the endometrium to form the endometrial cups, may stimulate the production of humoral antibodies which in turn bind to paternal transplantation antigens situated on trophoblast cells and thus mask them from further maternal recognition.

It is interesting to speculate that at least a proportion of the high rate of early, non-infective abortions which occur in mares (van Niekerk, 1965; Merkt, 1966; Bain, 1969 and others) may result from a breakdown of whatever the protective mechanism is which prevents the already-sensitized mare from immunologically recognizing and reacting against the paternal component of the foetal trophoblast. Many authors have presented evidence to suggest that a significant percentage of spontaneous abortions in women may result from specific immunization against paternal antigens. Habitual abortion occurs more frequently in marriages where incompatibility in the ABO blood group system exists between the partners (Matsunaga & Isoh, 1958; Chung & Morton, 1961; Wren & Voss, 1961; Bardawil, Mitchell, McKeogh & Marchant, 1962), and Javert (1962) quoted three instances of habitually aborting women who achieved successful pregnancies following a change of husband. A similar situation exists in thoroughbred horses where many instances are recorded of individual mares which have aborted on two or more consecutive occasions when mated to the same stallion, but have subsequently produced normal, healthy offspring when covered by a different stallion. Experiments designed to investigate this possibility are in progress at the present time, but in the interim, the equine endometrial cup reaction remains somewhat of an enigma which defies many of our preconceived notions of the immunobiological role of trophoblast in mammalian pregnancy.

SUMMARY

Equine endometrial cups are composed of specialized gonadotrophin-secreting trophoblast cells which invade the maternal endometrium between 36 and 38 days of gestation. These cells express paternally-derived foetal transplantation antigens and are ultimately destroyed by the maternal cell-mediated immune response which they invoke. Foetal genotype profoundly influences PMSG production by endometrial cup cells and the intensity and rate of success of the maternal leucocyte response directed against them.

The suggestion is advanced that PMSG may constitute an immunoprotective layer around endometrial cup cells which partially protects the cells from the attack of sensitized lymphocytes. It is further postulated that a process of immunological enhancement prevents maternal recognition of, and immunological reaction against, trophoblast cells of the chorio-allantois.

This work was supported by the Thoroughbred Breeders' Association. I wish to express sincere thanks to Dr R. V. Short FRS, Mr L. E. A. Rowson FRS and Professor T. R. R. Mann FRS for providing laboratory and other facilities and for constant encouragement and advice. I am also sincerely grateful to Dr R. M. Moor for much fruitful and enjoyable collaboration. Mr J. Taylor, Miss S. Cattling and Mrs Margaret Shams gave valuable technical assistance. Mr John Fuller prepared the diagrams and Miss Barbara Hawkes kindly typed the manuscript.

REFERENCES

ALLEN, W. E. (1971). The occurrence of ovulation during pregnancy in the mare. *Vet. Rec.*, **88**, 508.

ALLEN, W. R. (1969*a*). A quantitative immunological assay for pregnant mare serum gonadotrophin. *J. Endocrinol.*, **43**, 581.

ALLEN, W. R. (1969*b*). The immunological measurement of pregnant mare serum gonadotrophin. *J. Endocrinol.*, **43**, 593.

ALLEN, W. R. (1969*c*). Factors influencing pregnant mare serum gonadotrophin production. *Nature, Lond.*, **223**, 64.

ALLEN, W. R. (1970*a*). Endocrinology of early pregnancy in the mare. *Equ. Vet. J.*, **2**, 64.

ALLEN, W. R. (1970*b*). Equine gonadotrophins. Ph.D. Thesis, University of Cambridge.

ALLEN, W. R. & MOOR, R. M. (1972). The origin of the equine endometrial cups. I. Production of PMSG by foetal trophoblast cells. *J. Reprod. Fert.*, **29**, 313.

ALLEN, W. R., HAMILTON, D. W. & MOOR, R. M. (1973). The origin of the equine endometrial cups. II. Invasion of the endometrium by trophoblast. *Anat. Rec.*, **177**, 485.

ALLEN, W. R. & ROSSDALE, P. D. (1973). A preliminary study upon the use of prostaglandins for inducing oestrus in non-cycling thoroughbred mares. *Equ. Vet. J.*, **5**, 64.

AMOROSO, E. C. (1952). Placentation. In *Marshall's Physiology of Reproduction* (ed. A. S. Parkes), vol. XI, p. 127. Longmans' Green: London.

AMOROSO, E. C. (1955). Endocrinology of pregnancy. *Brit. Med. Bull.*, **11**, 117.

AMOROSO, E. C. & FINN, C. A. (1962). Ovarian activity during gestation, ovum transport and implantation. In *The Ovary* (ed. S. Zuckerman), vol. I, p. 451. Academic Press: London.

AMOROSO, E. C., HANCOCK, J. L. & ROWLANDS, I. W. (1948). Ovarian activity in the pregnant mare. *Nature, Lond.*, **161**, 355.

BAHL, O. P. (1969*a*). Human chorionic gonadotrophin. I. Purification and physicochemical properties. *J. biol. Chem.*, **244**, 567.

BAHL, O. P. (1969*b*). Human chorionic gonadotrophin. II. Nature of the carbohydrate units. *J. biol. Chem.*, **244**, 575–83.

BAHL, O. P. (1973). Chemistry of human chorionic gonadotrophin. In *Hormonal Proteins and Peptides* (ed. C. H. Li), vol. I, p. 171, Academic Press, London.

BAIN, A. M. (1967). The ovaries of the mare during early pregnancy. *Vet. Rec.*, **80**, 229.

BAIN, A. M. (1969). Foetal losses during pregnancy in the Thoroughbred mare: a record of 2562 pregnancies. *N. Z. Vet. J.*, **17**, 155.

BAINES, M. G., ANSELL, J. D., MCLAREN, A. & MICKLEM, H. S. (1974). Maternal immunization to paternal antigens during normal pregnancy in multiparous mice. *J. Reprod. Fert.* (in press).

BARDAWIL, W. A., MITCHELL, G. W., MCKEOGH, R. P. & MARCHANT, D. J. (1962). Behaviour of skin homografts in human pregnancy. *Am. J. Obstet. Gynec.*, **84**, 1283.

BEER, A. E. & BILLINGHAM, R. E. (1971). Immunobiology of mammalian reproduction. *Adv. Immunol.*, **14**, 1.

BIELANSKI, W., EWY, Z. & PIGONIOWA, H. (1956). Differences in the level of gonadotrophin in the serum of pregnant mares. *Third International Congress on Animal Reproduction*, 1956, Cambridge, p. 110.

BRADBURY, S., BILLINGTON, W. D. & KIRBY, D. R. S. (1965). A histochemical and electron microscopical study of the fibrinoid of the mouse placenta. *J. R. Microscop. Soc.*, **84**, 199.

BRADBURY, S., BILLINGTON, W. D., KIRBY, D. R. S. & WILLIAMS, F. A. (1969). Surface mucin of human trophoblast. *Am. J. Obstet. Gynec.*, **104**, 416.

BREYERE, E. J. & BARRETT, M. K. (1960). Prolonged survival of skin homografts in parous female mice. *J. natn. Cancer Inst.*, **25**, 1405.

CATCHPOLE, H. R. & LYONS, W. R. (1934). The gonad-stimulating hormone of pregnant mares. *Am. J. Anat.*, **55**, 167.

CHALLIS, J. R. G., HEAP, R. B. & ILLINGWORTH, D. V. (1971). The plasma concentrations of oestrogen and progesterone in non-pregnant, pregnant and lactating guinea pigs. *J. Endocrin.*, **51**, 333–45.

CHANDLEY, ANN, C., JONES, R., DOTT, H. M., ALLEN, W. R. & SHORT, R. V. (1974). Meiosis in interspecific equine hybrids. I. The mule (*E. assinus* × *E. caballus*) and hinny (*E. caballus* × *E. assinus*). *Cytogenetics* (in press).

CHUNG, C. S. & MORTON, N. E. (1961). Selection at the ABO locus. *Am. J. Hum. Genet.*, **13**, 9.

CLEGG, M. T., BODA, J. M. & COLE, H. H. (1954). The endometrial cups and allantochorionic pouches in the mare with emphasis on the source of equine gonadotrophin. *Endocrinology*, **54**, 448.

CLEGG, M. T., COLE, H. H., HOWARD, C. B. & PIGON, H. (1962). The influence of foetal genotype on equine gonadotrophin secretion. *J. Endocrin.*, **25**, 245.

COLE, H. H. & GOSS, H. T. (1943). The source of equine gonadotrophin. In *Essays in Biology in Honour of H. M. Evans*, p. 107. University of California Press: California.

COLE, H. H. & HART, G. H. (1930). The potency of blood serum of mares in progressive stages of pregnancy in effecting sexual maturity of the immature rat. *Am. J. Physiol.*, **93**, 57.

CURRIE, G. A. (1970). The conceptus as an allograft: Immunological reactivity of the mother. *Proc. R. Soc. Med.*, **63**, 61.

CURRIE, G. A. & BAGSHAWE, K. D. (1967). The masking of antigens on trophoblast and cancer cells. *Lancet*, **i**, 708.

DAY, F. T. & ROWLANDS, I. W. (1940). The time and rate of appearance of gonadotrophin in the serum of pregnant mares. *J. Endocrin.*, **2**, 255.

DE FEO, V. J. (1967). Decidualization. In *Cellular Biology of the Uterus* (ed. R. M. Wynn), p. 191. Meredith: New York.

EWART, J. C. (1897). *A critical period in the development of the horse.* Adam and Charles Black: London.

EWART, J. C. (1915). Studies on the development of the horse. I. Development during the third week. *Trans. R. Soc., Edinburgh*, **51**, 287.

FINN, C. A. (1971). The biology of decidual cells. In *Advances in Reproductive Physiology* (ed. M. W. H. Bishop), **5**, p. 1. Logos Press: London.

GOOD, W. (1967). Masking of antigens on trophoblast. *Lancet*, **i**, 1156.

GOSPODAROWICZ, D. & PAPKOFF, H. (1967). A simple method for the isolation of pregnant mare serum gonadotrophin. *Endocrinology*, **80**, 699.

GOSPODAROWICZ, D. (1972). Purification and physicochemical properties of the pregnant mare serum gonadotrophin (PMSG). *Endocrinology*, **91**, 101.

GRAY, A. P. (1972). *Mammalian hybrids*. Commonwealth Agriculture Bureaux: Buckinghamshire.

HAMILTON, D. W., ALLEN, W. R. & MOOR, R. M. (1973). The origin of the equine endometrial cups. III. Light and electron microscopic study of fully developed equine endometrial cups. *Anat. Rec.*, **177**, 503.

HAŠKOVÁ, V. (1962). Transplantation non-antigenicity of the placenta. *Nature, Lond.*, **193**, 278.

HAŠKOVÁ, V. (1963). Differences in the antigenic effectiveness of the foetal part of the mouse placenta depending on the strain combination employed. *Folio Biologica*, **9**, 99.

HEAP, R. B. & DEANESLY, RUTH, (1966). Progesterone in systemic blood and placentae of intact and ovariectomized pregnant guinea-pigs. *J. Endocrin.* **34**, 417.

HEAP, R. B., PERRY, J. S. & CHALLIS, J. R. G. (1973). Hormonal maintenance of pregnancy. In *American Handbook of Physiology*, section **7**, *Endocrinology*, vol. IX, *Female Reproductive System*, part 2, p. 217.

JAVERT, C. T. (1962). Further follow-up on habitual abortion patients. *Am. J Obstet. Gynec.*, **84**, 1149.

KALISS, N. (1958). Immunological enhancement of tumour homografts in mice. *Cancer Res.*, **18**, 992.

KING, J. M., SHORT, R. V., MUTTON, D. E. & HAMERTON, J. L. (1966). The reproductive physiology of male zebra–horse and zebra–donkey hybrids. *Symp. Zool. Soc., Lond.*, **15**, 511.

KIRBY, D. R. S. (1968). Immunological aspects of pregnancy. In *Advances in Reproductive Physiology* (ed. Anne McLaren), **3**, p. 33. Logos Press: London.

KIRBY, D. R. S., BILLINGTON, W. D., BRADBURY, S. & GOLDSTEIN, D. (1964). Antigen barrier of the mouse placenta. *Nature, Lond.*, **204**, 548.

KIRBY, D. R. S., BILLINGTON, W. D. & JAMES, D. W. (1966). Transplantation of eggs to the kidney and uterus of immunised mice. *Transplantation*, **4**, 713.

MARTINEK, J. J. (1970). Fibrinoid and the fetal–maternal interface of the rat placenta. *Anat. Rec.*, **166**, 587.

MATSUNAGA, E. & ITOH, S. (1958). Blood groups and fertility in a Japanese population, with special reference to intra-uterine selection due to maternal-foetal incompatibility. *Ann. Hum. Genet.*, **22**, 111.

MERKT, H. (1966). Foal heat and embryonic resorbtion. *Zuchthygiene*, **1**, 102.

MICKLEM, H. S., ASFI, C., STAINES, N. A. & ANDERSON, N. (1970). Quantitative study of cells reacting to skin allografts. *Nature, Lond.*, **227**, 947.

MITCHELL, D. (1971). Early foetal death and a serum gonadotrophin test for pregnancy in the mare. *Can. Vet. J.*, **12**, 41.

MORRIS, C. J. O. R. (1966). The chemistry of gonadotrophins. In *Marshall's Physiology of Reproduction* (ed. A. S. Parkes), vol. III, p. 379. Longman's Green: London.

ROWLANDS, I. W. (1949). Serum gonadotrophin and ovarian activity in the pregnant mare. *J. Endocrin.*, **6**, 184.

ROWLANDS, I. W. (1963). Levels of gonadotrophin in tissues and fluids with emphasis on domestic animals. In *Gonadotrophins: their chemical and biological properties and secretory control* (ed. H. H. Cole), p. 74. W. Freeman & Co.: San Francisco.

SCHAUDER, W. (1912). Untersuchungen uber die eihaute und embryotrophe des pferdes. *Arch. Anat. Physiol.*, 259.

SCHOMBERG, D. W., COUDERT, S. P. & SHORT, R. V. (1967). Effects of bovine luteinizing hormone and human chorionic gonadotrophin on the bovine corpus luteum *in vivo*. *J. Reprod. Fert.*, **14**, 277.

SHORT, R. V. (1959). Progesterone in blood: IV. Progesterone in the blood of mares. *J. Endocrin.*, **19**, 207.

SIMMONS, R. L. (1966). Mechanisms of trophoblast non-antigenicity. *Fedn Proc.*, **25**, 356.

SIMMONS, R. L. & RUSSELL, P. S. (1962). The antigenicity of mouse trophoblast. *Ann. N. Y. Acad. Sci.*, **99**, 717.

VAN NIEKERK, C. H. (1965). II. Early embryonic resorption in mares. *J. South African Vet. Med. Assoc.*, **36**, 61.

VAN NIEKERK, C. H. (1973). Morphological and physiological changes in the genital system of mares. D.V.Sc. Thesis, University of Pretoria.

VAN RENSBURG, S. J. & VAN NIEKERK, C. H. (1968). Ovarian function, follicular oestradiol-17B, and luteal progesterone and 20-hydroxypregn-4-en-3-one in cycling and pregnant equines. *Onderstepoort J. Vet. Res.*, **33**, 195.

WREN, B. G. & VOSS, G. H. (1961). Blood group incompatibility as a cause of spontaneous abortion. *J. Obstet. Gynaec. Br. Commonw.*, **68**, 637.

DISCUSSION

EDWARDS: Evidence from electron microscopy has shown the trophoblast and endometrial or epithelial cells in very close apposition, even including cytoplasmic fusion, in pregnant women or female animals (Wynn, 1967; Enders & Schlafke, 1971; Martinek, 1971; Tekelioglu-Uysal, Edwards & Kisnisci, 1975). The invasive capacity of the chorionic girdle cells in the horse may not be such a unique characteristic. I am unconvinced about the endocrinological differences between the mare and other species since claims have been made that HCG is produced initially in the human as a result of interaction between the syncytio- or cytotrophoblast and the endometrium. There is no evidence showing that human trophoblast isolated before implantation can synthesize HCG, and we have been unable to detect it in the medium from human blastocysts growing in culture (Edwards, 1973); this hormone is found in human plasma very shortly after implantation. Is the relationship between trophoblast and endometrium in the human very dissimilar to the mare?

ALLEN: I agree that striking similarities exist between primates and equidae in the endocrinology of pregnancy. The trophoblasts of both secrete large amounts of gonadotrophin, a feature not found in other species. Perhaps the most striking differences between the two are that in women, with a highly invasive haemochorial placenta, the HCG-secreting syncytiotrophoblast remains physically attached to the foetal placenta and elicits no maternal cell-mediated immune response, despite the intimate relationship with endometrial cells. In the mare, on the other hand, with a diffuse and non-invasive epitheliochorial placenta, the PMSG-secreting chorionic girdle cells sever their connection with the foetal placenta and exist in isolated colonies within the endometrium. In this situation, they elicit a strong maternal immune reaction. A maternal–foetal interaction may be necessary for gonadotrophin production by the placenta and I would certainly not dispute this possibility. However our experiments have demonstrated that equine chorionic girdle cells will manufacture PMSG *in vitro* in the absence of maternal influence and a great many people over the past 30 years or so have demonstrated HCG production by human syncytiotrophoblast, hydatidiform mole and chorionic tissues *in vitro* (Gey, Seegar & Hellman, 1938). Thus, any maternal interaction must have occurred prior to removal of these placental tissues and perhaps maternal influences act in terms of a one-time irreversible switch mechanism.

EDWARDS: Relevant data could be obtained by taking girdle tissue from a horse conceptus prior to 30 days of gestation, and culturing it beyond 30 days. Would it then produce PMSG?

ALLEN: Girdle tissue taken before 30 days and cultured for several days will not produce PMSG at any time, even though it grows well in culture. This may be evidence of the necessity for a maternal/foetal interaction to switch on the synthesis of PMSG by the girdle tissue.

McLaren: I am not completely convinced by your suggestion that the hormone levels are determined entirely by the foetal genotype. The reciprocal difference is so enormous that the maternal tissue may be stimulating the trophoblast. The horse could have a weak maternal stimulus but rather a responsive trophoblast, whereas the donkey has a strong maternal stimulus but a very unresponsive trophoblast. The F_1 hybrid trophoblast would be intermediate. This suggestion would in turn lead to a reconsideration of your findings on in-vitro culture. The maternal stimulus might act before 32 days, when I believe you removed the tissue for culture. This hypothesis is testable by reciprocal embryo transfers, where the horse embryo put into the donkey would give even higher hormone levels than are found with a hinny, and the donkey into the horse should give even lower levels than with mule embryos.

Allen: Yes, this is indeed an interesting possibility. However, we have found that chorionic girdle tissue recovered from mares prior to 30 days of gestation does not secrete any PMSG in culture, even though the tissue grows well and forms the same large, binucleated, decidual-like cells seen in the cultures of girdle tissue recovered after day 30. Thus, there seems to be some form of switch mechanism to stimulate the girdle cells to produce PMSG which acts at around the thirtieth day of pregnancy. However, we do not know if this is a maternal influence or is inherent in the development of the embryo.

Some three years ago now, I tried, without success, reciprocally to transfer donkey zygotes into mares and horse zygotes into donkeys. We have, however, had some success with transferring hybrid eggs and have managed to achieve two hinny conceptuses developing in mares (Allen & Rowson, 1972). With the first of these pregnancies, PMSG appeared in the blood of the mare at 38 days but had disappeared again by day 42. Despite this complete lack of gonadotrophin in the serum, the mare went on to produce a live foal which was later verified by tests using G-6-PDH measurements as a hinny (Hamerton *et al.*, 1969). In the second pregnancy, we attempted to recover and culture the chorionic girdle tissue prior to the development of endometrial cups but we discovered that hinny chorionic girdle tissue begins to invade the maternal endometrium as early as 32 days of gestation, some four or five days before invasion commences by horse, donkey or even mule girdle cells. Thus, in this one very valuable animal, endometrial cup development was already well under way when we operated on day 35, so we were unable to recover the chorionic girdle tissue. Histology of the young endometrial cups revealed an extensive infiltration of lymphocytes which appeared to be destroying the newly-developed cup cells.

Curiously, from both the endocrinological and immunological standpoints, everything seems to favour the survival of hinny pregnancies compared with mules, although the reverse is actually the case. Only 10% or less of female donkeys will conceive when mated to a stallion whereas both the conception and foaling rates of mares mated to a jack donkey are no different from those in mares mated to normal horse stallions.

ALLISON: In some situations, immune responses increase hormone release, such as the long-acting thyroid stimulator and this may happen in your system. In your cultures, antibodies such as horse anti-mule and vice versa might possibly increase release under certain circumstances.

ALLEN: We are hoping to do this type of experiment during the coming year. We have already added an anti-PMSG antiserum in the absence of complement to cultures of girdle cells but without any obvious effect.

HEAP: Your findings on the time when trophoblast becomes capable of hormone synthesis are rather similar to ours with pig blastocysts. On day 11, but not on day 9 or 10, the pig blastocyst has the enzymic capacity to produce a considerable amount of oestrogen (Perry, Heap & Amoroso, 1973). This synthetic activity occurs several days before attachment occurs. Do you have any information on the production of oestrogen by the girdle tissues?

ALLEN: We have not yet specifically looked for oestrogen production by the cultured chorionic girdle cells. Judged by measurements of progesterone levels in the culture fluid and by histochemistry, the girdle cells are evidently active in steroid as well as gonadotrophin biosynthesis, and they may make oestrogen in addition to progesterone. Does a sudden 'switch-on' of oestrogen production occur with pig blastocysts?

HEAP: By day 11 of pregnancy, when the blastocyst starts to elongate there is considerable synthetic activity as indicated by the incorporation of labelled steroid precursors into the oestrone and oestradiol-17β *in vitro*. Before day 11, the incorporation is negligible.

BODMER: Do you assay PMSG with an antibody? If so, the apparent differences in levels between horses, donkeys and their hybrids could be a measure of the differential specificity of the antibody.

ALLEN: We use antisera prepared in rabbits and turkeys, but I doubt your suggestion is true. Our antisera were checked against a range of PMSG standards with known biological potency and the results corresponded very well. We have also checked odd serum samples from mares and donkeys carrying hybrid pregnancies in both ovarian and uterine weight bioassays and again, the immunological and biological potencies correspond nicely.

BODMER: Although your evidence for foetal trophoblast production of PMSG looks fairly convincing, you have not proved beyond doubt that the right types of PMSG are produced, for the antibodies were not specific for donkey and horse gonadotrophins. Have you tried making a PMSG antibody against the donkey in the horse and vice versa, to obtain a species-specific antibody? You could then decide which type of PMSG is present.

ALLEN: No, we have not thought this necessary in view of the in-vitro results.

BODMER: Yes, but unless the gene product is identified, you cannot be sure just from the level what is the species origin of the hormone.

EDWARDS: There may be a contribution to PMSG production from both the foetal and maternal sides. I would not accept his point entirely but Bodmer suggests that this test should be done in order to make absolutely sure that the PMSG being produced can be classified according to its species of origin. Are there any other differences that you could look at in donkey and horse PMSG? Do they have any different electrophoretic mobilities or any other characteristics?

ALLISON: It is likely to be the same since one is working with the same genus.

BODMER: From the kind of work that we do on somatic cell hybrids we know that confusion can arise if the species of origin of the cellular product cannot be established.

HEAP: Your evidence for the probable role of foetal genotype rests also in the striking difference in the amount of progesterone produced in the two different hybrid pregnancies. Might this not be a considerably more specific method of measurement than PMSG?

ALLEN: No doubt I am being too complacent but I just cannot see the urgent need to ascertain the degree of specificity of donkey as compared with horse PMSG or indeed, of either of the reciprocal hybrids. Differences may exist between the various types of PMSG but is this relevant to the point in question? McLaren's earlier suggestion of the possibility of a stronger or weaker maternal stimulus acting upon a more or less responsive trophoblast to stimulate PMSG production is very pertinent but I would again emphasize that any maternal influence must occur before 30 days of pregnancy and must be in the form of a switch-on mechanism. I cannot believe that the mother actually contributes directly to the supply of PMSG either in the form of the finished product or in the form of initial precursors of the hormone in the manner in which the foetus supplies the necessary precursors for oestrogen production by the human placenta. Our culture studies have shown very clearly that only chorionic girdle tissue will secrete PMSG *in vitro*. Maternal endometrium, foetal skin and normal chorioallantois produce no PMSG in culture at all. Furthermore, not only will chorionic girdle tissue secrete PMSG over a prolonged period in the complete absence of any maternal influence, it will secrete gonadotrophin *in vitro* at exactly the same rate as endometrial cup cells do *in vivo*. Any possible maternal influence in the in-vivo situation must surely be sublimated by the genetic effect.

GARDNER: The enzyme G-6-PDH is very different in horses and donkeys, and could perhaps be useful in tracing cell lineages (Cohen & Ratazzi, 1971).

COOMBS: Loke's findings concerning HL-A on the trophoblast are important here (see Discussion after Johnson's paper, p. 101). You presumably find histocompatibility antigens on endometrial cup cells and not on the remaining trophoblast cells. If, in fact, the cup cells are derived from the trophoblastic tissue then you have to postulate the masking of these antigens on trophoblast. Or could the cup cells be derived from some other cell which is not trophoblast?

ALLEN: All our evidence strongly indicates that the endometrial cup cells are derived from the specialized layer of trophoblast known as the chorionic girdle. Major differences must exist between trophoblast cells of the chorionic girdle and those of the remaining chorio-allantois, either in the expression of HL-A equivalents and other histocompatibility antigens or in the relative degree of masking of these antigens. Paternally-derived antigens could be expressed on both types of trophoblast cells, those exposed on the girdle cells during the very first stages of cup development being then partially masked by the PMSG produced by the cells. The initial exposure could stimulate the formation of humoral antibodies in the mare which would in turn bind to similar antigens expressed on normal trophoblast of the chorioallantois and thereby protect these cells by a process of immunological enhancement.

COOMBS: More precision would of course be needed in determining whether the antigens are expressed and the relevant part of the cell.

ALLEN: Attempts to detect antigenic differences between the surfaces of cup cells and normal trophoblast are being carried out now.

BAGSHAWE: If the production of PMSG has nothing to do with the abrogation of the immunological response, then other antigens involved in this situation may be important for rejection. Under these circumstances a high level of PMSG would be found when the rejection process has taken place rather slowly and low levels when the cup cells have been destroyed more rapidly. You are suggesting that PMSG abrogates the destruction of the endometrial cup cells but it could be the other way round.

ALLEN: I doubt it. From the histological evidence, the secretion of PMSG does not appear significantly to lower or mask the antigenicity of cup cells, but it does appear to provide some sort of protection for the cells from the maternal response. Hinny endometrial cup cells secreting large amounts of PMSG are destroyed more slowly by the leucocytes attracted to them than are mule cup cells secreting only small amounts of gonadotrophin.

BAGSHAWE: But suppose that the PMSG is not taking this role; then there might be some other mechanism involved leading to immunological rejection. Where rejection is taking place effectively there would be low levels of PMSG because of cell destruction and high levels where the cells are persisting. This would be a co-incident effect rather than a causally related effect.

ALLEN: Two years ago I would have agreed with your proposal but our more recent in-vitro findings argue against it. Mule girdle cells grown *in vitro*, and therefore absent from maternal immune pressures, secrete only small amounts of PMSG compared with hinny girdle cells in the same system. Therefore, one cannot argue that the low PMSG concentrations in the serum of mares carrying mule conceptuses arises through the premature destruction of the endometrial cups. An inverse correlation is apparent between the rate of gonadotrophin production by girdle cells and the rate of their rejection from the maternal tissues.

GARDNER: Are the observed differences in the invasiveness of hydatidiform moles in human pregnancies due to differences in the trophoblastic cells or is there a deficiency of the maternal decidual response? Secondly would Anne McLaren comment on the importance of carbon dioxide in inducing decidualization, especially with respect to the discovery by Wilson that metabolically active tumours fail to evoke decidualization in the mouse uterus.

BAGSHAWE: We have not observed a close correlation between the development of an invasive mole and the decidual reaction. Certainly a decidual reaction occurs in the presence of some invasive moles and some choriocarcinomas but the factors important in determining the response are not known. Most moles appear to be more invasive than normal trophoblast. It is very unusual for a mole to be completely evacuated at the time of primary evacuation because the trophoblast has penetrated beyond the endometrium and is inaccessible.

EDWARDS: Can deciduomata be induced in the human?

BILLINGTON: They occur spontaneously, and also following steroid administration (De Feo, 1967).

ALLEN: Some decidualization occurs spontaneously towards the end of the cycle, especially tissue near the spiral arteries (Wilson, 1969).

MCLAREN: I remain open-minded about carbon dioxide and decidualization. With tumours, an important point is that only a rather narrow range of carbon dioxide concentration was effective in inducing deciduomata (Hetherington, 1968), so the amount of carbon dioxide produced by the tumour would be critical. Some tumours are known to have a low carbon dioxide output.

EDWARDS: The endometrial cup cells are making large quantities of PMSG but this does not appear to be related to the amount of progesterone secreted. Hence, the PMSG must not be a luteotrophin and must have some other role.

ALLEN: I have doubts about the supposed luteotrophic role of PMSG. Follicular growth occurs in the mare as early as 17–20 days of pregnancy although these follicles do not normally ovulate until around day 40 when PMSG is being produced. Furthermore, luteal function declines quite rapidly in many mares following surgical removal of the conceptus from the uterus or abortion after the thirty-eighth day of pregnancy. This occurs despite the continued normal function of the endometrial cups and the presence of high concentrations of PMSG remaining in the blood. PMSG may play a minor role in ovulating and/or luteinizing the already well-developed follicles but other foetal factors are obviously involved in some manner.

ALLISON: Does immunization of a mare against donkey tissues affect the probability of obtaining a normal foetus?

ALLEN: We have recently begun a series of these experiments but a long time is needed to obtain data in animals with an 11-month gestation like the horse.

EDWARDS: You have already contributed to this type of experiment by successive mating of horses and donkeys to each other. Antibodies will have been produced as a result of the incompatible pregnancy.

ALLISON: No, the immunization by pregnancy may represent a different situation because Allen found that there were some animals which, if they produced one live foetus, would produce others, and this might indicate a state of immunological enhancement.

EDWARDS: Do you get normal pregnancies after successive matings?

ALLEN: We have not been able to look at them very closely as yet although the two mares which I showed on the slide (Fig. 7, p. 227) both produced normal foals after the second concurrent mule pregnancy, despite low levels of PMSG in the blood and the short life of the endometrial cups. Not so many years ago, large numbers of mules were bred for use by the army in India and other eastern countries. Large mule breeding studs were established for this purpose and the same mares were mated to jack donkeys year after year without any apparent decline in fertility (Hickman, personal communication). The hinny pregnancy is a different sort of problem. Only a small percentage of donkeys will conceive when mated to a stallion but the blockage seems to be right back at fertilization since those animals which do not conceive, show no alteration of cycle length. I do have one donkey which has conceived three or four hinny conceptuses, all of which have been removed from her uterus at various stages of early gestation. We may have reached an interesting stage with this animal since on three occasions during the past breeding season, she appeared to conceive when mated to a stallion only to undergo early embryonic death and resorbtion of the pregnancy around 40 days.

ALLISON: This animal may have a high titre of enhancing antibodies in her serum. Does her serum react with horse cells?

ALLEN: John Ansell, who has recently come to work with us from McLaren's laboratory will be looking at this during the coming season.

AMOROSO: Is there any evidence for the presence of PMSG on the surface of the endometrial cups or elsewhere on the endometrium, i.e. outside the endometrial cup area where the PMSG is produced?

ALLEN: Yes, as originally shown by Cole and yourself (Cole & Goss, 1943; Amoroso, 1955), very high concentrations of PMSG are present in the sticky, brown endometrial cup secretion which adheres to the overlying foetal membranes. However, we have no evidence at all for PMSG production by any maternal or foetal tissue other than the foetal-derived endometrial cup cells.

AMOROSO: Judged from the anatomical considerations of the placenta, it is quite possible that lateral seepage of PMSG from the endometrial cups could occur. A copious production of 'uterine milk' occurs from the endometrial glands during pregnancy in the mare and this material presumably contains PMSG. Because there is little degradation of the uterine tissues in the mare, and a layer of 'uterine milk' must be interposed between the surface of the chorion and the surface of the remainder of the endometrium.

You observed large numbers of lymphocytes and plasma cells in and around the endometrial cups but you avoided mention of eosinophils. If eosinophilic cells are present, what are their functions?

ALLEN: Large numbers of eosinophils are present during the degenerative phase of the cups.

AMOROSO: The lymphocytes appear to be largely confined to the base of the endometrial cups, where the cup is sitting on a mass of lymphocytic tissue. However, during degradation of the cups, eosinophils and plasma cells are found. We must remember that horses show an abundance of lymphocytes in their blood profile and I would like to know if there is any connection between the eosinophils and some of the immunological responses we have been discussing.

COOMBS: There is a certain association between eosinophils and certain types of reaginic antibody activity but no one knows the function of these eosinophils.

AMOROSO: Your demonstration of the derivation of the endometrial cup cells from the chorionic girdle is clear and elegant. However, I am still worried, despite the remarks of many people at this symposium, that no clear definition has been made of the production of protein hormones by the placenta of a variety of species nor of the possible role of these hormones in immuno-suppression. We must think in terms of a family of hormones and not content ourselves with only the protein hormones. The steroids may also have a role and since the early days of the century, the function of steroids in maintaining pregnancy has been stressed. In the absence of oestrogen and progesterone, the pregnancy is lost and I would like to call this 'rejection' rather than 'abortion'. The relationship between steroid and protein hormones and rejection may well be related to the accumulation of large amounts of these hormones during human pregnancy.

REFERENCES

ALLEN, W. R. & ROWSON, L. E. A. (1972). Transfer of ova between horses and donkeys. *Proc. 7th Int. Cong. Anim. Reprod. Munich, 1972*. P. 484.

AMOROSO, E. C. (1950). The source and composition of the uterine milk in the mare. *J. Anat.*, **84**, 78.

COHEN, M. M. & RATAZZI, M. C. (1971). Cytological and biochemical correlation of late-X chromosome replication and gene reactivation in the mule. *Proc. natn. Acad. Sci. USA*, **68**, 554.

COLE, H. H. & GOSS, H. (1943). The source of equine gonadotrophin. In *Essays in biology in honour of H. M. Evans*, p. 107. University of California Press: California.

DE FEO, V. J. (1967). Decidualization. In *Cellular Biology of the Uterus* (ed. R. M. Wynn), p. 191. North-Holland: Amsterdam.

EDWARDS, R. G. (1973). Studies on human conception. *Am. J. Obstet. Gynec.*, **117**, 587.

ENDERS, A. C. & SCHLAFKE, S. (1971). Penetration of the uterine epithelium during implantation in the rabbit. *Am. J. Anat.*, **132**, 219.

GEY, G. O., SEEGAR, G. E. & HELLMAN, L. M. (1938). The production of a gonadotrophic substance (prolan) by placental cells in tissue culture. *Science*, **88**, 306.

HAMERTON, J. L., RICHARDSON, B. J., GEE, P. A., ALLEN, W. R. & SHORT, R. V. (1971). Non-random late replication of X chromosomes in mules and hinnies. *Nature, Lond.*, **232**, 312.

HETHERINGTON, C. M. (1968). Induction of deciduomata in the mouse by carbon dioxide. *Nature, Lond.*, **219**, 863.

MARTINEK, J. J. (1971). Ultrastructure of the deciduotrophoblastic interface of the mouse placenta. *Am. J. Obstet. Gynec.*, **109**, 424.

PERRY, J. S., HEAP, R. B. & AMOROSO, E. C. (1973). Steroid hormone production by pig blastocysts. *Nature, Lond.*, **245**, 45.

TEKELIOGLU-UYSAL, M., EDWARDS, R. G. & KISNISCI, H. A. (1975). Ultrastructural relationships between decidua, trophoblast and lymphocytes at the beginning of human pregnancy. *J. Reprod. Fert.*, (in press).

WILSON, E. W. (1969). Alkaline phosphatase in pre-decidual cells of the human endometrium. *J. Reprod. Fert.*, **19**, 567.

WYNN, R. M. (1967). Comparative electron microscopy of the placental functional zone. *Obstet. Gynec.*, **29**, 644.

ANTIGENIC DISPARITY: DOES IT AFFECT PLACENTAL SIZE, IMPLANTATION OR POPULATION GENETICS?

By ANNE McLAREN

Early in 1966, David Kirby came to Edinburgh and discussed with Bryan Clarke certain observations recently made in his laboratory (Billington, 1964; James, 1965). It seemed that there could be implications for population genetics. Together they put forward a hypothesis (Clarke & Kirby, 1966) to explain one of the puzzling phenomena of immunology, the genetic diversity of transplantation antigens.

Clarke and Kirby suggested that a balanced polymorphism for transplantation antigens was maintained in mammals by a selective system in which antigenic disparity between mother and embryo was beneficial to the development of the young. The hypothesis was elegant, attractive and genetically plausible. In the present paper I shall show that its experimental foundation is no longer secure and that it is unlikely to be correct.

Although Clarke & Kirby's hypothesis has stimulated valuable experimental work and new thinking, the surviving author, B. Clarke, believes that it can no longer be sustained (Clarke, personal communication). However, once an appealing theory has become implanted in the scientific consciousness, it takes more than a few contrary facts to uproot it. The following recent examples, which could be multiplied many times, are typical of the uncritical and misleading views which are now prevalent:

> The fact that allogeneic fetuses in general appear to be favoured in utero (Billington, 1964; Clarke & Kirby, 1966; James, 1965) could be an indirect evolutionary consequence of the adaptation for maternal recognition of male antigenicity.
> Lappé & Schalk (1971)

> . . . the increasing evidence that the fetus is favoured by a maternal-fetal histoincompatibility supplies a very satisfactory explanation of histocompatibility polymorphisms . . . (Kirby, 1970; Clarke & Kirby, 1966; James, 1965, 1969; Billington, 1964).
> Prehn (1971)

> . . . the now well documented advantages of a mild maternal immunity against the fetus, can adequately explain the complex histocompatibility polymorphisms.
> Prehn (1971)

> . . . there is evidence that at least in some species the size of the placenta, and probably the extent of trophoblastic invasion, are affected by (1) the existence or otherwise of genetic disparity between an embryo and its mother and (2) by

the immunological status of the latter with respect to the alien antigens of her fetuses, i.e. whether normal, specifically immune, or tolerant. (Followed by references to Billington, 1964, 1965; McLaren, 1965*a*; James, 1965, 1967).

Beer & Billingham (1971)

It has been suggested that antigenic disparity between mother and foetus may actually favour the growth and survival of the foetus, rather than causing selection against it (Clarke & Kirby, 1966; Hull, 1966). This is based on data in the mouse showing that antigen differences can apparently increase the size of the placenta (Billington, 1964; James, 1965; McLaren, 1965*a*) and by observations in the mouse (Hull, 1969) and rat (Michie & Anderson, 1966; Palm, 1970), showing significant and sometimes striking increases in the relative survival of offspring which were antigenically incompatible with their mothers. Such heterozygote advantage could, in principle, account for the maintenance of histocompatibility polymorphism.

Bodmer (1972)

. . . there is little doubt that . . . immune dissimilarity between mother and fetus in mice is manifested by greater placental and fetal growth (Billington, 1964; James, 1965).

Wertelecki & Mantel (1973)

Antigenic disparity and placental weight

The first line of work considered by Clarke and Kirby (Billington, 1964) claimed to show that antigenic differences between mother and foetus led to increased placental weight. One circumstance in which antigenic differences between mother and foetus are almost always present is in matings between inbred strains. In every group of organisms in which inbreeding has been successfully practised (e.g. mice, chickens, *Drosophila*, maize), the F_1 progeny of matings between inbred strains have shown heterosis in respect of size, developmental rate, and general fitness. Heterosis is a genetic phenomenon, associated with increased heterozygosity. In mice, F_1 foetuses are usually larger than the corresponding inbred foetuses, and the placentae (which in late pregnancy are composed largely of embryonic rather than maternal tissue in the mouse) are usually heavier in pregnancies between than within strains (e.g. McLaren, 1965*a*). In certain strain combinations, however, placental weight may be similar or even lower in between-strain pregnancies than it is in within-strain pregnancies (Hetherington, 1971). Billington (1964) studied placental weight in A_2G and C57BL inbred mice: A_2G females had heavier placentae than C57BL females, and placental weight was increased when females of either strain were mated to males of the opposite strain. He proposed that the increased size of F_1 placentae was due in the main not to heterosis, but to the antigenic disparity between mother and foetus in matings between strains. As evidence for this heterodox view, he reported that A_2G embryos produced heavier placentae than C57BL embryos when both were transferred to C57BL foster-mothers.

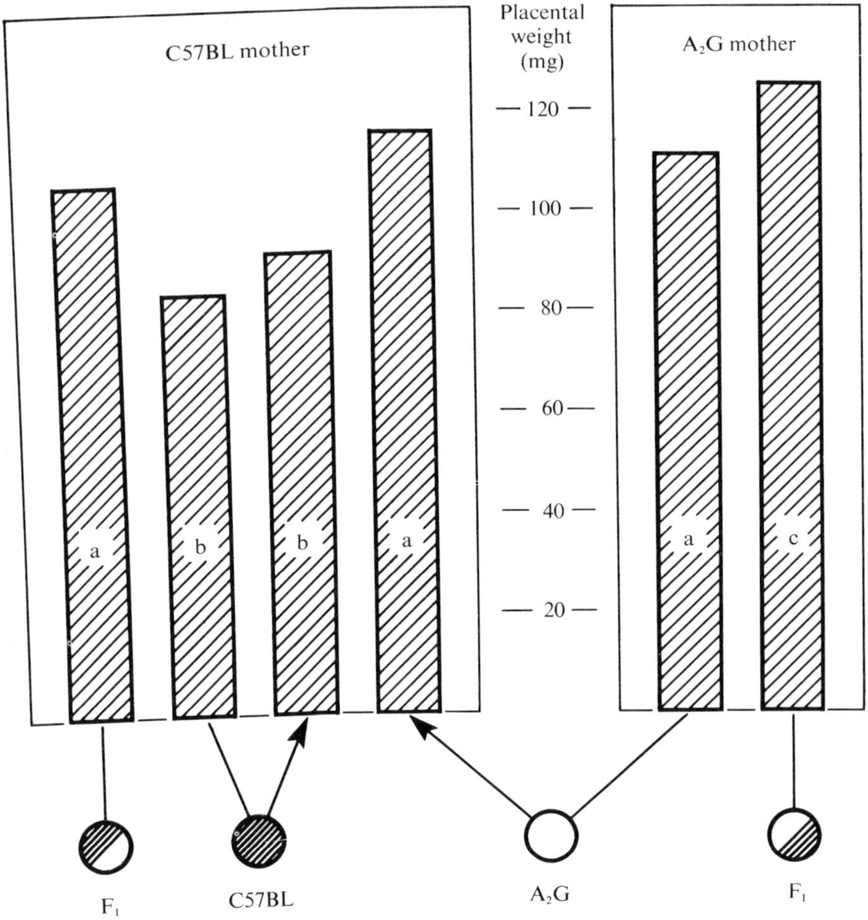

Fig. 1. Placental weight in relation to genotype of mother and foetus (data from Billington, 1964). Arrows indicate egg transfer, lines indicate normal pregnancy. Blocks carrying the same letter do not differ significantly when allowance is made for litter size.

Billington's data are shown in Fig. 1. Both C57BL and A_2G embryos have slightly larger placentae on transfer to C57BL females than in normal pregnancies, presumably because the transfer litters were smaller and placental weight is known to be inversely related to litter size (McLaren, 1965*a*). The A_2G embryos do indeed have heavier placentae than the C57BL embryos, on transfer to C57BL females as well as in normal pregnancies. This could be a consequence of antigenic disparity between mother and foetus in the between-strain transfers; alternatively, it may merely indicate that A_2G embryos have larger placentae than C57BL embryos for genetic reasons, since the genotype of the embryo is known to influence placental size (e.g. Hetherington, 1971). The greater size of F_1 placentae in A_2G than in C57BL mothers could be due to sex linkage, since sexes were not distinguished, or to a cytoplasmic effect, or to an effect of the maternal environment on the F_1 embryos.

These results can hardly be said to provide unequivocal evidence for an effect of antigenic disparity upon placental weight. As far as I know, no further evidence has been published on placental weight of transferred embryos, nor on the crucial reciprocal transfer, of C57BL embryos to A_2G mothers. Demonstrations that F_1 placentae are heavier than inbred placentae do not in themselves provide evidence that antigenic disparity increases placental weight.

An attempt to distinguish by purely genetical means the effects of heterosis from those of antigenic disparity was made by Hetherington (1971). The progeny of F_1 females will have a high level of heterozygosity whatever the genotype of the sire. However, backcross matings of F_1 females to either parental strain will provide no antigenic disparity between mother and young, whereas outcross matings should provide considerable antigenic disparity. Three strains of mice (JU, CBA, C57BL) were intercrossed to give the three possible types of F_1 female, and these were either backcrossed to the parental strains, or outcrossed to the third strain. No significant differences were observed between backcross and outcross matings in placental weight; in contrast, decidual weight (on the seventh day of pregnancy) was consistently lower in outcross matings than in either of the corresponding backcrosses, as though depressed by antigenic disparity. Foetal weight showed a positive association with decidual weight, an effect confirmed by Hetherington (1972).

In matings of congenic or allogeneic inbred strains of mice differing at the major H-2 locus, Finkel & Lilly (1971) were unable to detect any relationship between antigenic disparity and placental weight. The effect of H-2 and H-3 differences in the absence of other antigenic variation was examined by Hetherington (1973) using congenic strains. Neither decidual, placental nor foetal weight was affected by antigenic disparity at either locus. In crosses between strains, all three parameters tended to be intermediate between the parental strain values, supporting the view that the increase in placental and foetal weight associated with matings between conventional inbred strains is the result of heterosis.

No clear-cut effect of antigenic disparity between mother and foetus has been reported in man. For the ABO locus, foetal–maternal incompatibility (in the sense of incompatibility of foetal antigens to maternal isoagglutinins) has been reported to decrease placental weight (Jones, 1968), to increase placental weight but only in first-parity pregnancies and then only for male conceptuses (Toivanen & Hirvonen, 1970), and to have no effect on placental weight (Seppälä & Tolonen, 1970; Hohler, Bardawil & Mitchell, 1972). A recent report indicates that pregnancies in which at term the leucocytes of the mother and baby show *in vitro* a positive 'mixed lymphocyte reaction' have heavier placentae than those in which the in vitro-reaction is negative (Jenkins & Good, 1972). It is not clear whether the positive mixed lymphocyte reaction indicates some prior sensitization of the maternal lymphocytes, and if so whether the increased placental size might be a cause of the maternal sensitization, or an effect.

The most extreme example of antigenic disparity between mother and young is seen in interspecific crosses. Rogers & Dawson (1970) crossed two species of *Peromyscus* that had placentae of similar weight. In one direction, the hybrid placentae were half the normal weight; in the reciprocal cross, the placentae were three times as heavy as normal. It seems likely that factors other than antigenic disparity were influencing the size of the hybrid placentae.

Maternal immunity and placental weight

The other line of work considered by Clarke and Kirby claimed to show that the immunological status of a female affected her response to an antigenically disparate foetus by modifying placental weight (James, 1965, 1967). Three groups of C57BL female mice were used: an untreated control group, an 'immune' group that had recently received skin grafts and/or spleen cell injections from A_2G mice, and a 'tolerant' group that had at birth received spleen cell injections from A_2G mice. On mating to A_2G males, the 'immune' females proved to have heavier placentae, heavier foetuses, and shorter gestation periods than the controls, while the 'tolerant' females had lighter placentae, lighter foetuses, and longer gestation periods than the controls. Gestation period is known to be influenced by total foetal mass (McLaren, 1967), and foetal weight and placental weight are closely correlated; it was therefore presumed that the immunological status of the mother affected placental weight primarily.

The differences in placental weight were found consistently throughout the latter part of pregnancy, though the significance levels claimed in the text of James (1967) bear little relation to the standard errors given in his Table 1: for example, 113 ± 4.5 is claimed to differ at the 0.01 level of significance from 107 ± 4.4. However, the results appeared to lend support both to the 'antigenic disparity' interpretation of Billington's work and to the hypothesis of Clarke and Kirby. A further investigation was therefore undertaken by Ann Clarke, again using C57BL and A_2G mice.

The attempt to repeat James's results on 'tolerant' females failed, since more than 95% of C57BL mice injected at birth with A_2G spleen cells developed graft-versus-host disease (Clarke, 1971). This is not unexpected, since the A_2G strain is derived from the A strain and still resembles it in respect of histocompatibility antigens (Clarke, 1973), and Billingham & Brent's (1959) observation that the A $\rightarrow$ C57BL combination gives a high incidence of graft-versus-host disease and a minimal yield of tolerant animals has since been amply confirmed. James (1967) gives no information on the incidence of graft-versus-host disease in his neonatally injected C57BL mice, nor on the proportion that were proved tolerant by skin graft tests. Since body weights of mated females are not given, one cannot eliminate the possibility that some or all of his 'tolerant' animals were suffering from subclinical graft-versus-host disease, and that the reduced size of their foetuses and placentae arose from this cause rather than from a diminution of antigenic stimulation.

In Clarke's 'immune' group, placental weight was similar to that of the controls, and foetal weight was if anything somewhat reduced (Clarke, 1971). Immunization resulted in a decreased litter size, with more early embryonic mortality.

Thus Clarke's (1971) findings are in direct contradiction to those of James (1965, 1967). The reason for this lack of agreement is not obvious. A possible clue may lie in James's comment that he omitted from his analysis all litters of less than five young, and all litters containing dead embryos. If James's starting population contained some larger or reproductively more efficient females that had larger-than-average foetuses and placentae, and if these same females proved relatively resistant to any tendency of immunization to reduce litter size and to increase embryonic mortality, an apparent increase in foetal and placental size in the immunized group might have resulted from these omissions. Unfortunately it has not been possible to examine the complete records of James's experiment.

Clarke (1971) tested further the effect on foetal and placental weight of pre-immunization against paternal antigens, using a different strain combination (C57BL females immunized against, and mated to, CBA males). Again placental weight was similar in immunized and control groups, while foetal weight was slightly depressed by immunization.

Foetal weight was found by Hetherington (1971, 1972) to show a positive association with decidual weight. The reduction in decidual weight in the presence of antigenic differences between mother and embryo (Hetherington, 1971) might indicate some immunological effect upon the decidual response, although in the opposite direction to that which the Clarke–Kirby hypothesis would suggest. There is direct evidence for such an effect, since Clarke & Hetherington (1971) succeeded in reducing decidual weight by immunization of female mice to the antigens of the sire before mating. Passive immunization also proved effective in depressing the decidual response (Clarke, 1973). The depression of foetal weight observed by Clarke (1971) after maternal immunization may therefore have been a consequence of reduced decidual response. The reduction reported by Breyere & Springer (1969) in live young born to female mice immunized against the antigens of the sire could result from a similar cause. Parmiani & Della Porta (1973) also found a reduction in pregnancy rate and litter size in female rats immunized against embryonic antigens.

Further evidence against James's claim that maternal immunization benefits reproductive performance comes from the finding that, in sheep, immunization of the ewe to antigens of the ram before mating (Boshier & Moriarty, 1970) had no effect on foetal or placental weight.

Female mice made immune or tolerant to Y antigen by neonatal injection of male spleen cells showed no disturbance of sex ratio among their progeny, and immunization produced no alteration in the weight of the male foetuses (McLaren, 1962, 1965*b*). On the other hand, Lappé & Schalk (1971) found that female mice immunized against Y antigen and also splenectomized produced a higher proportion of male progeny than did controls. They postulate that embryonic mortality among male young is reduced by a maternal immune response, and that this selective advantage is normally neutralized by 'blocking' antibody produced by the spleen. Jackson, Mann & Schull (1969) found a significant deficit of female births to women negative for the X-linked Xg^a antigen, whose husbands were positive for the antigen, and whose daughters (but not sons) would therefore all have been histoincompatible with the mother. The effect was confined to births after the first daughter, suggesting an adverse effect of maternal immunization to Xg^a antigen. Unlike the analogous Rhesus situation, no increased abortion rate or neonatal jaundice was seen in the histoincompatible matings, so the effect presumably operates very early in pregnancy.

No effect on gestation length was detected in female rats made tolerant to their mates (Husain and Ketchel, 1965). In man, the presence of leucocyte agglutinins in the mother during pregnancy was reported by Brain (1968) to have no effect on birth weight, while Jensen (1962) found in such pregnancies a marginally significant increase in the number of babies weighing less than 2500 g at birth. Placental weight was not examined.

In summary, the evidence suggests that immunization of the mother against antigens present in the foetus does not in general increase foetal or placental weight, but may on the contrary decrease foetal weight.

Parity effects on placental, decidual and foetal weights

It has been suggested (Jones, 1968; Warburton & Naylor, 1971) that the well-known effects of parity on placental weight and birth weight (for references, see Warburton & Naylor, 1971; Clarke & Hetherington, 1972) may have an immunological basis. In mice, parous females show a specific unresponsiveness to the transplantation antigens of their mate; this unresponsiveness may be associated with humoral and cellular immunity to the male strain antigens (for references, see Hetherington, 1972).

An obvious test of the possibility that parity effects on reproductive performance are due to immunological factors is to see whether the parity effects persist when successive litters are fathered by antigenically different males. Hetherington (1972) examined decidual weight, foetal weight and placental weight in first and second pregnancies of females belonging to three inbred strains of mice, mated on each occasion to males of either their own strain or one of the other two strains. Placental weight in second pregnancies tended to be greater for a given genotype than in first pregnancies, but was unaffected by the genotype of the sire used for the first pregnancy. In contrast, the sire of the first pregnancy influenced decidual weight and foetal weight in the second pregnancy. Both tended to be greater in within-strain second pregnancies if the earlier pregnancy had also been within-strain rather than between-strain. It is hard to devise any immunological explanation for this apparent example of telegony. On the other hand, the tendency for decidual weight to be different in between-strain second pregnancies (greater in some strain combinations, less in others) if the first pregnancy had been between-strain rather than within-strain could have an immunological basis.

A similar study was carried out on women by Warburton & Naylor (1971). Placental weight and birth weight showed a tendency to increase with parity in both white and negro women. Successive pregnancies by the same or different fathers were compared. The increments in placental weight and birth weight were usually smaller (sometimes significant so) when the father changed between pregnancies. These results might suggest an immunological involvement, though as the authors prudently point out, 'it is quite likely that other factors differ between women who did and did not change mates between Study pregnancies'.

Antigenic disparity and trophoblast growth

Another line work of which seemed relevant to the question of placental size and antigenic disparity concerned the growth of trophoblast transplanted to an extra-uterine site in an isogeneic or an antigenically disparate allogeneic host. If trophoblast growth were increased by an antigenic difference, it might explain an effect of antigenic disparity on placental weight, if indeed there were such an effect.

Billington (1965) transplanted ectoplacental cones from within-strain matings of BALB/c and C57BL inbred mice to the right testes of BALB/c, C57BL and CBA inbred males. After eight days, trophoblast growth was assessed by comparing the weight of right and left testes. Preliminary findings had shown on control material that 'in the mouse the left testis is almost invariably within the range 0 to 5 mg heavier than the right testis';

in Billington's transplantation experiment, the cases in which the left testis weighed more at autopsy than the right (host) testis were excluded from the data. The mean increase in testis weight was greater for transplants between strains than within strains, but none of the differences were significant at the 5% level. For the C57BL/BALB/c combination, where reciprocal transplants were performed, it is possible to calculate the amount by which the results for the two between-strain transplants combined exceed those for the combined within-strain transplants; the difference is still not statistically significant (6.18 ± 4.07 mg).

In an attempt to elucidate the basis for Billington's claims, Clarke (1969) transplanted ectoplacental cones from CBA females pregnant by either CBA or C57BL males, to the testes of CBA, C57BL or (CBA × C57BL)F_1 recipients. Some of the CBA recipients were preimmunized against C57BL tissue. No significant differences were found between the extent of growth of CBA and F_1 trophoblast in any of the recipient types used, suggesting that in these experiments trophoblast growth was not enhanced by antigenic difference between donor and recipient. Immunization of the recipients also failed to affect trophoblast growth. Significant differences were found among recipient types in the ability of the testis to support trophoblast growth, with CBA testes providing the most and C57BL the least favourable environment.

The amount of trophoblast growth was found to be associated with the size of the ectoplacental cone transplanted, with mean ectoplacental cone size varying significantly between females. Since Billington (personal communication, cited by Clarke, 1969) neither measured ectoplacental cone size nor distributed cones from individual females randomly between experimental groups, so that the cones in each group were derived from a different two to three females, Clarke (1969) suggested that his results might be attributable to accidents of donor sampling.

Clarke's failure to detect any effect of antigenic disparity on trophoblast growth in the mouse testis has been confirmed more recently by James, Acierto & Murphy (1972). Koren, Abrams & Behrman (1968) found no association between trophoblast growth and antigenic differences in transfers of trophoblast to the mouse kidney, and Skreb, Svajger & Levak-Svajger (1971) report a similar finding in the rat.

All the above studies on trophoblast growth involve periods of not more than one to two weeks. When longer periods are involved, as with the growth of malignant trophoblast in human choriocarcinoma, there is extensive evidence that antigenic disparity reduces trophoblast proliferation, so that choriocarcinoma is more prevalent in inbreeding communities (Azar, 1962; Iliya, Williamson & Azar, 1967) and when mother and foetus are compatible on the ABO or HL-A systems (Scott, 1962; Llewellyn-Jones, 1965; Mogensen, Kissmeye-Nielsen & Hauge, 1969).

Disturbed segregation ratios

If Clarke & Kirby's hypothesis were correct, disturbances would be expected in the segregation of histocompatibility genes, and genes linked to them. A few such reports have indeed appeared. Phenotypic ratios which depart from Mendelian expectation may be due, in mammals, to phenotypic overlapping (incomplete penetrance) or to disturbances of gametogenesis, fertilization, implantation, pre- or post-implantational viability or postnatal survival. Progeny of different genotypes must be differentially affected, and the genotype of the mother may also be relevant. Antigenic disparity between mother and embryo could produce disturbed segregation ratios as a result either of selective implantation or of differential embryonic viability.

Selective fertilization seems to be responsible for disturbed segregation in at least two mammalian systems. Bateman (1960) showed in mice that the deficiency relative to wild-type progeny of short-tailed $T+$ progeny when $+t^e$ males were mated to tail-less Tt^e females, and of Tt^e progeny when Tt^e males were mated to $T+$ females, could be explained by a tendency for t^e spermatozoa to unite more frequently with + than with T eggs, and more frequently with T than with t^e eggs, when a choice of eggs was presented by heterozygous females. The reciprocal crosses showed no evidence of disturbed segregation. The T-locus is closely linked to the H-2 histocompatibility locus.

Secondly, Canham, Birdsall & Cameron (1970) found anomalous frequencies of transferrin phenotypes among the offspring of *Peromyscus* heterozygous for the two most common forms of transferrin, J and M. There was a highly significant deficiency of JM and excess of M young, no anomalies of segregation were seen in other mating types, and the possibility that selection against J and JM embryos was involved seemed remote. Their data are most readily explicable on the basis of selective fertilization, with M spermatozoa able to fertilize a majority of M eggs, an advantage apparently manifesting in the reproductive tract of heterozygous females only.

In rats, two examples of heterozygous advantage have been reported. Michie & Anderson (1966) investigated a case in which heterozygosity at a histocompatibility locus was still maintained after more than seventy generations of close inbreeding. One homozygous class was detected in the population at low frequency, the other was never recovered. These results are now thought to be explicable in terms of heterosis leading to differential prenatal survival without need of the selective fertilization hypothesis originally preferred by the authors (Michie, personal communication). Since most matings were between heterozygotes, antigenic disparity between mother and young is not involved.

In the second example (Palm, 1969, 1970), an excess of rats heterozygous for the major histocompatibility locus Ag-B was observed in the progeny of backcross matings, but only when the female was inbred and the favoured progeny therefore antigenically incompatible with their mother. In reciprocal backcrosses of F_1 females to inbred males, there was no disturbance of the expected 1:1 segregation ratio. In some series, the excess of heterozygotes was only seen among male progeny. Analysis revealed a high level of postnatal mortality, that showed many of the features of graft-versus-host disease ('runting' syndrome), and that bore most heavily on homozygous young born to homozygous mothers, thus producing the observed excess of heterozygotes. Palm (1970) concludes that the graft-versus-host reaction is due to maternal–foetal incompatibility at some non-Ag-B locus, perhaps involving transfer of sensitized cells from mother to foetus, and that the Ag-B incompatibility of Ag-B heterozygotes in homozygous mothers is exerting a protective effect on the young. A similar graft-versus-host reaction due to transfer of sensitized cells from mother to foetus in rats has been produced experimentally by Beer & Billingham (1973). Another example of the protective effect of maternal–foetal incompatibility at one histocompatibility locus against the adverse effects of incompatibility at a different locus is seen in our own species, where incompatibility between mother and baby at the ABO locus protects against Rh immunization (for references see Finn *et al.*, 1961). For 'protective' loci such as these, antigenic disparity between mother and young undoubtedly has a beneficial effect on the outcome of pregnancy, and heterozygosity will be favoured.

A deficiency rather than an excess of heterozygotes is seen in backcross matings involving the recessive *hair-loss* gene in mice, when heterozygous males are mated to homozygous recessive females (Hollander & Gowen, 1959). Segregation ratios in the reciprocal backcross are normal. The heterozygous young usually die within a few days of birth, with multiple abnormalities that suggest collagen deficiency (McKenzie, 1967). Immunological studies indicate that antigenic incompatibility is not involved (McLaren, unpublished), so presumably in this example the interaction between mother and foetus is biochemical in nature, perhaps relating to collagen metabolism.

Hull (1964) carried out crossing experiments with two inbred strains of mice differing at the agouti (*A,a*) coat-colour locus. In backcrosses of F_1 males, he found a significant excess of heterozygotes among the progeny of females of one strain (*aa*) but not the other (*AA*). In the two reciprocal backcrosses, and in the F_1 intercross, i.e. in all those matings where heterozygous females were used, he found a significant deficiency of heterozygotes among the progeny. He concluded that young whose coat-colour genotype resembled that of the mother were at a disadvantage prenatally. No sex or parity effect was observed. Since the agouti locus is closely linked to the H-3 histocompatibility locus, he carried out a subsequent series of backcrosses (Hull, 1969) involving two congenic strains differing only for the chromosome segment bearing these two loci. When heterozygous females were used, no disturbed segregation ratios were observed; in one of the reciprocal backcrosses, there was a deficiency of homozygous young (i.e. young resembling the mother in genotype), but only in third and fourth litters; in the other backcross there was a deficiency of homozygous young at all parities, but only affecting male progeny. No weight differences between homozygous and heterozygous young were detected at birth or at weaning.

Since the same loci were involved in both series, it is tempting to seek a common explanation of the observed segregation disturbances, but none has been proposed. In the first series, three out of four disturbances were in heterozygous mothers, which rules out any conventional incompatibility system depending on maternal recognition of foetal antigens. In the second series no segregation disturbance was observed in heterozygous mothers, and an immunological involvement is perhaps suggested both by the parity effect and by the restriction of the disturbance in one cross to male progeny which possess an additional antigenic difference from the mother. Toivanen & Hirvonen (1970) claimed that placental weight was increased in ABO-incompatible human pregnancies for male conceptuses, but in their case the effect was apparently confined to first parity pregnancies while the segregation distortion observed by Hull (1969) increased with parity. Using the same congenic strains as Hull (1969), Hetherington (1973) found no effect of the H-3 locus on decidual, placental or foetal weight, but he only examined first parity litters and did not distinguish the sex of the young.

It seems unlikely that any of the above examples of disturbed segregation, with the possible exception of Hull's observations, involve selective implantation. Where heterozygotes show an advantage in heterozygous as well as homozygous mothers (Michie & Anderson, 1966; Hull, 1964), the effect may be due to heterosis at or even before implantation, and the heterozygous embryos may implant more successfully than the slower-developing homozygous embryos. Selective implantation, however, implies interaction between the genotype of the mother and that of the embryo, for example by antigenic incompatibility influencing implantation for better or worse. There is as yet little evidence for such a phenomenon. One possible example is described by Fekete (1947), who found that inbred C57BL and DBA females showed a higher incidence of pregnancy when embryos of the other strain rather than their own were transferred to their uteri ($0.01 < P < 0.02$). In sheep, Boshier & Moriarty (1970) observed that more placentomes per foetus were formed at implantation in between-breed than in within-breed pregnancies. This may have been due to heterosis rather than to antigenic disparity, since pre-immunization of the ewes against the antigens of the sire failed to induce a significant increase in placentome number.

5. Antigenic disparity can increase trophoblast proliferation	(Billington, 1965)	Clarke, 1969 James *et al.*, 1972 Koren *et al.*, 1968 Skreb *et al.*, 1971	Incorrect
6. Maternal immunization can decrease decidual weight	Hetherington, 1971 Clarke & Hetherington, 1971 Clarke, 1973		Correct
7. Maternal immunization can decrease foetal survival	Clarke, 1971 Breyere & Springer, 1969 Beer & Billingham, 1973 Rh sensitization		Correct
8. Antigenic incompatibility at certain loci can protect against immunization at other loci, and so increase survival	Palm, 1969; 1970 Rh (ABO)		Correct
9. Maternal–foetal genetic differences can affect implantation, placental weight, foetal survival, not necessarily by immunological means	Fekete, 1947 Rogers & Dawson, 1970 Hull, 1964; 1969 Hollander & Gowen, 1959		Correct

to include among them the hypothesis that antigenic disparity between mother and embryo is beneficial to the development of the embryo.

SUMMARY

Evidence is reviewed bearing on the hypothesis (Clarke & Kirby, 1966) that antigenic disparity between mother and embryo is beneficial to the development of the embryo, and that this effect contributes to the maintenance of histocompatibility polymorphisms. The experimental foundations on which the hypothesis was originally based are no longer secure, and much recent evidence points in the opposite direction. It is concluded that the hypothesis is no longer tenable.

I am grateful to David Kirby who stimulated the interest of so many of us in this field, to David Billington, Ann and Bryan Clarke, and Colin Hetherington for valuable discussions, and to the Ford Foundation for financial support.

REFERENCES

AZAR, H. A. (1962). Cancer in Lebanon and the Near East. *Cancer*, **15**, 66.

BATEMAN, N. (1960). Selective fertilization at the *T*-locus of the mouse. *Genet. Res.*, **1**, 226.

BEER, A. E. & BILLINGHAM, R. E. (1971). Immunobiology of mammalian reproduction. *Adv. Immunol.*, **14**, 1.

BEER, A. E. & BILLINGHAM, R. E. (1973). Maternally acquired runt disease. *Science*, **179**, 240.

BILLINGHAM, R. E. & BRENT, L. (1959). Quantitative studies on tissue transplantation immunity. *Phil. Trans. R. Soc. B*, **242**, 439.

BILLINGTON, W. D. (1964). Influence of immunological disparity of mother and foetus on size of placenta in mice. *Nature, Lond.*, **202**, 317.

BILLINGTON, W. D. (1965). The invasiveness of transplanted mouse trophoblast and the influence of immunological factors. *J. Reprod. Fert.*, **10**, 343.

BODMER, W. F. (1972). Evolutionary significance of the *HL-A* system. *Nature, Lond.*, **237**, 139.

BOSHIER, D. P. & MORIARTY, K. M. (1970). Some effects on the conceptus of prior immunological sensitization of ewes to the sire. *J. Reprod. Fert.*, **21**, 495.

BRAIN, P. (1968). Leucocyte agglutinins and birth weight. *Med. J. Aust.*, **1**, 1046.

BREYERE, E. J. & SPRINGER, W. W. (1969). Evidence of allograft rejection of the conceptus. *Transplant. Proc.*, **1**, 71.

CANHAM, R. P., BIRDSALL, D. A. & CAMERON, D. G. (1970). Disturbed segregation at the transferrin locus of the deer mouse. *Genet. Res.*, **16**, 355.

CLARKE, A. G. (1969). Factors affecting the growth of trophoblast transplanted to the testes. *J. Reprod. Fert.*, **18**, 539.

CLARKE, A. G. (1971). The effects of maternal pre-immunization on pregnancy in the mouse. *J. Reprod. Fert.*, **24**, 369.

CLARKE, A. G. (1973). Studies on the immunological interaction between mother and foetus. Ph.D. Thesis, University of Edinburgh.

CLARKE, A. G. & HETHERINGTON, C. M. (1971). Effects of maternal pre-immunization on the decidual cell reaction in mice. *Nature, Lond.*, **230**, 114.

CLARKE, A. G. & HETHERINGTON, C. M. (1972). Immunogenetic aspects of maternal/fetal relations. *J. Reprod. Fert., Suppl.*, **15**, 99.

CLARKE, B. C. & KIRBY, D. R. S. (1966). Maintenance of histocompatibility polymorphisms. *Nature, Lond.*, **211**, 999.

FEKETE, E. (1947). Differences in the effect of uterine environment upon development in the DBA and C57 black strains of mice. *Anat. Rec.*, **98**, 409.

FINKEL, S. I. & LILLY, F. (1971). Influence of histocompatibility between mother and foetus on placental size in mice. *Nature, Lond.*, **234**, 102.

FINN, R., CLARKE, C. A., DONOHOE, W. T. A., MCCONNELL, R. B., SHEPPARD, P. M., LEHANE, D. & KULKE, W. (1961). Experimental studies on the prevention of Rh haemolytic disease. *Brit. Med. J.*, **1**, 1486.

HARPER, M. J. K., SKARNES, R. C., CHANG, M. C. & MCGAUGHEY, R. W. (1970). Effect of antilymphocytic serum (ALS) on pregnancy in rodents. *Biol. Reprod.*, **2**, 335.

HETHERINGTON, C. M. (1971). The decidual cell reaction, placental weight, foetal weight and placental morphology in the mouse. *J. Reprod. Fert.*, **25**, 417.

HETHERINGTON, C. M. (1972). The effect of paternal genotype on parity-dependent changes in the decidual cell reaction, placental weight and fetal weight in the mouse. *J. Reprod. Fert.*, **31**, 83.

HETHERINGTON, C. M. (1973). The absence of any effect of maternal/fetal incompatibility at the *H-2* and *H-3* loci on pregnancy in the mouse. *J. Reprod. Fert.*, **33**, 135.

HOHLER, C. W., BARDAWIL, W. A. & MITCHELL, G. W. (1972). Placental weight and water content relative to blood types of human mothers and their offspring. *Gynec. Obstet.*, **40**, 799.

HOLLANDER, W. F. & GOWEN, J. W. (1959). A single-gene antagonism between mother and fetus in the mouse. *Proc. Soc. exp. Biol. Med.*, **101**, 425.

HULL, P. (1964). Partial incompatibility not affecting total litter size in the mouse. *Genetics*, **50**, 563.

HULL, P. (1966). Possible stability of polymorphism in a multiallele incompatibility system in the mouse. *Genetics*, **54**, 1049.

HULL, P. (1969). Maternal–foetal incompatibility associated with the H-3 locus in the mouse. *Heredity*, **24**, 203.

HUSAIN, R. & KETCHEL, M. M. (1965). Normal pregnancy and parturition in rats with acquired immunological tolerance to their mates. *Nature, Lond.*, **206**, 522.

ILIYA, F. A., WILLIAMSON, S. & AZAR, H. A. (1967). Choriocarcinoma in the Near East. *Cancer*, **20**, 144.

JACKSON, C. E., MANN, J. D. & SCHULL, W. J. (1969). *Xg*^{a} blood group system and the sex ratio in man. *Nature, Lond.*, **222**, 445.

JAMES, D. A. (1965). Effects of antigenic dissimilarity between mother and foetus on placental size in mice. *Nature, Lond.*, **205**, 613.

JAMES, D. A. (1967). Some effects of immunological factors on gestation in mice. *J. Reprod. Fert.*, **14**, 265.

JAMES, D. A. (1969). Antigenicity of the blastocyst masked by the zona pellucida. *Transplantation*, **8**, 846.

JAMES, D. A., ACIERTO, S. & MURPHY, B. D. (1972). Growth of mouse trophoblast transplanted to syngeneic and allogeneic testes. *J. exp. Zool.* **180**, 209.

JENKINS, D. M. & GOOD, S. (1972). Mixed lymphocyte reaction and placentation. *Nature New Biol.*, **240**, 211.

JENSEN, K. G. (1962). Leucocyte antibodies in serums of pregnant women: serology and clinic. *Vox Sang.*, **7**, 454.

JONES, W. R. (1968). Immunological factors in human placentation. *Nature, Lond.*, **218**, 480.

JOOSTE, S. V. (1968). The effects of heterologous antilymphocytic antiserum in the late prenatal and early postnatal period in mice. *Transplantation*, **6**, 277.

KIRBY, D. R. S. (1967). Inhibition of egg implantation and induction of abortion in mice by heterologous immune serum. *Nature, Lond.*, **216**, 1220.

KIRBY, D. R. S. (1970). The egg and immunology. *Proc. R. Soc. Med.*, **63**, 59.

KOREN, Z., ABRAMS, G. & BEHRMAN, S. J. (1968). The role of host factors in mouse trophoblastic tissue growth. *Am. J. Obstet. Gynec.*, **100**, 570.

LAPPÉ, M. & SCHALK, J. (1971). Necessity of the spleen for balanced secondary sex ratios following maternal immunization with male antigen. *Transplantation*, **11**, 491.

LLEWELLYN-JONES, D. (1965). Trophoblastic tumours. Geographical variations in incidence and possible aetiological factors. *J. Obstet. Gynaec. Brit. Commonw.*, **72**, 242.

MCKENZIE, C. (1967). An investigation of the *hair-loss* mutant in the mouse with particular reference to the problem of maternal-foetal interactions. Ph. D. Thesis, University of Edinburgh.

MCLAREN, A. (1962). Does immunity to male antigen affect female reproductive performance? *Nature, Lond.*, **195**, 1323.

MCLAREN, A. (1965*a*). Genetic and environmental effects on foetal and placental growth in mice. *J. Reprod. Fert.*, **9**, 79.

MCLAREN, A. (1965*b*). Growth of male young in mothers immunized against Y antigen. *Transplantation*, **3**, 260.

MCLAREN, A. (1967). Effect of foetal mass on gestation period in mice. *J. Reprod. Fert.*, **13**, 349.

MEDAWAR, P. B. M. (1953). Some immunological and endocrinological problems raised by the evolution of viviparity in vertebrates. *Symp. Soc. Exp. Biol.*, **7**, 320.

MICHIE, D. & ANDERSON, N. F. (1966). A strong selective effect associated with a histocompatibility gene in the rat. *Ann. N.Y. Acad. Sci.*, **129**, 88.

MOGENSEN, B., KISSMEYER-NIELSEN, F. & HAUGE, M. (1969). Histocompatibility antigens on the HL-A locus in gestational choriocarcinoma. *Transplant. Proc.*, **1**, 76.

MORTON, J. R., GILMOUR, D. G., MCDERMID, E. M. & OGDEN, A. L. (1965). Association of blood-group and protein polymorphism with embryonic mortality in the chick. *Genetics*, **51**, 97.

PALM, J. (1969). Association of maternal genotype and excess heterozygosity for Ag-B histocompatibility antigens among male rats. *Transplant. Proc.*, **1**, 82.

PALM, J. (1970). Maternal–fetal interactions and histocompatibility antigen polymorphisms. *Transplant. Proc.*, **2**, 162.

PARMIANI, G. & DELLA PORTA, G. (1973). Effects of antitumour immunity on pregnancy in mouse. *Nature New Biol.*, **241**, 26.

PREHN, R. T. (1971). Perspectives on oncogenesis: does immunity stimulate or inhibit neoplasia? *J. Reticuloendothel. Soc.*, **10**, 1.

ROGERS, J. F. & DAWSON, W. D. (1970). Foetal and placental size in a *Peromyscus* species cross. *J. Reprod. Fert.*, **21**, 255.

SCOTT, J. S. (1962). Choriocarcinoma: observations on the etiology. *Am. J. Obstet. Gynec.*, **83**, 185.

SEPPÄLÄ, M. & TOLONEN, M. (1970). Histocompatibility and human placentation. *Nature, Lond.*, **225**, 950.

SKREB, N., SVAJGER, A. & LEVAK-SVAJGER, B. (1971). Growth and differentiation of rat egg-cylinders under the kidney capsule. *J. Embryol. exp. Morph.*, **25**, 47.

SNELL, G. D. (1968). The *H-2* locus of the mouse: observations and speculations concerning its comparative genetics and its polymorphism. *Folia Biol., Praha*, **14**, 335.

TOIVANEN, P. & HIRVONEN, T. (1970). Placental weight in human foeto–maternal incompatibility. *Clin. exp. Immun.*, **7**, 533.

WARBURTON, D. & NAYLOR, A. F. (1971). The effect of parity on placental weight and birth weight. An immunological phenomenon? A report of the collaborative study of cerebral palsy. *Am. J. Hum. Genet.*, **23**, 41.

WERTELECKI, W. & MANTEL, N. (1973). Increased birth weight in leukemia. *Pediat. Res.*, **7**, 132.

DISCUSSION

EDWARDS: Did Palm ever cross animals heterozygous for AgB? This would be rather an essential control of her system in order to ensure that heterozygous foetuses developing in a heterozygous mother were not affected, and help to decide that no genes other than AgB were involved. I had not realized before that this particular control was missing in the experiment until you described her work.

McLAREN: One would expect a 1:2:1 ratio in the progeny of intercrosses of heterozygotes. As far as I know, Palm has only published data from backcrosses.

JENKINS: Hancock and I (Jenkins & Hancock, 1972) have worked with lymphocyte reactions in different parity groups. With increasing parity there appears to be a decrease in the degree of lymphocyte reactivity in autologous serum and yet high parity groups of women have larger placentae. We were concerned that the large placental-weight group (Jenkins & Good, 1972) included more multiparous women than in the low placental-weight group, but this was evidently not the case because women of high parity had a lower reactivity in mixed lymphocyte reactions.

I wish to add one more point. We have more than one hundred mother–baby pairs referenced to HL-A types (Jenkins, 1973) and there is an apparent association between increasing antigenic disparity between mother and foetus and heavier placentae. The data are not yet significant.

BODMER: Many published studies show no convincing evidence that antigenic disparity, the presence of antibodies or similar effects exert any significant effect on most of the parameters of fertility. Studies in humans are very difficult because of the lack of controls, and many factors, such as socio-economic status, illegitimacy and so on are involved too. Perhaps the best approach is to examine segregation distortion using statistical approaches and the HL-A system. Pooled data from a large number of HL-A types have indicated that the answer, once again, is clearly negative. Fairly large bodies of data have, for example, revealed no evidence of a distortion with respect to the last birth in comparison with the first in highly multiparous women (Mattiuz *et al.*, 1970).

EDWARDS: I agree with all the comments and conclusions made by Anne McLaren. Recently, however, Lappé and Schalk (1971) immunized mice that had been splenectomized against the Y antigen, in an attempt to select against male foetuses by withdrawing any protection through enhancement. The sex ratio obtained was exactly the opposite to that expected, for they found an excess of male offspring. There was evidently no selective loss of female foetuses to explain the distortion in the sex ratio, as would be suspected if, for example, a decline occurred in litter size. Lappé and Schalk conclude that a selective form of implantation occurs in their mice, which favours males. I cannot understand why or how, especially if the transplantation and other antigens are not expressed on the blastocyst. I wonder if you can shed any light on their observations?

McLAREN: I found this paper difficult to interpret, too. The authors suggested that the removal of the spleen was having a protective effect on the male blastocysts, perhaps by eliminating the production of some blocking factor which, in intact animals, prevented the male blastocysts from implanting preferentially. I would like to see the work repeated before basing too many conclusions on it.

EDWARDS: They include their results in a wider theory suggesting that different tissues stimulate each other mitotically, resulting in a large growth of one or other types of tissue when they make contact. In this case, of course, the male foetuses are stimulated and therefore implant preferentially.

HOWE: These investigators attempted to interpret the maternal–foetal relationship on the basis of their observations of immunostimulation of certain animal tumours (Prehn & Lappé, 1971).

McLAREN: Yes, and Prehn (1971, 1972) has similar data on the effects of immunity on tumour growth.

LAWLER: A point in relation to the HL-A system is that the antibodies undoubtedly cross the placenta yet they do not harm the foetus. Neonatal neutropenia due to maternal anti-leucocyte antibodies has been recorded in relation to an antigenic system confined to neutrophils, but this is genetically distinct from the HL-A system (Lalezari *et al.*, 1960). As far as the HL-A system is concerned, a simple mechanism for inducing foetal morbidity apparently does not exist.

BILLINGTON: I would like to add that in general the points made by Anne McLaren are valid and acceptable. There could be disagreement with her interpretation only in minor points of detail, although some of the quoted evidence requires final verification.

REFERENCES

JENKINS, D. M. (1972). Immunological studies of the lymphocyte in human pregnancy. M.D. Thesis, Queens University, Belfast.

JENKINS, D. M. & GOOD, SHEILA (1972). Mixed lymphocyte reaction and placentation. *Nature New Biol.*, **240**, 211.

JENKINS, D. M. & HANCOCK, K. W. (1972). Maternal unresponsiveness to paternal histocompatibility antigens in human pregnancy. *Transplantation*, **13**, 619.

LALEZARI, P., NUSSBAUM, S., GELMAN, S. & SPAET, T. H. (1960). Neonatal neutropenia due to maternal iso-immunization. *Blood*, **15**, 236.

LAPPÉ, M. A. & SCHALK, J. (1971). Necessity of the spleen for balanced secondary sex ratios following maternal immunization with male antigen. *Transplantation*, **11**, 491.

MATTIUZ, P. L., IHDE, D., PIAZZA, A., CEPPELLINI, R. & BODMER, W. F. (1970). New approaches to the population genetics and segregation analysis of the

HL-A system. In *Histocompatibility Testing 1970*, (ed. P.I. Terasaki), p. 193. Munksgaard: Copenhagen.

Prehn, R. T. (1971). Perspectives on oncogenesis: does immunity stimulate or inhibit neoplasia? *J. Reticuloendothel. Soc.*, **10**, 1.

Prehn, R. T. (1972). The immune reaction as a stimulator of tumor growth. *Science*, **176**, 170.

Prehn, R. T. & Lappé, M. A. (1971). Immunostimulation theory of tumour development. *Transplant. Rev.*, **7**, 26.

INDEX